BIOLOGICAL THREATS AND TERRORISM

Assessing the Science and Response Capabilities

Workshop Summary

Stacey L. Knobler, Adel A.F. Mahmoud, and Leslie A. Pray, *Editors*

Based on a Workshop of the
Forum on Emerging Infections

Board on Global Health

INSTITUTE OF MEDICINE

NATIONAL ACADEMY PRESS
Washington, DC

NATIONAL ACADEMY PRESS • 2101 Constitution Avenue, N.W. • Washington, DC 20418

NOTICE: The project that is the subject of this workshop summary was approved by the Governing Board of the National Research Council, whose members are drawn from the councils of the National Academy of Sciences, the National Academy of Engineering, and the Institute of Medicine.

Support for this project was provided by the U.S. Department of Health and Human Services' National Institutes of Health, Centers for Disease Control and Prevention, the U.S. Agency for International Development, and the U.S. Food and Drug Administration; U.S. Department of Defense; U.S. Department of State; U.S. Department of Veterans Affairs; U.S. Department of Agriculture; American Society for Microbiology; Bristol-Myers Squibb Company; Burroughs Wellcome Fund; Eli Lilly & Company; Pfizer; GlaxoSmithKline; and Wyeth-Ayerst Laboratories.

This report is based on the proceedings of a workshop that was sponsored by the Forum on Emerging Infections. It is prepared in the form of a workshop summary by and in the name of the editors, with the assistance of staff and consultants, as an individually authored document. Sections of the workshop summary not specifically attributed to an individual reflect the views of the editors and not those of the Forum on Emerging Infections. The content of those sections is based on the presentations and the discussions that took place during the workshop.

International Standard Book Number: 0-309-08253-6
Library of Congress Control Number: 2002101411

Additional copies of this report are available for sale from the National Academy Press, 2101 Constitution Avenue, N.W., Box 285, Washington, D.C. 20055. Call (800) 624-6242 or (202) 334-3313 (in the Washington metropolitan area), or visit the NAP's home page at www.nap.edu. The full text of this report is available at www.nap.edu.

For more information about the Institute of Medicine, visit the IOM home page at: www.iom.edu.

Printed in the United States of America.

The serpent has been a symbol of long life, healing, and knowledge among almost all cultures and religions since the beginning of recorded history. The serpent adopted as a logotype by the Institute of Medicine is a relief carving from ancient Greece, now held by the Staatliche Museen in Berlin.

COVER: The background for the cover of this workshop summary is a photograph of a batik designed and printed specifically for the Malaysian Society of Parasitology and Tropical Medicine. The print contains drawings of various parasites and insects; it is used with the kind permission of the Society.

"Knowing is not enough; we must apply.
Willing is not enough; we must do."
—Goethe

INSTITUTE OF MEDICINE

Shaping the Future for Health

THE NATIONAL ACADEMIES

National Academy of Sciences
National Academy of Engineering
Institute of Medicine
National Research Council

The **National Academy of Sciences** is a private, nonprofit, self-perpetuating society of distinguished scholars engaged in scientific and engineering research, dedicated to the furtherance of science and technology and to their use for the general welfare. Upon the authority of the charter granted to it by the Congress in 1863, the Academy has a mandate that requires it to advise the federal government on scientific and technical matters. Dr. Bruce M. Alberts is president of the National Academy of Sciences.

The **National Academy of Engineering** was established in 1964, under the charter of the National Academy of Sciences, as a parallel organization of outstanding engineers. It is autonomous in its administration and in the selection of its members, sharing with the National Academy of Sciences the responsibility for advising the federal government. The National Academy of Engineering also sponsors engineering programs aimed at meeting national needs, encourages education and research, and recognizes the superior achievements of engineers. Dr. Wm. A. Wulf is president of the National Academy of Engineering.

The **Institute of Medicine** was established in 1970 by the National Academy of Sciences to secure the services of eminent members of appropriate professions in the examination of policy matters pertaining to the health of the public. The Institute acts under the responsibility given to the National Academy of Sciences by its congressional charter to be an adviser to the federal government and, upon its own initiative, to identify issues of medical care, research, and education. Dr. Kenneth I. Shine is president of the Institute of Medicine.

The **National Research Council** was organized by the National Academy of Sciences in 1916 to associate the broad community of science and technology with the Academy's purposes of furthering knowledge and advising the federal government. Functioning in accordance with general policies determined by the Academy, the Council has become the principal operating agency of both the National Academy of Sciences and the National Academy of Engineering in providing services to the government, the public, and the scientific and engineering communities. The Council is administered jointly by both Academies and the Institute of Medicine. Dr. Bruce M. Alberts and Dr. Wm. A. Wulf are chairman and vice chairman, respectively, of the National Research Council.

FORUM ON EMERGING INFECTIONS

GARY ROSELLE, Program Director for Infectious Diseases, VA Central Office, Veterans Health Administration, Department of Veterans Affairs, Washington, DC

DAVID SHLAES, Vice President, Infectious Disease Research, Wyeth, Pearl River, New York

JANET SHOEMAKER, Director, Office of Public Affairs, American Society for Microbiology, Washington, DC

P. FREDRICK SPARLING, J. Herbert Bate Professor Emeritus of Medicine, Microbiology, and Immunology, University of North Carolina, Chapel Hill, North Carolina

I. KAYE WACHSMUTH, Deputy Administrator, Office of Public Health and Science, United States Department of Agriculture, Washington, DC

C. DOUGLAS WEBB, Senior Medical Director, Bristol-Myers Squibb Company, Princeton, New Jersey

MICHAEL ZEILINGER, Infectious Disease Team Leader, Office of Health and Nutrition, U.S. Agency for International Development, Washington, DC

Liaisons

ENRIQUETA BOND, President, Burroughs Wellcome Fund, Research Triangle Park, North Carolina

NANCY CARTER-FOSTER, Director, Program for Emerging Infections and HIV/AIDS, U.S. Department of State, Washington, DC

EDWARD McSWEEGAN, National Institute of Allergy and Infectious Diseases, National Institutes of Health, Bethesda, Maryland

STEPHEN OSTROFF, Associate Director for Epidemiologic Science, National Center for Infectious Diseases, Centers for Disease Control and Prevention, Atlanta, Georgia

Staff

STACEY KNOBLER, Director, Forum on Emerging Infections

MARJAN NAJAFI, Research Associate

LAURIE SPINELLI, Project Assistant

BOARD ON GLOBAL HEALTH

DEAN JAMISON, *(Chair)*, Director, Program on International Health, Education, and Environment, University of California at Los Angeles
YVES BERGEVIN, Chief, Health Section, UNICEF, New York, New York
PATRICIA DANZON, Professor, Health Care Systems Development, University of Pennsylvania, Philadelphia
RICHARD FEACHEM, Institute for Global Health, UC San Francisco/UC Berkeley, San Francisco, California
NOREEN GOLDMAN, Professor, Woodrow Wilson School of Public and International Affairs, Princeton University, Princeton, New Jersey
ARTHUR KLEINMAN, Maude and Lillian Presley Professor of Medical Anthropology/Professor of Psychiatry and Social Medicine, Harvard Medical School, Boston, Massachusetts
ADEL MAHMOUD, President, Merck Vaccines, Whitehouse Station, New Jersey
ALLAN ROSENFIELD, Dean, Mailman School of Public Health, Columbia University, New York, New York
SUSAN SCRIMSHAW, Dean, School of Public Health, University of Illinois at Chicago
JOHN WYN OWEN, Secretary, Nuffield Trust, London, United Kingdom
GERALD KEUSCH, *(Liaison)*, Director, Fogarty International Center, National Institutes of Health, Bethesda, Maryland
DAVID CHALLONER, (*IOM Foreign Secretary*), Vice President for Health Affairs, University of Florida, Gainesville

Staff
JUDITH BALE, Director
PATRICIA CUFF, Research Associate
STACEY KNOBLER, Study Director
MARJAN NAJAFI, Research Associate
KATHERINE OBERHOLTZER, Project Assistant
JASON PELLMAR, Project Assistant
MARK SMOLINSKI, Study Director
LAURIE SPINELLI, Project Assistant

REVIEWERS

All presenters at the workshop have reviewed and approved their respective sections of this report for accuracy. In addition, this workshop summary has been reviewed in draft form by independent reviewers chosen for their diverse perspectives and technical expertise, in accordance with procedures approved by the National Research Council's Report Review Committee. The purpose of this independent review is to provide candid and critical comments that will assist the Institute of Medicine (IOM) in making the published workshop summary as sound as possible and to ensure that the workshop summary meets institutional standards. The review comments and draft manuscript remain confidential to protect the integrity of the deliberative process.

The Forum and IOM thank the following individuals for their participation in the review process:

Roger Breeze, Agricultural Research Service, Washington, DC
Richard Johnson, The Johns Hopkins University, Baltimore, Maryland
Stuart Nightingale, U.S. Department of Health and Human Services, Washington, DC
Gerald Parker, U.S. Army Medical Research and Material Command
M. Patricia Quinlisk, Iowa State Department of Health, Des Moines, Iowa
Robert Ryder, University of North Carolina at Chapel Hill, North Carolina

The review of this report was overseen by Melvin Worth, Scholar-in-Residence, National Academy of Sciences, who was responsible for making certain that an independent examination of this report was carried out in accordance with institutional procedures and that all review comments were carefully considered. Responsibility for the final content of this report rests entirely with the editors and individual authors.

Preface

The Forum on Emerging Infections was created in 1996 in response to a request from the Centers for Disease Control and Prevention and the National Institutes of Health. The goal of the Forum is to provide structured opportunities for representatives from academia, industry, professional and interest groups, and government* to examine and discuss scientific and policy issues that are of shared interest and that are specifically related to research and prevention, detection, and management of emerging infectious diseases. In accomplishing this task, the Forum provides the opportunity to foster the exchange of information and ideas, identify areas in need of greater attention, clarify policy issues by enhancing knowledge and identifying points of agreement, and inform decision makers about science and policy issues. The Forum seeks to illuminate issues rather than resolve them directly; hence, it does not provide advice or recommendations on any specific policy initiative pending before any agency or organization. Its strengths are the diversity of its membership and the contributions of individual members expressed throughout the activities of the Forum.

ABOUT THE WORKSHOP

In the wake of the events of September 11, already mounting concerns about bioterrorism became imminent priorities for policymakers, researchers,

---

*Representatives of federal agencies serve in an ex officio capacity. An ex officio member of a group is one who is a member automatically by virtue of holding a particular office or membership in another body.

public health officials, and private industry. These communities continue to grapple with ways to better understand the potential threats and ensure the country's ability to preempt an attack or respond to the consequences.

The Forum on Emerging Infections was uniquely positioned through its representation of multi-sector science and policy expertise to convene a working group discussion on the next steps for responding to bioterrorism.

Much has been written and discussed over the last decade about the potential use of biological agents in warfare or in terrorist attacks. Initiatives to prevent and respond to such events have been developed and implemented within areas of federal, state, and local government. The scientific, healthcare, policy, and law enforcement communities have also created specific agendas to address these threats.

The November 27–29, 2001, workshop of the Forum explored the current scientific understanding of threatening pathogens and what measures have been put in place to better monitor, prevent, and respond to their emergence. To determine where progress has been made and where gaps remain, Forum presentations and discussions reviewed existing policies, infrastructure, and research and scientific tools.

Additionally, Forum presentations and discussions sought to identify the obstacles to preparing an optimal response, particularly as it relates to the complexities of interaction among private industry, research and public health agencies, regulatory agencies, policymakers, academic researchers, and the public.

During this three-day workshop, Forum members and invited guests explored the issues surrounding emerging opportunities for more effective collaboration as well as the scientific and programmatic needs for responding to bioterrorism.

ORGANIZATION OF WORKSHOP SUMMARY

This workshop summary report is prepared for the Forum membership in the name of the editors, with the assistance of staff and consultants, as an individually authored document. Sections of the workshop summary not specifically attributed to an individual reflect the views of the editors and not those of the Forum on Emerging Infections sponsors or the Institute of Medicine (IOM). The contents of the unattributed sections are based on the presentations and discussions that took place during the workshop.

The workshop summary is organized within chapters as a topic-by-topic description of the presentations and discussions. Its purpose is to present lessons from relevant experience, delineate a range of pivotal issues and their respective problems, and put forth some potential responses as described by the workshop participants. The Summary and Assessment chapter discusses the core messages that emerged from the speakers' presentations and the ensuing discussions.

Although this workshop summary provides an account of the individual presentations, it also reflects an important aspect of the Forum philosophy. The workshop functions as a dialogue among representatives from different sectors and presents their beliefs on which areas may merit further attention. However, the reader should be aware that the material presented here expresses the views and opinions of those participating in the workshop and not the deliberations of a formally constituted IOM study committee. These proceedings summarize only what participants stated in the workshop and are not intended to be an exhaustive exploration of the subject matter.

ACKNOWLEDGMENTS

The Forum on Emerging Infections and the IOM wish to express their warmest appreciation to the individuals and organizations who gave valuable time to provide information and advice to the Forum through participation in the workshop.

The Forum is indebted to the IOM staff who contributed during the course of the workshop and the production of this workshop summary. On behalf of the Forum, we gratefully acknowledge the efforts led by Stacey Knobler, director of the Forum and coeditor of this report, who dedicated much effort and time to developing this workshop's agenda, and for her thoughtful and insightful approach and skill in translating the workshop proceedings and discussion into this workshop summary. We would also like to thank the following IOM staff and consultants for their valuable contributions to this activity: Leslie Pray, Rob Coppock, Marjan Najafi, Laurie Spinelli, Judith Bale, Mark Smolinski, Katherine Oberholtzer, Patricia Cuff, Paige Baldwin, Jennifer Otten, Clyde Behney, Bronwyn Schrecker, Sally Stanfield, Sally Groom, Michele de la Menardiere, Francesca Moghari, and Beth Gyorgy.

Finally, the Forum also thanks sponsors that supported this activity. Financial support for this project was provided by the U.S. Department of Health and Human Services' National Institutes of Health, Centers for Disease Control and Prevention, and the U.S. Food and Drug Administration; U.S. Department of Defense; U.S. Department of State; U.S. Department of Veterans Affairs; U.S. Department of Agriculture; American Society for Microbiology; Bristol-Myers Squibb Company; Burroughs Wellcome Fund; Eli Lilly & Company; Pfizer; GlaxoSmithKline; and Wyeth-Ayerst Laboratories. The views presented in this workshop summary are those of the editors and workshop participants and are not necessarily those of the funding organizations.

Adel Mahmoud, Chair
Stanley Lemon, Vice-Chair
Forum on Emerging Infections

Contents

APPENDIXES

BIOLOGICAL THREATS AND TERRORISM

Summary and Assessment

In the wake of September 11 and recent anthrax events, our nation's bioterrorism response capability has become an imminent priority for policymakers, researchers, public health officials, academia, and the private sector. Experts from each of these communities and the Forum on Emerging Infections convened for a three-day workshop discussion—the subject of this summary—to identify, clarify, and prioritize the next steps that need to be taken in order to prepare and strengthen bioterrorism response capabilities.

From the discussions, it became clear that of utmost urgency is the need to cast the issue of a response in an appropriate framework that captures the attention and understanding of policymakers and the public to garner sufficient and sustainable support for response initiatives. Such understanding would recognize that the protection of the nation's health is essential to ensuring national and global security. There was much debate, however, on what constitutes an appropriate framework to deliver this message.

No matter how the issue is cast, numerous workshop participants agreed that there are many gaps in the public health infrastructure and countermeasure capabilities that must be filled in order to assure a rapid and effective response to another bioterrorist attack. Many priorities for action—from encouraging antibiotic and vaccine research and development to educating first responders—were identified and discussed. Throughout the workshop, there were repeated calls for partnerships, including interagency and interdepartmental partnerships in government; and public-private partnerships that could harness the considerable power and knowledge of the academic community and industry at large.

FRAMING THE ISSUE

Bioterrorism is no longer a hypothetical event. A bioterrorist attack has occurred and could occur again at any time, under any circumstances, and at a magnitude far greater than we have thus far witnessed. U.S. bioterrorism preparedness efforts have so far focused on a number of potential agents, in particular anthrax, smallpox, plague, botulinum toxin, tularemia, and viral hemorrhagic fevers. Details of each of these threats were reviewed on the first day of the workshop. Anthrax is a proven risk and of most immediate concern, although smallpox, because it is capable of person-to-person transmission, engendered an equivalent sense of urgency. However, there is a plethora of potential, credible bioterrorist agents. One workshop participant noted that the Soviet Union is known to have weaponized some 30 different biological agents, including drug- and vaccine-resistant strains.

It is impossible for us as a nation to provide a specific defense against each of these many agents within a reasonable time frame: the diversity of readily available potential bioterrorist agents is great; the technology and knowledge that make it possible to bioengineer drug- and vaccine-resistant antimicrobial strains are becoming increasingly accessible; there are many crucial gaps in our countermeasures and public health response capabilities; and it was noted that one can develop a new bioweapon within only two to three years compared to the eight to ten years that it typically takes to bring a new vaccine or antimicrobial product into the market.

It is possible, however, to bolster our nation's general biodefense to a level at which we can at least minimize, if not prevent, the potentially catastrophic consequences of a large-scale bioterrorist attack. Workshop participants asked the question, how do we convince policymakers and those who allocate funds that new and substantial resources are needed at all levels, from local surge capacities for clinical care to research facilities expansion? The public health, scientific, and private industrial communities involved with bioterrorism defense must present their needs in an appropriate framework and present a vision to which the country can respond.

The most powerful strategy may be to cast bioterrorism defense as a national security issue first and foremost. Indeed, it was suggested that the only way to acquire the resources needed to develop the capacity that bioterrorism defense requires is to equate these tools with other weapons defense tools. It must be made clear that the nation's capability to respond to a bioterrorist attack is, in essence, a weapons defense system. Although most people do not know the details of how much money or research and development are required to sustain our country's armed forces, nonetheless they are able to express the essential role of such capabilities in our national security. Several workshop participants suggested that bioterrorism defense requires the same attention and understanding.

There was some concern though, that the priorities of a bioterrorism public health initiative not be pitted against those directed at more general control of emerging infectious diseases. In fact, bioterrorism defense is intimately tied to emerging infectious disease preparedness. The current situation was compared to the emergence of HIV/AIDS two decades ago—it is illuminating of many of the existing problems within the health care system about which we have become complacent. For some time now, it has been apparent to many that the public health infrastructure needs to be strengthened. Workshop participants recognized that now might be an opportune moment to foster better understanding of what has been in the past decade an often unheard call in the interest of public health, from increasing infectious disease surveillance to confronting antimicrobial resistance.

It was proposed that an alternative strategy would be to address bioterrorism response preparedness in a coordinated fashion with broad emerging infectious disease issues. For example, propose this effort as a defense system with concurrent substantial benefits for the public health system, as well as microbial and biomedical science. Recent efforts to put together a bioterrorism initiative in New York City, for example, were as much helped by the threat of naturally occurring pneumonic plague as by the threat of plague being used as a biological weapon. However, the question remained for many—is it possible to convey such a textured message that includes both bioterrorism and infectious disease preparedness needs? Many present felt it best to cast the issue as one of national defense.

Terrorism, by its nature, often evokes a more urgent response from the government and the public than does the emergence of other infectious disease or public health crises. It was noted that the 4 recent deaths from anthrax, but not the ongoing global AIDS crisis (which causes thousands of deaths daily), had the temporary effect of largely incapacitating the U.S. Congress and the U.S. Postal System. Several discussants observed that there is a national security component to bioterrorism response preparedness that makes it a very different issue than what they have been dealing with in public health up until this point. The question is, how do they attract the nation's attention so that they can acquire the necessary funds and resources to prepare in the event of another bioterrorist attack? It was pointed out that past efforts to obtain resources for emerging infectious disease preparedness have been only moderately successful at best. Strategically and tactically, it was suggested that a more compelling story than that of emerging infectious disease was needed, especially during a time when there are many competing immediate national security issues also on the table. Even the story that has unfolded over the last several months has not garnered the attention and focus that this effort requires. Again, it was suggested that bioterrorism defense be put in the context of a national security emergency that requires a new level of funding and a new way of operating.

Other questions raised were: In the absence of further attacks, how do we keep this issue before Congress? How do we prevent a lapse back into a false sense of security? How do we ensure the sustainability of our response to bioterrorism? If another act of terrorism occurs targeting an element other than the public's health, how do we ensure sustained funds for a bioterrorism preparedness effort? Thus, many agreed that the time is ripe, while the events of September 11 and the subsequent anthrax attack are fresh in the public's memory, to at least procure initial funds with which we can begin to develop the capacity that we need to defend against a future bioterrorist attack.

To better assess the amount of funding that is required, it was suggested that a prioritized list be developed setting out actions that should commence now and others that can be developed over time. If the funding is spread out over a longer period of time in this manner, the amount of funding necessary for building the capacity that is needed might be more digestible for the people who appropriate these funds. In order to do this, however, several participants noted that a governing structure would need to be in place for prioritizing and implementing such a plan.

Although the best framework for bioterrorism defense may be in terms of national security, there was some admonition that we not overreact. Some believed that it was important not to inflate the threat and create a challenge that cannot be met. And, whether the issue is framed solely as a national security threat or as a national security threat with concurrent benefits for the public health infrastructure, it must be stressed that in no way should efforts toward bioterrorism preparedness diminish other specific public health programs.

Finally, as one workshop participant indicated, framing the issue involves more than wrapping words around it. It was noted that allotting only a small percentage of the Department of Health and Human Services (DHHS) budget for bioterrorism defense sends a strong message to the nation and Congress that the public health community does not really consider the threat to be as serious as they claim. In order to be credible, these funds need to be reallocated in a manner that sends the message that *yes, this is important*.

FILLING THE GAPS

Several presenters and participants observed that the recent anthrax attack stretched the response network to its limit. However, it was a relatively small event and, as such, has raised many questions about our capacity and resources to cope with a larger attack. Even though there has been progress toward strengthening our response capabilities in recent years, there are still many gaps that leave us vulnerable to a potentially catastrophic event. But where are the gaps, how should they be filled, and which gaps should be filled first?

Prioritizing bioterrorist defense needs is, in essence, a very complex systems problem that involves many different people and groups from both the public and private sectors. As a first step toward strengthening our nation's response capabilities, there were repeated calls throughout the workshop for partnerships within and among government, private industry, academe, the health care system, and the intelligence community. The absence of multiple experts from the intelligence community and hospital care facilities in the workshop discussions did not allow robust discussion on their related issues and priorities.

It was mentioned that significant amounts of untapped laboratory capacity and scientific expertise exist within the global pharmaceutical industry in particular. It was suggested that a better way to leverage that capacity be devised by recognizing that industry truly is a partner in public health and able to contribute in very substantial ways to sustainable vaccine and drug research and development. To this end, workshop participants considered how to enlist the appropriate level of industry management and government decisionmakers in the right forums to discuss what needs to be done, who will do it, and how it will be done. It was recognized that the magnitude of the problem is clearly such that no single company can take on this challenge. Industry representatives noted that if there were plans in place, however, laid out by those who know best what the priorities are from a research and intelligence standpoint, it is likely that the pharmaceutical industry would respond very positively. One possible strategy suggested by a Forum member was to build a pharma-based industrial consortium backed with government spending to drive vaccine and antimicrobial development to where the market clearly is not going to take it.

As central to the effort as it is, it was pointed out that big Pharma is not the only industry in the private sector with which partnerships should be sought. The biotechnology industry is another important resource for new technologies that could be applied in a wide variety of ways to the bioterrorism defense effort, from improved rapid diagnostics and vaccine and therapeutics development to faster information transfer. One discussant pointed out that the challenge might not be so much that we need to develop more new technologies but that we need to optimize current applications. Users need to make their needs known and partnerships formed such that the demands can be met with the available supply.

It was suggested that new mechanisms and channels for strengthening all of these various partnerships be crafted. After September 11 it became apparent just how many interested companies and applicable products exist, as well as how many individual researchers are interested in the type of research that biodefense demands. But the question for some workshop participants remained, how do we coordinate all of these efforts? As one participant noted, the challenge before us is that everyone calls for coordination, but no one likes to be coordinated. In fact, this issue may require a new coordinating structure that has yet to emerge. One suggestion was that it be modeled after the Office of Scientific Research and Development that was set up during World War II and involved the efforts of some 6,000 scientists. With both federal agencies and industry actively engaged in this debate, now may be the time to move toward the establishment of such a structure.

The recent establishment of the Office of Public Health Preparedness (OPHP) was described during the workshop. The office is expected to oversee activities related to bioterrorism and bioterrorism preparedness within DHHS. The office will likely serve as public health's primary liaison with the Department of Veterans Affairs (VA), Federal Emergency Management Agency (FEMA), Office of Homeland Security, the Department of Defense (DoD), and other agencies involved with bioterrorism defense, including intelligence. That such an office has been created at a very high governmental level within DHHS was acknowledged as a very important first step toward centralizing power in a constructive way. However, because there are many other departments besides DHHS involved in this effort, some workshop participants expressed concern that there is still a need for a government-wide change in the way the prevention, control, and elimination of bioterrorist attacks are approached. Several participants called for broad changes in the communication and coordination both within and among these various other federal agencies.

Partnerships, coordination, and centralized leadership may be the necessary first steps. Still, the range of problems that were described and detailed throughout this workshop comprises a daunting portfolio. Several participants described the need to wrap our hands around this issue and try in some rational way to say, "There is the set of bioterrorism issues that we must address." Because of the enormity of the problem at hand, coupled with the reality of budget constraints, we must set priorities both in terms of research and public health preparedness. Which gaps should be filled first? It was suggested that moving such an agenda forward would require attention to the following issues:

- Prioritize response measures. Determine how to allocate present funds and decide which components of biodefense need to be strengthened first.
- Invest in real-time response role-playing exercises based on probable biological attack scenarios to assist in identifying appropriate infrastructure and training needs for an optimal response.
- Improve and sustain the understanding of the public, policymakers, and political leaders about the risks of bioterrorism and the capacity reqired to counter the threats. Convince those who allocate resources that new and substantial resources are needed at all levels.
- Evaluate the components of the National Stockpile Inventories. Determine the capacity of existing assets and establish priorities for developing and procuring additional needs.
- Consider the development of a peer review system for screening new bioterrorism defense research ideas.
- Craft innovative mechanisms and channels for forming and strengthening partnerships required for an effective response.
- Evaluate the necessity of a legal strategy for bioterrorism response.
- Consider the role for and responsibilities of a civilian biodefense program.

The many conspicuous gaps in our nation's biodefense science and response capabilities, from the striking insufficiency of vaccines and therapeutics to local public health departments already struggling with limited resources were described by workshop participants. A number of workshop participants suggested the urgent need to prioritize actions to be taken in order to strengthen our response capabilities. The following issues were identified by different individuals as priorities for action or consideration during the presentations and roundtable and panel discussions. The three major components are vaccines and therapeutics; research needs; and response infrastructure. Response infrastructure involves communications and information, laboratory capacity, disease detection and surveillance, and local response. Additional individual priorities are identified in the authored papers that follow in subsequent chapters of this report. The ordering and statement of priorities is a reflection of the workshop agenda. It is again emphasized that this summary document reflects only the statements and opinions of individual participants expressed during the workshop and is not intended to be an exhaustive exploration of the subject matter.

VACCINES AND THERAPEUTICS

Vaccines and antimicrobial therapeutics are vital to bioterrorist defense. Even if the best possible public health infrastructure existed and responses were as rapid as they need be, we might still be faced with a disease that could not be countered. Participants noted that the current biodefense arsenal is sparse and very little progress has been made in recent years in the development of new products.

The hurdles in vaccine discovery and production in particular were described as high for private industry to pursue. Although this may change following September 11, still the potential market is too risky to encourage the vaccine industry to make the large investments needed for the type of research, development, and manufacturing facilities that will be necessary to bolster our biodefense vaccine supply to an adequate level. Several participants proposed the need for incentives in the vaccine industry, such as expedited regulatory pathways; direct financial awards or contracts; tax credits; guaranteed liability indemnification; and partnerships in product development. One participant urgently called for clear vaccine production priorities, the development and procurement of top-priority vaccines, and the designation of somebody to be responsible for ensuring that these actions are accomplished.

It was suggested that development of biodefense vaccines must be a collaborative effort. With regard to the development of several potential new vaccines and therapeutics, efficacy is an important issue that often requires the use of monkey models that can only be studied in select laboratories. Likewise, a recent independent review of DoD's vaccine acquisition program recommended that DoD consider a collaborative approach to vaccine research and development with industry partners. The review also recommended a dedicated facility that would allow maximal flexibility and expandable manufacturing capability for the production of various types of vaccines. Workshop participants debated whether this would be government-owned and contractor-operated or contractor-owned and contractor-operated.

Discussion of other priorities related to vaccine development and production included the consideration of alternate methods of vaccine administration that would be more amenable to rapid dispersion, such as oral or inhalational vaccines; applications of genomics and high throughput technology in the identification of genomic markers for vaccine efficacy; technological applications that would improve vaccines; the use of combination vaccines; the potential applications of the new DNA vaccination technology; and, the application of military data on vaccine use to civilian populations.

One participant noted that is was important to recognize the inherent differences between military and civilian bioweapons defense vaccine usage and develop a specific vaccination policy for bioweapons defense in the civilian population.

Presenters stated that just as important as vaccine research and development is the research and development of new drug therapeutics. As with vaccines, there has been very little market incentive to produce these products. There was some concern that the FDA's new delta rules—the statistical requirements for clinical trials for antibiotics—add yet another major disincentive due to the large size of clinical trials that the new rule could demand. Several participants emphasized that there has been only one new class of antibiotic developed in the past two decades, and resistance to it emerged even before it entered the commercial market.

Presenters described that very little information on antivirals for potential bioterrorist agents is available, and none have thus far proven to be of clinical utility in humans. Accumulating evidence is beginning to suggest that different viruses may have common molecular targets. In light of this, one proposed strategy was to consider the development and production of family-specific antivirals, which could be tested against viral relatives in areas where disease naturally occurs. However, some participants highlighted concerns that antivirals have several disadvantages, including their toxic side effects and limited ability to reverse the effects of disease.

Antitoxins are another type of therapeutic agent discussed by presenters that could be developed for the biodefense arsenal. Basic research on the anthrax toxin system has led to some exciting prospects for antitoxin targeting. The most promising are the dominant negative inhibitors (DNIs), mutant forms of the protective antigen that block translocation of the virulence factors across the plasma membrane. Currently, DNIs are in the late stages of product development. Several participants suggested that if DNIs can be proven efficacious in infected animal models, they could be produced and deployed very rapidly. There are several other approaches in much earlier stages of development.

Finally, monoclonal antibodies provide yet another possible post-exposure therapeutic. Workshop presenters described the use of recombinant monoclonal antibodies that have been implicated for several different biothreat agents, including anthrax, smallpox, and botulinum neurotoxins. One of their primary advantages is that their route through the discovery and approval process may be faster than that of any other biodefense therapeutic. Thus, with the appropriate funding, they could serve as a very important short-term biodefense measure.

Again it was suggested that following September 11, the pharmaceutical and biotechnology industries have expressed a willingness to participate in the effort to build up the nation's biodefense arsenal of vaccines and therapeutics, but it seems that the direction, authority, and incentives are lacking. Considerations suggested by workshop participants for moving these issues forward include:

- Initiate stakeholder dialogue. Facilitate the role of the major vaccine and drug manufacturers in the development and production of biodefense countermeasures with clear direction and effective collaboration between industry and government.
- Provide incentives, priorities, and leadership to ensure that antimicrobial development and production are sufficiently scaled up.
- Decide to what extent our research, development, and production efforts should be directed toward agent-specific countermeasures versus broad-spectrum antimicrobials that could be used to defend against a wide range of agents.
- Direct more effort into developing better vaccine delivery technologies (e.g., aerosolized vaccines that can be dispensed much more quickly than injectable vaccines).
- Improve the usability of DNA vaccination platform technology.
- Explore the potential applications of genomics research to vaccine development and production.
- Encourage the research and development of new antitoxin treatments.
- Explore the potential for monoclonal and polyclonal human antibodies as countermeasures to bioterrorist agents and provide incentives to encourage development.
- Direct more effort toward alternative antiviral therapeutics, such as immunomodulators.
- Consider ways to accelerate vaccine and drug FDA review and licensure without compromising the independent and rigorous assessment of product safety and effectiveness.

RESEARCH NEEDS

The public health response to the anthrax attack was based on decades-old data. Clearly, our knowledge base needs to be revitalized. The country must develop a bioterrorism defense research agenda in the context of current and emerging understandings of the threats. Indeed, research is crucial for developing new and effective vaccines and therapeutics. It must be determined how much effort should be addressed to developing generic antivirals or broad-spectrum antibiotics versus target-specific agents. The issue is complicated by the spectra of genetic engineering and the increasing ease with which new, therapeutic-resistant strains of particular agents can be developed. Participants asked, do we become more and more specific, or do we develop an arsenal of weapons that can be used in any situation?

The implications of the long-term research on anthrax lethal factor which was presented in this workshop is an excellent example of what are sometimes unforeseen beneficial applications of basic scientific research. Examples of the type of knowledge that is needed and that can be gleaned from basic research include a better understanding of microbial biology; the human body's innate immunity; the potential applications of computational techniques and infectious disease modeling; and aerosol biology.

Funding for basic scientific research, as well as for research associated with the development of new vaccines and therapeutics, must extend beyond the actual experimental work. Discussants noted that in order to accommodate the increased need for safely contained laboratory facilities, where some of this research must be conducted, laboratory capacity needs to be expanded. One possibility is the construction of new BSL-3 or BSL-4 laboratories and animal facilities in order to validate new vaccines and therapeutics. It would also be necessary to train a larger cadre of individuals who are skilled and knowledgeable about these particular infectious diseases and trained to conduct research in these highly specialized laboratories.

Some participants considered it important to determine not only what science should be done but also how it should be done. Who should have access to materials, equipment and information? Do we take extreme measures such as requiring every individual who uses a centrifuge to be logged in and out by security agents? Although it was recognized that security is a very serious issue, we must be very careful that we not impose undo restrictions in the laboratory sector that could ultimately diminish scientific discovery by slowing or preventing important advances.

The research needs for responding to biological threats extend well beyond the biological and medical issues at hand. It was noted that more operational systems research was needed—particularly as it relates to all of the various aspects of the public health infrastructure and how it operates. The question was asked: *are the plans that are being implemented working?*

Presenters representing public health organizations identified a broad range of short-term applied research needs. Some suggested innovative, automated surveillance systems. Currently, many of the surveillance systems are "drop in" systems that are "dropped into" the Super Bowl or other high profile events. But it is impossible to predict when or where a bioterrorist attack is going to occur. It was also noted that broad-spectrum diagnostics for both environmental and clinical detection are needed. Currently, there are reasonably good assays available for only a limited range of specific agents. As with research and development of vaccines and therapeutics, one of the major challenges to rapid and accurate detection and diagnosis is the immense diversity of microorganisms. There can be considerable variability even within a strain, let alone a species, that can greatly complicate diagnosis and detection by making it difficult to separate out the causal agent from related agents as well as other microbes that are naturally occurring in the environment.

Whether basic or applied, short or long term, several participants recognized that all research could benefit from collaboration and federal incentives. Otherwise, applied research in particular tends to fall through the cracks if it is not profitable for the investigator. The following priorities were suggested by participants as important in developing a research agenda that responds to bioterrorism threats:

- Increase basic research on pathogenesis of disease caused by potential bioterrorist agents and on the human immune system response.
- Recruit, educate, and train more people in the fields of microbiology, aerosol biology, forensic epidemiology, and environmental microbiology.
- Consider how the scientific community can take a proactive role in helping to increase laboratory security and the safe transfer of knowledge while at the same time not restricting the advancement of science.
- Increase operational systems research that addresses whether preparedness plans are working.
- Consider the role of computational modeling in bioterrorism response preparedness efforts.
- Improve rapid molecular diagnostics.
- Improve environmental detection capability.

THE RESPONSE INFRASTRUCTURE

Numerous workshop participants agreed that the public health infrastructure must be strengthened in order to ensure a rapid, effective response in the event of another bioterrorist attack. But given obvious budget constraints, much concern was expressed that there is currently not enough money or human resources to strengthen public health services to an ideal capacity. It was suggested, however, that at least those specific components that warrant the most immediate attention should be reinforced. Indeed, if efforts are spread too thin and focus is lost, members of Congress and other sources of funding will ask, have you really delivered a product or has there been measurable improvement? Thus, workshop participants noted the imperative to set clear priorities. Public health infrastructure issues were addressed in the workshop through four categories, the order of which is a reflection of the workshop agenda sequence: communication and information; laboratory capacity; surveillance, detection, and diagnosis; and strengthening the local response.

Communication and Information

The issue of communication was front and center during the recent anthrax events. Several speakers observed that the way in which we generate and communicate information is extremely important. During the course of the anthrax events, most of the communications were backdoor connections instead of dialogue in a set forum. The result was that it was very difficult for the government to provide a single credible source about what was happening. Several presenters and participants expressed the need for a better forum for health and medical communications across all levels of government as well as the academic community and private sector.

It was suggested that appropriate and effective risk communication during and immediately following an attack is crucial. Bioterrorism is intended to generate panic in a population. Risk communication is an important first step toward diffusing public panic and assuaging people's fears. Some participants emphasized that when authorities speak to the general population, it is extremely important that they provide hard, accurate information that people can actually use and which gives them a sense of their own control. Public health professionals, from the national to local levels, should be able to offer an informed perspective on pertinent issues and effectively communicate the real risks in ways which are both meaningful to and appropriate for diverse populations.

Considered equally important by some workshop participants was the sharing of information among partners who are involved in the response and investigation. Ideally, it was proposed that every individual, whether they are in health care, law enforcement, or another sector involved in the response, have real-time high-speed access to current and consistent information. To this end, several participants proposed the ongoing need to upgrade, secure, and back up the communication systems that support information systems.

Finally, having a single database of information that is shared among labs was identified as another essential priority. A direct linkage and distribution of data between public health labs and the Centers for Disease Control and Prevention (CDC) in particular was suggested in order to ensure rapid dissemination of crucial information. Some specific priorities for improving public communication and information policy were identified by workshop participants.

- Identify a single authority with appropriate scientific expertise who can serve as the primary spokesperson for the government during a bioterrorist emergency.

• Develop policies and public information tools for the appropriate use of vaccines and therapeutics as a bioweapons defense measure in civilian populations.

• Improve communication systems, both in terms of providing information to the public and sharing information with partners who are involved in the investigation.

• Develop ways to disseminate crucial information and response technology quickly.

Public Health Laboratory Capacity

A workshop speaker suggested that adequate response capabilities include sufficient laboratory resources for environmental detection and clinical diagnosis. As has been shown with recent events, harnessing these resources in the midst of a crisis can be very difficult. Advance planning is essential. Increased laboratory capacity is needed at all levels, including local, state, and federal. The roles of the different levels of laboratories (i.e., A, B, C, and D) that make up the Laboratory Response Network need to be clarified. For example, CDC laboratories were stretched to the limit with recent events, building a temporary level A laboratory to do ground-level screening when such work might have been delegated to other established level A facilities.

Workshop speakers called for more serious consideration of the extent to which the public health laboratory response network can and should interact with other laboratory systems around the country, especially veterinary diagnostic labs and particularly the National Veterinary Services Laboratory, which is generally considered to be the CDC of the veterinary world.

Several workshop participants noted a lack of regulatory oversight of products being marketed for detection as a growing concern because of the unnecessary increased laboratory workload that stems from false positives. Other laboratory capacity issues that were identified by workshop participants included specimen transport; worker safety; security; information management; and having nearby, accessible chemical expertise to help deal with cases where a white powder, for example, is not necessarily biological. To address such issues the following measures were suggested by individuals during roundtable discussions:

• Increase laboratory capacity, aerosol biology and non-human primate testing capability in particular.

• Strengthen and clarify the roles of the various components of the laboratory response network. Establish ongoing interactions with other laboratory systems, especially the veterinary diagnostic labs and most importantly the National Veterinary Service Laboratory.

• Evaluate laboratory security issues.

• Ensure adjacent chemical and biological expertise in diagnostic and detection laboratories.

Disease Detection and Surveillance

Rapid detection of an outbreak is crucial to setting an appropriate response in motion, especially in the case of person-to-person transmission. It was noted that the single most important defensive measure that can be taken in this regard is to educate front line healthcare providers, including nurses, doctors, clinicians, and others who are in positions to detect an unusual signal and respond appropriately. It is they who are going to see that first rash, lesion or other symptom, and it is they who will sound the alarm. In order to train front line clinicians appropriately, one speaker described the need to consider the content of the educational materials that need to be provided as well as how to disseminate these materials.

Workshop participants suggested that surveillance capacity be evaluated, strengthened, and accelerated. The need for a high-speed surveillance system that connects hospitals, emergency rooms, laboratories, and public health departments was described by several presenters. The technology is available, and there are many different types of surveillance systems currently being used. But none of these systems in and of itself is an ideal choice for broad-based application. It is likely that our best strategy will be to combine and integrate various components of each of these systems.

A workshop presenter described the need for capacity to sequence microorganisms rapidly in order to identify foreign genetic elements and determine if the outbreak is naturally occurring or intentional and to identify where and when the infectious pathogen originated. Another presenter noted that although the development of rapid diagnostics is generally moving forward, new products and methods usually do not make it into public health labs because of financial and other constraints.

There was a call from several participants to establish a standard protocol for specimen collection so that unidentified agents are still viable after having been transported as well the need to establish protocol for collecting samples from unusual places such as the inside of a computer. It was noted that radiation of mail poses an additional challenge to specimen collection and transport, because now diagnostic specimens cannot survive normal shipping. Consequently, alternative routes for medical diagnostic specimens will need to be established. Individual participants urged the following steps to action:

- Educate front line healthcare providers so that the astute laboratory clinician, nurse, or doctor who sees the first patient or wave of patients can recognize an attack early and sound the alarm.
- Develop a sustainable, standardized information-gathering database which is shared among labs and which the CDC can access.
- Develop consistency among different agencies with regards to environmental sampling; develop standardized sampling protocols.
- Develop a protocol for processing and transporting clinical aerosol samples to laboratories where they can be examined.

- Improve CDC's access to consultations with systems engineering expertise, building engineering expertise, etc., during a response.

Local Response

A strong public health infrastructure that can detect cases and deliver the appropriate therapeutics in a timely manner requires resources and organization at the community level. However, workshop presenters representing these organizations noted that there are many significant gaps in our local response capabilities. For example, delivering the stockpile to where it is needed is likely the least of our worries. Rather the challenge will be in distributing its contents. There is a need at both the state and local levels to identify emergency authorities and delegate responsibilities.

A strong local response involves not only local public health agencies, but also hospitals, the law enforcement community, and the community at large. It was observed that there is a striking disparity in public health capacity not only among states but also among jurisdictions within states.

Strengthening local and state public health agencies will require an infusion of resources, including both trained personnel and financial resources. Workshop participants emphasized that these resources should flow into an organized system. It was proposed that every state evaluate its own system, including its legal system, and implement its own plan of action for organizing and strengthening its response capabilities. A discussant added that this effort could be aided by the collection and dissemination of best practices information about what works and what does not work.

A workshop discussant described that current median size of the approximately 3,000 local health departments in the U.S. is 13 employees. Thus, most local health departments are small and of limited capacity. Part of the problem is that local health departments suffer a dismal salary structure, a situation that needs to be rectified in order to attract experienced, trained professionals. Furthermore, only about 20% of the current public health workforce has any academic training in public health. This capacity needs to be strengthened. One suggestion, for example, is to implement training programs such as the competency-based curricula developed by the Columbia University School of Public Health to meet the local emergency preparedness needs of the New York City Department of Health. The Columbia University–New York City Department of Health program is one of several CDC public health preparedness centers that links academe and public health practice.

Other suggestions for building greater local public health capacity include devising better volunteer management programs so that during a crisis the volunteers do not become a part of the problem rather than the solution; strengthening epidemiology and surveillance systems, which are typically sparse and lacking within local public health agencies; and improving relationships among local public health departments, state health departments, and national CDC laboratories.

Hospitals are already struggling with extremely limited resources and fiscal challenges. Thus the question was raised, what would they do if they suddenly have 100,000 anthrax patients or thousands of patients with botulism requiring mechanical ventilation? A speaker suggested that hospitals engage in evaluating their ability to care for mass casualties, which will require more than simply counting beds. Evaluating, revising, and implementing responsive plans should be based on probable bioterrorist attack scenarios and include the multiple components of the hospital health care system. Revised disaster plans should also ensure that the plans are not only chemical, but also biological event-appropriate. For example, patients from a biological event may be contagious and unable to be moved or require containment (unlike a response to chemical exposures). It was suggested that hospitals participate in joint planning exercises with their local and state public health agencies.

- Establish liaisons between CDC and state and local agencies to build a base for a cooperative effort.
- Encourage state and local health departments to assess their own needs and resources and to then move toward implementing these requirements.
- Strengthen local medical care surge capacity, including personnel, training, space, supplies, and equipment.
- Consider the usefulness of joint training and preparedness exercises among local public health departments and hospitals.
- Develop clear plans for how state and federal resources will be mobilized to support local agencies in response to a bioterrorist attack.
- Develop clear plans with regard to how local and community level agencies will distribute drugs, vaccinations, or other interventions that would be needed to be rapidly mobilized in a mass casualty situation.
- Enable the exchange of best practice information among local jurisdictions.

BIOTERRORISM AS AN INTERNATIONAL ISSUE

There were many references throughout the workshop to the international aspects of bioterrorism response preparedness. Many also noted that this is a very important but seemingly underdeveloped concept. Various efforts on the part of different governmental and public health agencies to discuss bioterrorism response preparedness with international partners, such as the World Health Organization were described. The need to continue these discussions and open the doors to a broader international dialogue was identified as essential by several workshop participants.

Not only is bioterrorism a potential global threat, such that a smallpox scare anywhere in the world should be considered a global public health emergency, but there is a broad range of other issues that should also be considered. These include global usage of our very limited vaccine resources; food-borne disease tracking; international surveillance (e.g., it was suggested that international surveillance officers be strategically placed in other countries; it was also suggested that we enlist the help of Department of Agriculture veterinarians in labs and embassies throughout the world); and the role of the international community in the enforcement of the biological weapons convention.

The involvement of the international community in the research and development of new vaccines and therapeutics is crucial. It was suggested, for example, that we reevaluate live vaccine research that has been conducted in Russia and China. Israeli scientists are also known to have conducted animal studies of engineered experimental vaccines, but no data are available. The possibility of conducting clinical testing of new products in disease-endemic areas needs to be seriously considered. For example, it was suggested that scientific research pursue family-specific antiviral agents that could readily be tested against viral relatives in areas where the diseases naturally occur.

The need to open international doors to build bridges and allow for a broader international dialogue regarding bioterrorism and civilian defense measures was proposed by several participants. For example, one speaker noted that it was extraordinarily unfortunate that initial information about what actually happened during the unintentional release of anthrax at the Soviet biological research complex in Sverdlovsk in 1979 was not readily available when officials first began to address the recent anthrax outbreak in the United States. Follow up by U.S. officials with the treating Russian physician from the 1979 event eventually proved useful in managing the recent attack. Another participant noted that recognizing and integrating the capacity of international scientists and facilities could prove enormously beneficial in maximizing resources and fostering responsible scientific research and use of dangerous pathogens.

Deterrence and prevention of a biological attack is another important international issue. For example, one speaker suggested that we consider a molecular forensics laboratory where we would maintain molecular fingerprints (i.e., nucleotide sequences) of all bioterrorist agents worldwide so that the origins of samples could be identified. Making it known that we have the means to identify perpetrators could serve as a form of deterrence. Even if the perpetrators were not associated with the lab of origin, having identified the lab would at least provide a starting point for the investigation. However, other participants expressed concern that the complex molecular genealogies of these pathogens, combined with what would inevitably be an incomplete database, would make this kind of effort far too confusing.

In addressing the global aspect of bioterrorism, these two issues stood out as important to many participants:

• Expand existing partnerships and develop new partnerships with former Soviet scientists who were once part of the bioweapons program but are now under- or unemployed.

• Develop targeted strategies related to disease surveillance and vaccine and drug development that include both the international communities resources and concerns.

THE ROLE OF THE INTELLIGENCE COMMUNITY

At several points during the course of the discussion, reference was made to the involvement of the intelligence community in bioterrorism preparedness. Of utmost concern: *to what extent will prioritization require input from the intelligence community with regard to identifying which agents are most likely to be used in a bioterrorist attack and therefore which specific defensive measures should be strengthened?* The general impression is that intelligence information could appropriately direct decision makers towards certain preventive strategies, including a targeted research agenda, against the most likely bioterrorist threats.

However, it is unclear how this collaboration with the intelligence community should be initiated. Several participants argued not to wait for that information before making critical decisions about allocating resources. Some discussants observed that many people who have been privy to this kind of information would likely agree that while the information is extremely useful, it is not as useful to the public health community as one would expect. On the other hand another participant noted, if the former Soviet Union's bioweapons program involved thirty agents and we are only dealing with a list of five or six, there is a large gap between what could happen and for what we are prepared.

The question was also raised, would it be possible to clear a select group of people to access the Department of Defense's classified library of our former offensive biological warfare program? Perhaps having the kind of firsthand knowledge that one would acquire by reading through that material would help us understand some of the threats that we face, such as an aerosolized attack. In that library there is likely a great deal of information about aerosol biology, a subject about which we have very little current knowledge.

PUBLIC AWARENESS AND CIVILIAN BIODEFENSE

The consequences of a bioterrorist attack extend far beyond public health. A bioterrorist event invokes terror and panic—psychological effects that also have serious economic implications. Indeed, one participant judged that one of the most successful elements of the September 11 terrorist act was the unforeseen damage to the airline, transportation, tourism, and restaurant industries.

Public awareness entails more than risk communication during a crisis. It is also a critical part of the more general bioterrorist defense capacity-building effort. Now that the recent anthrax release has entered a recovery stage, several workshop participants suggested that it is essential that we continue educating the public about the risks of bioterrorism. There is concern that many people do not realize how much damage a bioterrorist attack larger than what we have seen to date could inflict. In fact, we may be doing a disservice to the American public by not finding a way to educate them in a non-alarmist fashion about the serious nature of this problem.

An increased public awareness may help overcome the challenge of convincing those who allocate resources that new and substantial resources are needed at all levels of capacity-building, from multi-sector collaborative vaccine production to informed local first responders. Several workshop participants expressed the belief that increased public awareness will indirectly send the message to Congress that this is a serious issue that demands immediate attention. Currently, it is unclear whether policymakers and those who are in positions to lead the effort to fill the gaps in our response capabilities have fully realized the reality of what looms ahead.

It was described by one participant that a major challenge to public awareness, however, is the chance that this issue could diminish in importance if we are fortunate enough not to experience another attack in the near future. As such, it is crucial that we keep this issue in front of the nation and that we continue to develop our response to it.

There was much discussion about whether public awareness could somehow be manifest as civilian biodefense and whether civilian biodefense could someday serve to empower local communities and decentralize the response to an attack. That is, on the one hand, we do everything that we need to do to strengthen the infrastructure of public health. But on the other hand, civilians could be educated about what they can do as individuals in terms of protective measures. With the awareness and proper training, civilian biodefense could become an important part of the local response. Indeed, as one workshop participant envisioned, is it not imaginable that in the far, far future we might be able to treat ourselves?

Adel Mahmoud, M.D., Ph.D.
Chair, Forum on Emerging Infections
President, Merck Vaccines

Stanley Lemon, M.D.
Vice-Chair, Forum on Emerging Infections
Dean of Medicine, University of Texas, Galveston

1

Introduction

FRAMING THE DEBATE WITH REAL-TIME CONSIDERATIONS

Bioterrorism is no longer a hypothetical event. A bioterrorist attack has occurred and could occur again at any time. The recent anthrax attacks not only caused five tragic deaths but significantly altered governmental operations in Washington, D.C., and substantially impacted our mail system. Importantly, we now have a window of opportunity to examine how we responded to that initial first hit and ask what must be done to better prepare for another attack. There is much concern that if we do not focus now with a clear framework for action, we will have lost a critical opportunity to capture the often short attention span of the nation.

We cannot assume that either the public or the policy makers truly understand the threat that still looms before us. The recent anthrax attack was as close to a traditional HAZMAT type of event as a biological event could be in terms of a defined source; teams could arrive at the site, define a perimeter, and identify who needed care. However, there are many potential biological scenarios that could unfold in very different ways that would require different strategies and investments. For example, there are many imaginable scenarios in which we would not know who had been exposed, nor would we even recognize the attack until cases started appearing in health care centers and hospitals across the country.

From a political perspective, since September 11, our awareness of our vulnerability to a bioterrorist attack is much greater than it was just three years ago, when hearings began with the Health, Education, Labor, and Pension Committee on whether we were vulnerable to such an assault. Today, we are much more aware of the holes and gaps in our system. The recently introduced Bioterrorism

Preparedness Act of 2001 is an attempt to fill those gaps. The leadership in Congress and the President of the United States are both committed to addressing bioterrorism in an appropriate, mature, and sophisticated way. However, it is imperative that experts in the field continue to emphasize to political leaders that the problem is much more complex than simply stockpiling. We must communicate in a way that is educational but not alarmist.

In order to better prepare, we must evaluate our financial support for this effort. There is not nearly enough money to strengthen across-the-board basic public health services. However, it is possible to strengthen certain components of the public health infrastructure. Thus, we need to prioritize what can and should be done.

Although the military has done an extraordinarily good job over the decades of preparing and planning for biowarfare, biowarfare is very different from bioterrorism. The issue is much more complex for the civilian population. For example, the civilian population is more diverse in terms of age and health which poses unique challenges in terms of which vaccines and antibiotics are most appropriate for different populations (i.e., children, immunocompromised individuals, geriatric patients, etc.). In fact, it was suggested that we may even need to pursue an antiviral approach to treating smallpox, given the risk of vaccinia in such a diverse civilian population.

In our effort to build up our biodefense arsenal, we must decide whether it is wiser to develop vaccines and therapeutics that target specific agents or a much broader spectrum of antivirals and antibiotics that can be used more generally. Unfortunately, we do not know enough about the immune system response to develop broad-spectrum agents. Does this mean that, given the increasing accessibility of the technology and knowledge needed for bioengineering drug- and vaccine-resistant microbial strains, we must continually develop more and more specific antiterrorist agents? One option is to continue research on broad-based agents, while in the meantime continuing to develop target-specific agents.

It was suggested that the very small fraction of the National Institutes of Health's (NIH) total budget that is spent on bioterrorism be reevaluated. Some of the ground-breaking scientific findings that were presented during this workshop, for example on the mechanisms of the pathogenesis of anthrax, is testament to the long-term benefits that can result from basic research.

Because of limited national capabilities, especially with regard to containment conditions required for efficacy studies with aerosolized pathogens, research questions must be prioritized. We must lay out a clear research agenda and invest appropriately to pursue that agenda in order to build the knowledge base that is necessary for developing better drugs and vaccines. The scientific community must mobilize to help reduce real risks in a way that will not be overly cumbersome to legitimate science.

Finally, we must evaluate and strengthen the public health infrastructure to ensure that it is fully capable of rapidly delivering countermeasures in the event of another attack. Again, several priorities for action that were identified during this session of the workshop were reiterated during the roundtable discussion and are summarized in the Summary and Assessment.

Several workshop participants expressed that the level of cooperative effort under such great stress and tremendous pressure during recent events has been very heartening. However, it also became clear that there is a strong need for more direction and coordination. Indeed, this has led to the recent establishment of the Office of Public Health Preparedness (OPHP) , which will coordinate Department of Human and Health Services (DHHS) efforts in bioterrorism and bioterrorism preparedness. This new office will be the primary liaison with the Office of Homeland Security, the Department of Defense (DoD), the Department of Veterans Affairs (VA), the Federal Emergency Management Agency (FEMA), intelligence, and other governmental sectors that play a role in the national security team. The goals of the office are to address a broad range of issues, from expediting smallpox vaccine production to funding hospital planning programs.

THE ROLE OF RESEARCH IN COUNTERING BIOTERRORISM

Anthony S. Fauci,[*] M.D.
Director
National Institute for Allergy and Infectious Diseases

Civilian biodefense preparedness requires a multifaceted and comprehensive approach within DHHS, involving first and foremost the Centers for Disease Control and Prevention (CDC). Other involved agencies include NIH, which plays an important role in basic research and developing medical interventions; the Food and Drug Administration (FDA), which plays an important role in the regulatory approval of vaccines, therapeutics, and diagnostics; and the Office of Emergency Preparedness (OEP), which is responsible for mobilizing resources to coordinate state and local responses.

The most likely bioterrorist agents are the Category A agents which include smallpox, anthrax, plague, botulinum toxin, tularemia, and the viral hemorrhagic fevers. Given a limited resource pool, we must prioritize what we can and should do regarding therapies, diagnostics, and prevention for each of these agents.

NIH's total bioterrorism research funding from 1998 through 2002 is shown in Figure 1-1. Although there have been considerable increases over this timeframe, as indicated in the piechart, only 0.4% of the entire NIH budget is spent

[*] This statement reflects the professional view of the author and should not be construed as an official position of the National Institute for Allergy and Infectious Diseases.

on bioterrorism. This will expand substantially. As shown in Figure 1-2, most of the $93 million allotted to bioterrorism research funding in 2002 is for vaccine (53.2%) and basic (38.5%) research. The remainder goes toward diagnostics (1.3%) and antibiotic/antiviral (7.0%) research.

From a medical and biomedical standpoint, the military has done an extraordinarily good job over the past decades in preparing and planning for biowarfare. However, there are several important differences between biowarfare and bioterrorism. In terms of protection, the military thinks primarily in terms of the tactical and strategic use of bioweapons against them. Protection of the civilian population is a much more complex issue, the components of which are not considered front-burner issues for the military, nor should they be. First, the civilian population is significantly more diverse than the military population in terms of age and health and, as such, poses unique challenges such as knowing which vaccinations and antibiotics should be administered to children, pregnant women, the aged, and individuals with medical conditions. Second, military preparedness emphasizes vaccine protection; however, there are many more bioterrorist agents than there are bioweapons, and it is neither feasible nor desirable to vaccinate the entire civilian population against all microbes on every list. Civilian attacks will be sudden and unexpected, requiring rapid diagnostics and antimicrobial treatments.

Vaccines need to be developed for all groups of civilians, not just healthy young men and women between the ages of 18 and 40 years (i.e., the bulk of the military). This will require perfecting those vaccines that already exist and developing new vaccines. For example, DoD is collaborating with NIH to develop a recombinant protective antigen as the immunogen for a new anthrax vaccine. Other ongoing studies are addressing the development of a preventive vaccine for Ebola.

Smallpox vaccine research is based on a three-pronged plan: immediate, intermediate, and long-term. For immediate use, dilutional studies are being conducted to evaluate whether diluted vaccines can be used to "stretch" the current smallpox vaccine stockpile. Preliminary results look gratifying, and it appears that the stock is in fact quite potent. For intermediate use, we are negotiating a second generation of cell culture-based vaccines. The long-term research goal is to develop a better third-generation smallpox vaccine that has fewer side effects and can be used to vaccinate everyone. The well-known toxicities of the smallpox vaccine are described in Table 1-1.

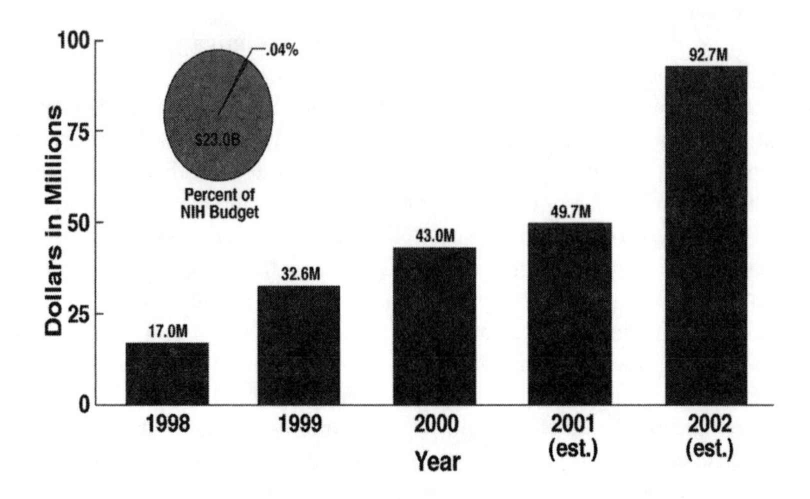

FIGURE 1-1 NIH Bioterrorism Research Funding, FY 1998–2002

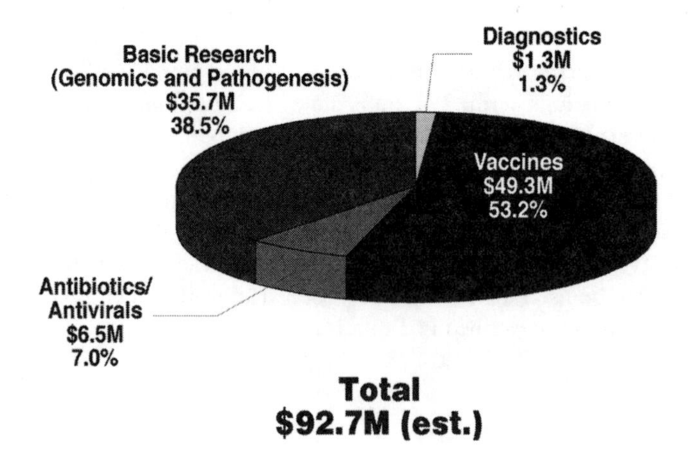

FIGURE 1-2 NIH Bioterrorism Research Funding FY 2002.

Vaccination Status	Estimated No. of Vaccinations	No. of Cases						
		Postvaccinial Encephalitis	Progressive Vaccinia	Eczema Vaccinatum	Generalized Vaccinia	Accidental Infection	Other	Total
Primary vaccination	5,594,00	16(4)	5(2)	58	131	142	66	**418**
Revaccination	8,574,00	0	6(2)	8	10	7	9	**40**
Contacts	–	0	0	60(1)	2	44	8	**114**
Total	**14,168,000**	**16(4)**	**11(4)**	**126(1)**	**143**	**193**	**83**	**572**

TABLE 1-1 Complications of smallpox vaccination, United States, 1968
SOURCE: JAMA 1999; 281:2127.

Creative techniques can be used to develop a chimeric vaccine from an already existent vaccine. For example, genes of the West Nile virus are being inserted into an attenuated Yellow Fever virus vaccine to create a "chimeric" West Nile virus vaccine. Theoretically, this can be done with any microbe for which we have identified and cloned the appropriate genes. To this end, we are collaborating with the Department of Energy (DOE) in cloning all the important pathogenic microbes. Selected examples of dangerous pathogens whose genome we have sequenced or are in the process of sequencing include: *Bacillus anthracis* (Anthrax), *Brucella suis* (Brucellosis), *Burkholderia mallei* (Glanders), *Clostridium perfringes* (Epsilon toxin), *Coxiella burnetii* (Q fever), *Staphylococcus aureus* (Enterotoxin B), *Yersinia pestis* (Plague), *Variola major* (Smallpox), and *Vibrio cholerae* (Cholera).

Basic research can be a tough sell, especially when people want immediate gratification. True, we need immediate gratification because of the threat of an immediate bioterrorist attack, but we also need to prepare for the long term. The extraordinary and elegant work that John Collier and his colleagues have done on the mechanisms of the pathogenesis of anthrax has revealed multiple targets for intervention and serves as an excellent example of the long-term benefits that stem from basic research. We must continue investing in these types of studies if we, as a scientific research community, are truly going to address adequately the long-term threat of bioterrorism.

The goal of diagnostics research is to apply the available technologies in the development of rapid, sensitive, easy-to-use tools that can be used to identify cases in civilian settings and assist in case management. DoD has already begun to address this issue. But because molecular biology is such an important component of diagnostics, DoD needs to collaborate with agencies that are conducting relevant molecular biology research. Such collaboration also serves as a way to learn from each other's experiences and even mistakes.

There are many existing antimicrobial agents, such as cidofovir, that could be screened for their activity against potential bioterrorism agents. Cidofovir was originally developed for the treatment of cytomegalovirus (CMV) infections in HIV-infected individuals and has shown to be highly effective against a number of pox viruses in an animal model. The goal of antimicrobial research, however, is not just to identify new therapies but also to determine how they should be used in diverse populations. Pediatric populations, for example, pose a major challenge. Many drugs are not used because of their unknown or adverse effects in children.

In conclusion, we have been facing emerging and re-emerging diseases throughout history (see Figure 1-3). From an infectious disease perspective, the only difference between bioterrorism and any of these other naturally occurring diseases is that bioterrorism is deliberate. Thus, if we are to appropriately address this issue in a sustained way, we must also make a sustained effort toward addressing emerging and re-emerging diseases in general.

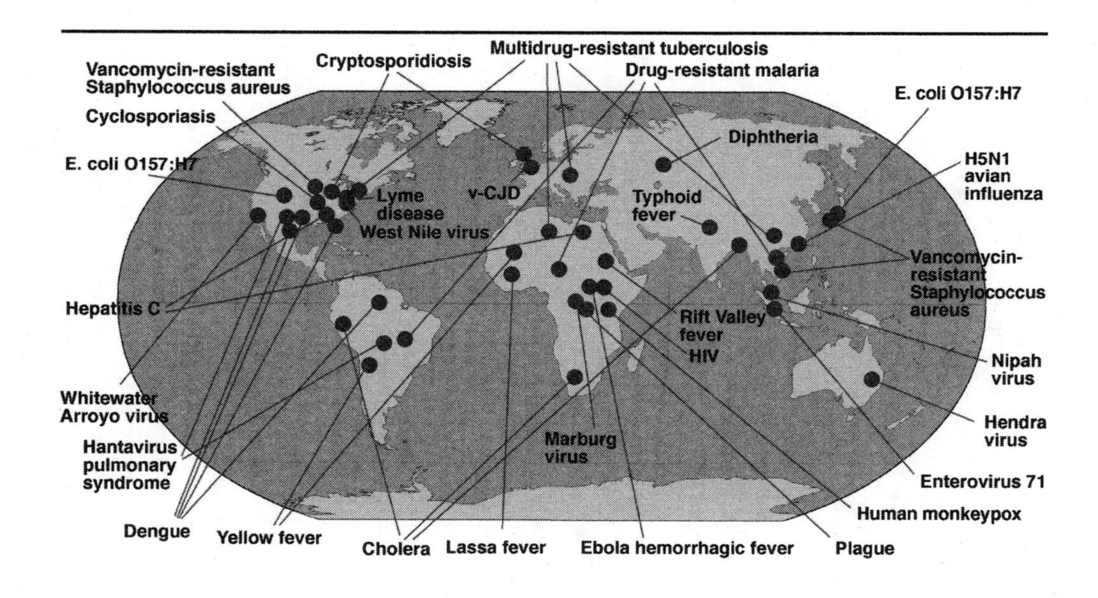

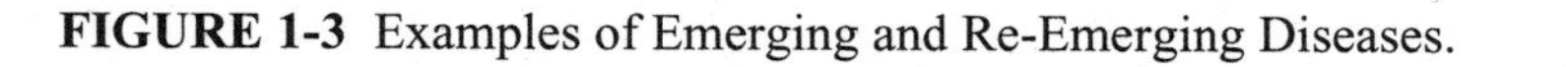
FIGURE 1-3 Examples of Emerging and Re-Emerging Diseases.

THE POLITICAL PERSPECTIVE OF THE BIOTERRORISM THREAT

William Frist, M.D.
United States Senator from Tennessee
United States Congress

In the public eye and public image, September 11 and the ensuing several weeks have been a benchmark for future generations. I hope that they will say that we may not have had all the answers, but at least we responded appropriately and moved forward. This challenge provides the opportunity to bring out the very best of our universities, our academies, the U.S. Congress, our nation's government, pharmaceutical companies, the public sector, and the partnerships within and among all of these various sectors of society.

Three years ago, hearings began within the Health, Education, Labor, and Pension Committee on whether we were vulnerable to a biological assault. The answer was yes, it was very clear that there was a threat and that we were highly vulnerable. The risk had risen as a result of the progression of science, the end of the Cold War, the lack of balance and counterbalance in power, the increase in terrorist activity, and many other dynamics. The question then became, when would such an event occur and in what shape and in what form?

As the chair and ranking member of the Subcommittee of Public Health, I worked with Senator Kennedy to write the "Public Health Threats and Emergencies Act of 2000." The bill had eleven co-sponsors, was passed by unanimous consent, and was signed by the President of the United States one year ago. It has provided the framework for prevention, preparedness, and consequence management that Senator Kennedy and myself are building on today.

Now, since September 11, our awareness of our vulnerability to a bioterrorist attack is obviously much greater. It has been spelled out, demarcated, and underlined. We are much more aware of what the holes and gaps are in our system. We were aware of the potential holes and gaps before September 11; now, we have even more proof. We may not know the perpetrators of recent events, but there is increasing hard evidence that formal efforts are being made by many countries to acquire biological weapons. We also know that there is a religious duty among terrorists to use germs in a way that can terrorize a nation or the world.

It is imperative that experts in this field continue to communicate with policymakers in a way that is educational but not alarmist. People need to be informed about what the appropriate response is at the local, state, and national levels. For example, the response across much of the country is that everything will be okay if we just make enough vaccines. This response needs to be altered to reflect the advantage of vaccines as well as the need to examine cost-effective prevention and preparedness activities. Because "public health infrastructure" is an unfamiliar concept to most political leaders, we need to figure out how to better articulate our plans for educating people in government about it.

Last week, we introduced a bill—the "Bioterrorism Preparedness Act of 2001"—written by Senator Kennedy and myself (see Appendix D). The bill attempts to build on the already existing framework, fill the gaps, and better prepare the nation, government, and society. It provides a framework that will keep the nation focused as we fill these gaps. It includes recommendations from the President of the United States regarding improving the National Pharmaceutical Stockpile including an authorization for a certain amount of money. However, it does not commit the money; that is a separate issue that will be playing out over the next several weeks. It addresses research and development of new vaccines and treatments. To a certain extent, it also addresses training initiatives, outreach, response capabilities, and epidemiologic capacity. It supports core capacities for our laboratories. It discusses the dual purpose of investing in preparedness for a bioterrorist attack, i.e., that this investment will spill over into basic public health needs that physicians and families deal with every day, whether it be the flu or other infections. The bill also addresses food and agricultural safety.

When that letter was sent to Senator Daschle's office, during the following four days, we witnessed an outbreak. Initially, things were under control. This despite the fact that we didn't know much about anthrax as a bioterrorist agent. We were able to conclude that fewer than thirty people had been exposed, and antibiotics were distributed as needed. Then, when we were made aware that somebody outside of the Hart Building had possibly been exposed to anthrax spores and developed inhalational anthrax from a piece of mail, there suddenly was a chance that our postal system would be shut down, just like the airlines had been shut down and our transportation system turned on its end on September 11. More people were admitted to the hospital; more people died; and our laboratories were stressed to their limit. We saw firsthand what our response was and what it should have been. If we pull together and form partnerships, the communication and laboratory capacity problems that we witnessed can be addressed in a positive manner.

I, Senator Kennedy, our leadership in Congress, and the President of the United States—we are all committed to addressing bioterrorism in an appropriate, mature, and sophisticated way. We would all like to be able to go back to our districts, families, and homes and say that we are prepared—not underprepared, but prepared—in the event of any future bioterrorist attack.

UPDATE ON THE IMPLICATIONS OF ANTHRAX BIOTERRORISM

James M. Hughes,* M.D.
Assistant Surgeon General and Director
National Center for Infectious Diseases
Centers for Disease Control and Prevention

The recent anthrax events represent an unprecedented biological attack on our nation. On October 3, the CDC received an initial report of the index case in Florida, a 63-year old male photo editor employed by American Media, Incorporated. His illness started on September 30 and was characterized by fever and an altered mental status. He was admitted on October 2 and seen by an infectious disease clinician, Dr. Larry Bush, who was very concerned about him. A lumbar puncture was performed and blood and cerebral spinal fluid (CSF) cultures sent to a local clinical laboratory. The laboratory very rapidly isolated a suspicious organism and promptly referred it to a member of the Laboratory Response Network (LRN), a branch of the state public health laboratory located in Jacksonville. The diagnosis was confirmed by the CDC on October 4.

Shortly thereafter, the CDC was notified of a 38-year old woman employed by NBC and who handled mail sent to Tom Brokaw. Onset of her illness, which was characterized by a typical skin lesion, occurred on September 25. On October 12, the diagnosis was confirmed at the CDC using immunohistochemical staining. Although it was not initially clear, it rapidly became evident that the individual had handled a letter that contained a suspicious powder.

To date, there have been twenty-two cases, including eighteen confirmed and four suspected. A confirmed case is defined as a clinically compatible illness confirmed either by isolation of the organism or other evidence based on two supportive laboratory tests. A suspected case is a clinically compatible illness that is either linked to a confirmed environmental exposure or supported by one supportive laboratory test.

Importantly, the epidemic curve for the first two phases of the outbreak is bimodal. The first cluster cases are associated with letters mailed from Trenton, New Jersey, on September 18. A larger cluster of cases followed mailings of letters to Senator Daschle and Senator Leahy. It remains to be seen whether the last two inhalational cases represent a third wave.

There have been a total of six cases in New Jersey, including four cutaneous and two inhalational. Six of those cases involved mail handlers, the seventh a bookkeeper. Fortunately, there have been no deaths. As part of the exposure assessment, more than 1,200 nasal swabs were collected, none of which were

* This statement reflects the professional view of the author and should not be construed as an official position of the Centers for Disease Control and Prevention or the U.S. Department of Health and Human Services.

positive. This assessment led to a clear identification of widespread contamination in the Hamilton mail processing facility, where defined groups have received antimicrobial prophylaxis.

There have been five inhalational cases in Washington, D.C., all in mail handlers. There have been two deaths. As part of the exposure assessment, nasal swabbing was conducted in the Hart Building soon after Senator Daschle's office received the letter. Twenty-eight nasal swabs were positive; those individuals are receiving antimicrobial prophylaxis and are being closely monitored. The environmental assessment in Washington, D.C., is ongoing.

In Connecticut, there has been one inhalational case which was fatal. It is not clear how the individual acquired her infection. Many cultures have been taken and many more are in progress. Antimicrobial prophylaxis has been administered to groups judged to be at increased risk, pending the results of the ongoing investigation.

Five geographic areas—Palm Beach County, Florida, Washington, D.C., Trenton, NJ, New York City, and Oxford, CT—have been the major focus of CDC assistance to state and local health departments, in collaboration with the Federal Bureau of Investigation (FBI) and others. But this is not just a state or local problem. It is a national problem and has overwhelmed public health laboratories in all fifty states.

There are several key issues that need to be addressed, whether the threat is anthrax, smallpox, plague, tularemia, botulism, or any other bioterrorism candidate:

- Rapid identification of the source and routes of exposure.
- Consideration of the possibility of new modes of transmission, as is being done now in New York City and Connecticut.
- Post-exposure prophylaxis issues related to effectiveness, adherence, and safety.
- Decontamination strategies.
- Detection and differential diagnosis (e.g., the increasing number of alerts over the last few weeks to possible smallpox cases highlights the need for adequate varicella diagnostic capacity).
- Optimal therapy.
- The research agenda.
- Local, state, and national capacity and preparedness planning.
- Communications.
- Partnerships.

These issues need to be addressed immediately.

FRAMING THE ISSUES

Edward M. Eitzen, Jr.,[*] M.D., M.P.H.
Commander, United States Army Medical Research Institute of Infectious Diseases

I am very honored to be part of this opening panel today, especially considering the fact that the other panelists are all leaders in the fields of government, infectious diseases, public health, and epidemiology. I hope that I will have a few points to add to the discussion from the perspective of the Army's biological medical defense program at the United States Army Medical Research and Materiel Command (USAMRMC) under MG John Parker, and at the United States Army Medical Research Institute of Infectious Diseases (USAMRIID).

The hypothetical has become real when considering biological terrorism. And although we collectively have been able to deal with the threats we have faced recently with the anthrax mail attacks, there are certainly areas where we can improve our ability to respond. I would like to outline a few personal thoughts in that regard.

Laboratory Diagnostic Capacity

The first area is laboratory diagnostic capacity. USAMRIID, along with CDC, is one of the two level D laboratories (see Appendix K: Glossary and Acronyms) in the national LRN. In that regard, since September 11th, USAMRIID has processed and analyzed over 8,200 samples for multiple threat agents including anthrax. We have run over 30,000 individual assays. Volume of samples has ranged from 20 per day to over 700 per day at peak. To say we have been running at capacity or over capacity would be an understatement. Most of our samples have been environmental as opposed to clinical. It is my perception that the nation needs considerably more diagnostic capacity, and the LRN needs to work better. The A, B, and C laboratories should be able to handle most samples, and the D laboratories should be used for confirmation and difficult or high priority analyses. If the D laboratories are going to function at this level of capacity, then significantly greater resources are needed. The average funding for USAMRIID's level D Special Pathogens laboratories over the past 5 years has been only about $750,000 per year. Our core diagnostics research missions have been essentially put on hold during this crisis.

---

[*] This statement reflects the professional view of the author and should not be construed as an official position of the United States Army Medical Research Institute of Infectious Diseases or the U.S. Army Medical Research and Material Command.

Research Capacity

More research capacity is needed. The choke point for biological defense research is the ability to perform challenge studies for efficacy with aerosolized pathogens or toxins in animal models; these studies normally require containment conditions to perform them safely. There are only a few organizations in the United States who can do this type of work, and all of them, including USAMRIID, have limited capacity. The large anthrax post-exposure antibiotic study which was performed in non-human primates (NHPs) at USAMRIID in 1990 and 1991 (and is the basis for current treatment regimens) took over 70 people to perform, took several months, cost nearly a million dollars, and was very labor intensive—all other bacteriology research essentially stopped. Laboratory space, aerobiology capability, people, funding, and time are all issues. USAMRIID's total yearly budget is about 50 million dollars, and one half to two thirds of this goes for maintenance and upkeep, and for salaries. That doesn't leave a lot of excess capacity. As one of the key national research assets in this area of expertise, this speaks to a national level issue in terms of research capacity. There are research questions that need to be answered in the short, middle, and long term, and a need to prioritize those questions nationally due to the limited capabilities.

Security

Preventing further attacks must be a high priority. Protecting air supplies of key facilities, and other assets such as subways and metro systems would seem to me to be very important when we have a perpetrator or perpetrators who have shown the capability to create a concentrated, pure preparation of aerosolizable anthrax spores.

Knowledge Assets

The expertise on issues surrounding terrorist use of biological agents is limited in this country, and recent events have shown that there are some gaps in our knowledge. We need to do what we can to capture and augment the expertise that we have in a defensive direction. But we have to be able to do realistic threat assessment. There are some constraints on what is considered appropriate in the context of a defensive biological program.

Education and Training

Education of our healthcare providers is probably the most important defensive measure we can take, so that the "astute clinician" can recognize the medical consequences of an attack early, and sound the alarm. USAMRIID has put on an Office of the Army Surgeon General (OTSG) sponsored satellite distance learning course in concert with the FDA and CDC partners for the past four years, a course which has educated over 52,000 military and civilian healthcare providers at a cost of only $55 per student. I am happy to announce that on Wednesday through Friday this week, in partnership with the VA's Emergency Management Group and U.S. Army Medical Research Institute of Chemical Diseases, we will put on the fifth course in this series from the FDA studios in Gaithersburg, MD. This program will give physicians and nurses throughout the United States the tools they need to recognize, treat, and sound the appropriate alarm if they see one of these bioterrorism related diseases. We need more education like this for our biological first responders—who are not the classical first responders (EMTs and paramedics), but rather are doctors and nurses in emergency departments and in primary care offices and settings.

As we move forward and face these new threats, there are many issues we must face together. One of the very gratifying aspects to me of the last two months has been the fantastic cooperation and great working relationships with our colleagues in other federal agencies (DHHS, CDC, FBI, Environmental Protection Agency [EPA], U.S. Postal Service) and state and local agencies as well. The level of cooperative effort under great stress and tremendous pressures has been very heartening to me personally. The relationships that have been developed over the last several years as we prepared have stood us well in this time of crisis. I am confident that with this spirit of cooperative effort that we can face the biological threats of the future, and protect and defend our nation against these threats.

FRAMING THE DEBATE: APPLYING THE LESSONS LEARNED

Michael T. Osterholm, Ph.D., M.P.H.
Director, Center for Infectious Disease Research and Policy and Professor, School of Public Health University of Minnesota

Very few of us should have been surprised by what happened on September 11. We were definitely shocked by the manner in which this catastrophic terrorism event happened on our shore, but the fact that it happened should not have surprised us. Many in our society understood and recognized that it was only a matter of time before terrorists would attack our Homeland; nonetheless most were in denial that it would actually happen. Understanding this phenomenon will be important as we move forward to better understand and prepare for what lies ahead.

The post-September 11 anthrax situation was the first real test of our country's response system to a potential bioterrorism attack. Many citizens consider the post-September 11 anthrax situation just a "scare", not an anthrax crisis.

While this limited hit (i.e., in terms of infectious disease) resulted in "only" five tragic deaths, it did alter governmental operations in Washington, D.C., substantially impacted our mail system, and caused fear and panic among much of our country. Given these facts, I believe this event should really be classified as a tragic dry run.

What are some of the lessons learned? Following September 11, the national media tended to congratulate the response by the emergency response personnel of New York City to the World Trade Center Towers disaster. Indeed, it was a very heroic effort. But there was a misunderstanding with regard to the conclusion that the emergency medical services of New York City were prepared to handle this situation. I would offer that this was not an adequate test of what could potentially happen during a terrorist attack involving biological agents. There were less than 4,000 people in the World Trade Center area that presented to hospitals for any type of medical treatment. If this had been a smallpox or anthrax situation resulting in tens of thousands of victims, many needing hospital beds, this same system would not have been able to respond.

It does us no good to speculate as to who were the perpetrator(s) of the anthrax situation. If a foreign entity is involved, this simply supports the notion that there are groups out there that can provide and deliver this sophisticated bioterrorism material. The challenge may even be greater if the perpetrator(s) is a domestic terrorist, as it would indicate that there is greater expertise out there than many had recognized and understood. Regardless, someone has created a very powerful bullet. However, to date they have used a very ineffective gun. Remember, the gun was not the technological hurdle—the bullet was the hurdle. The reality is, no matter who the perpetrators are, the capability exists to take this already prepared material and use it in a much more effective gun. Unfortunately there are many ways that this bullet can be used much more effectively. Such use will result in catastrophic numbers of cases that will far exceed the casualties of the World Trade Center tragedy. Given this, we need to focus not on the anthrax scare, but on the potential future anthrax crisis.

We have a window of opportunity to thoughtfully examine how we responded to this initial attack. How did the federal, state, and local agencies respond? What can we learn from our successes, and what can we learn from where our response was inadequate? The point is not to blame or give credit, but to review what happened and learn how we can be better prepared when the other shoe drops.

It is essential that we evaluate our short and long term financial support for the public health system. I believe Americans have a very short attention span. A Lexis/Nexis search of all major English newspapers and TV news transcripts over the past few months illustrates this. From September 11 through September 14, there were only 14 news stories on public health scares or crises (see Figure 1-4). During the week of October 23 to October 30—the height of the post-September 11 anthrax events—there were 558 such stories. But last week, there were only 104, and that number is dropping precipitously (see Figure 1-5). In a similar search for anthrax stories in particular, there were six from September 11 to September 16, 1,487 from October 16 to October 22, and only 146 last week, despite the new case in Connecticut. We will need to constantly remind our elected leaders and the public that it is only with ongoing and substantial resource support, that our country's public health system will be able to rebuild and arm itself for future attacks.

Our thinking now must consider longer-term threats and consequences. Nobody should be surprised next time if a biological agent is involved and consequences are much more substantial than what our nation has tragically experienced to date.

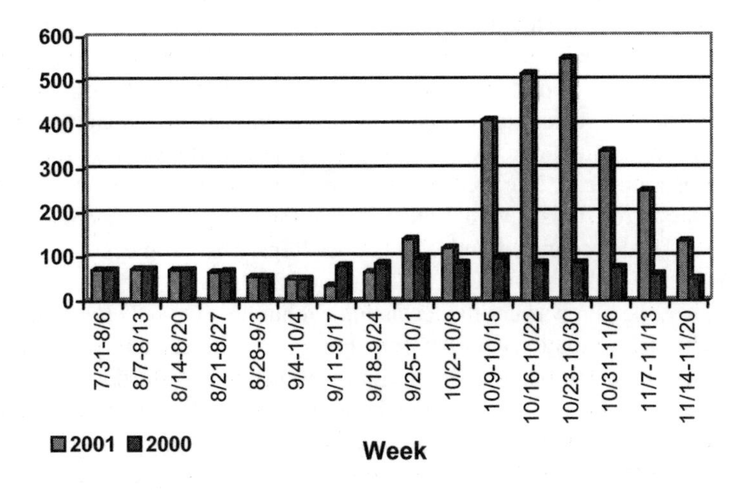

FIGURE 1-4 Newspapers, TV stories, re: public health threat, scare, crisis, 7/31-11/20, 2001 and 2000.

SOURCE: Lexis-Nexis Academic: Major world English language newspapers and TV transcripts.

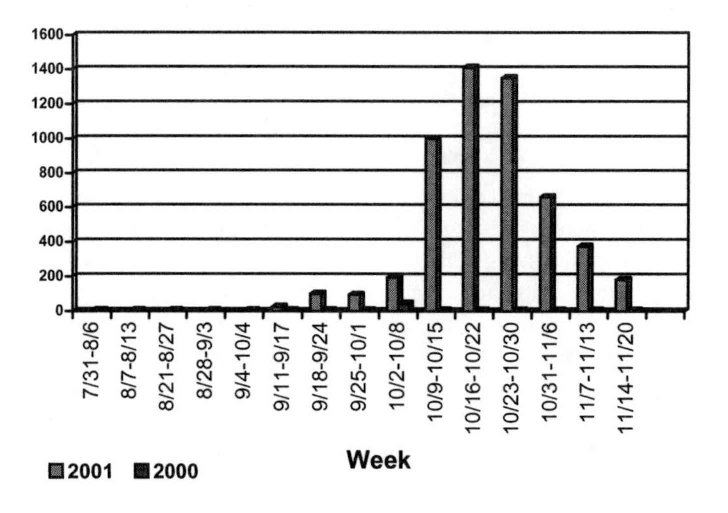

FIGURE 1-5 Newspaper, TV stories re: Anthrax, 7/31 to 11/13, 2001 and 2000.

SOURCE: Lexis-Nexis Academic: Major world English language newspapers and TV transcripts.

BUILDING CAPACITY TO PREVENT AND RESPOND TO BIOTERRORISM

Margaret A. Hamburg, M.D.
Vice President of Biological Programs
Nuclear Threat Initiative

Now is the time to define an agenda and move forward toward countering the threat of bioterrorism—to focus now with a clear framework for action. Even though our nation has experienced its first lethal bioterrorism attack, we cannot assume that the public and critical policy makers truly understand the threat that still looms before us. We need to continue to clearly define the threat. The recent anthrax attack was as close to a traditional HAZMAT type of event as a biological event could be in terms of a defined source and in the sense that teams could arrive at the site, define a perimeter, and identify who needed care. But we need to emphasize that there are many potential biological scenarios that could unfold in a very different way that would require a different focus, different strategies, and different investments. In this attack, the anthrax was delivered through the mail. But there are many other modalities that would lead to an unfolding disease epidemic with an unknown source. We would not know who had been exposed, nor would we even recognize the attack until cases started to appear in health care centers and hospitals across the country. We must continue to define and communicate to people in critical decision-making positions what needs to be done with respect to the public health infrastructure and disease surveillance so that we can rapidly detect, investigate, and respond to disease outbreaks.

We must recognize that the response will begin at the local level, thus we must ensure capacity at that level. This capacity must be augmented with clear plans for how state and federal resources will be mobilized to support local agencies. We need trained personnel, stronger laboratories, and better communication systems across all levels of government and in the private sector. We must invest in systems and activities that will be utilized as regularly as possible so that we are not testing new plans and strategies in the event of a crisis.

We need to recognize that the bioterrorism threat is embedded in a set of infectious disease concerns for which we should also be better preparing our nation. At the same time, there are some unique preparedness programs that pertain specifically to the bioterrorism threat, for example the National Pharmaceutical Stockpile. As the nation moves forward with its plan to expand the National Pharmaceutical Stockpile, our efforts should be linked with the best possible intelligence about what the real and credible threats are. Furthermore, the stockpile must be linked to a real time distribution system. We need to make much more concrete plans with regards to how we are going to distribute the drugs, vaccination, or other interventions that would need to be rapidly mobilized in a mass casualty situation involving very large numbers of individuals.

We also need to consider how to best prepare the medical care system to surge rapidly in the event of a mass casualty situation. This will involve careful advance planning, and a systematic look at regional capacities, how they can be mobilized, and how they can be augmented by state and federal resources.

In order to build our knowledge base and become a better prepared nation in both the short and long term, we need to lay out a clear research agenda and invest appropriately to pursue that agenda. This involves research and development for new drugs and vaccines; improved diagnostics for human samples; improved environmental detection capability; and basic research on how these organisms cause disease and how the human immune system responds.

We also need the type of systems research that will help us better understand the issues that have been vexing us so much over the past couple of months, such as environmental decontamination and personal protection.

Public health has always been an important form of public safety. Now it is a critical pillar in our national security framework as well. As such, public health expertise should be an important component of the new Office of Homeland Security, and a public health official should become part of the White House national security team.

Finally, we must consider prevention, of which a key element is intelligence. Recent events have led to a commitment to improve overall intelligence collection, and the scientific community must play an important role in this. Scientists are informed in critical ways, including fundamental knowledge of biological threat agents and the effects they cause, knowing what information in the scientific literature could be misused or misapplied by those who want to do harm, and insights into what is going on at various laboratories around the world. As such, scientists will be crucial to building new expertise in this complex area within the intelligence community.

The scientific medical community will also need to engage on the issue of improving biosecurity in terms of reducing access to dangerous pathogens. Steps have been taken in recent years through the select agent rule at the CDC and some of the new procedures implemented at so-called germ banks, such as the American Type Culture Collection. But the anthrax situation has demonstrated that we still don't have an adequate handle on whether dangerous pathogens are secured, who is using them, and why. The scientific community needs to mobilize now to help reduce real risks in a way that will not be overly cumbersome to legitimate science.

We must recognize that while advances in science and technology hold enormous promise for improving health, they also present many opportunities for misapplication or inadvertent harm. The Australian mousepox study is one example of an inadvertent finding that has laid out a road map for others to make an already dangerous pathogen more lethal.

Finally, we need to recognize that there is a great deal that can be done to further secure or destroy dangerous biological materials in the former Soviet Union. We need to expand existing partnerships and develop new partnerships with former Soviet scientists who were once part of the bioweapons program but are now under- or unemployed. We have an opportunity through those collaborations to address critical public health and medical issues of mutual concern and reduce the possibility of further development or spread of biological weapons.

BIOTERRORISM: COMMUNICATING AN EFFECTIVE RESPONSE

Scott R. Lillibridge,* M.D.
Special Assistant to the Secretary for National Security and Emergency Management
Office of the Secretary
Department of Health and Human Services

The DHHS bioterrorism preparedness began in earnest in FY 1999 with more than 155 cooperative grants from the CDC covering all fifty states for some component of laboratory science, surveillance, planning and preparedness, communications and information technology, and training. As we begin to examine lessons learned from the recent anthrax events, we can take stock of these

---

* This statement reflects the professional view of the author and should not be construed as an official position of the U.S. Department of Health and Human Services.

preparedness measures and imagine what might have unfolded if this infrastructure had not been in place.

Tribute should be paid to the many walks of public health, particularly to many present today who worked diligently to provide emergency services during the recent anthrax releases. Within DHHS, many accolades also go to our CDC, the Public Health Service Commissioned Corps, and other agencies within the Department that provided vital services during this act of bioterrorism. I would like to comment briefly on a few of the many lessons learned that will help shape the upcoming bioterrorism program within DHHS during the upcoming year.

First, the initial event in Florida and thereafter during the emergency, the issue of communications was arguably one of the most important areas in our public health response. The way in which we generate and communicate information during an emergency is incredibly important to effectively deal with the problem at hand and to reassure the public. It was clear that the media was able to gather information from a wide range of sources. For example, test results, from laboratories that may be initially operating outside of the federal response were moved into the public consciousness and the political arena at a remarkable speed. There was often little distinction made between a false positive, a presumptive positive, and a confirmed positive test result by the media. This greatly complicated the dissemination of processed information by health officials operating at the state and local level.

Clearly, in the future, our bioterrorism preparedness efforts will need to pay more attention to the infrastructure associated with health communications and information dissemination. A few common strategies for example will pay dividends in dealing with such an emergency from the public health perspective. First, having a state or local plan that provides for a single, consistent, authoritative spokesperson will be extremely important in addressing the public. Consideration for prepackaged infectious disease information, local spokespersons to disseminate a clear message, and the ability to rapidly investigate rumors will also be vital to tackling the communications issues associated with a fast developing epidemic response.

With respect to the recent anthrax emergency, it was clear when public health experts took a prominent role in the national news media and began to promote their message, it had a tremendous positive effect in helping the public understand what was being done on their behalf. For example, when state and local public health professionals began talking about the meaning of specific laboratory tests, information about exposure, or the need for medical prophylaxis the media was far more supportive of the response than when conflicting information from different and sometimes dubious sources were communicated. Consequently, it is essential that our public health infrastructure have a strong communications capacity that can be activated during an emergency. This will greatly assist the public health community in providing an informed perspective on science, public health, epidemic control measures, and other pertinent response issues.

A second issue to be learned from the recent anthrax attack is the importance of investing in laboratory infrastructure. It was very reassuring to know that the laboratory staff who confirmed the initial Florida case had been trained by CDC and were using standardized CDC reagents. Through the combined work of the Association of Public Health Laboratories (APHL), American Society of Microbiology (ASM), CDC, DoD, DOE and others, 81 such network laboratories are currently in place across the nation.

From the events associated with the recent anthrax releases, it is clear that a greater investment in linking these laboratories and in ensuring vital surge capacity will be needed in the future. In addition, the distinction between laboratories of the network and other laboratories outside of the network were often unimportant to the media and some response organizations. As a consequence, proper tracking of laboratory samples from various response sites became an important issue. It was clear that expanded laboratory sample tracking and coordination of results information will continue to be very important issues for future preparedness highlighting the importance of developing systems of electronic laboratory reporting. Another aspect of laboratory readiness that was not anticipated in the scope of the current response laboratory network was the need for a large environmental sampling component and some method for redistributing work throughout the network.

A third major lesson learned from recent events is the extreme importance of training front line healthcare providers, including nurses, doctors, clinicians, and infectious disease consultants in disease recognition. It was an alert clinician who initially identified many of these cases and then notified public health and law enforcement officials to trigger an investigation. Increasing our efforts to educate such providers will be extremely important to an ongoing bioterrorism preparedness effort.

A fourth lesson learned is that the public health infrastructure is strategically important to the national security of the United States. Before these events, during discussions that took place in FY 1999, the belief that the public health infrastructure needed to be enhanced in key areas was not widely shared within the interagency emergency response community. However, as the current emergency unfolded, the interagency community as well as the various other sectors of government began to view the public health infrastructure as being strategically important to the national security of the United States. I hope these beliefs will continue to drive an investment into our increasingly important public health infrastructure for some time to come.

Finally, recent events tested the adequacy of our stockpile and our ability to deploy it. The reality of deploying push packages and therapeutic prophylaxis into populations is very different from planning such activities in an office environment. During this response, it was apparent that stockpiling is more than simply a warehouse activity and many of our efforts to ready stocks work very well. Issues to be addressed in the future include, how much of the stockpile should be immediately ready to go, how much stock should be vendor-managed for delayed arrival, what drugs need to included, and how it should be best packaged. This reinforced our notion that stockpiling and the policy that directs its deployment will be an extremely important component of an overall national bioterrorism preparedness program.

In conclusion, Winston Churchill, who was involved in the dawn of chemical warfare in WWI and who also served as prime minister of England during WWII when the radiation age appeared, would probably tell us today that our investment in selected public health infrastructure and readiness for epidemics remains one of the best ways to prepare our population for the current threat of bioterrorism.

2
Assessing Our Understanding of the Threats

OVERVIEW

The focus of this session of the workshop was an assessment of known threats. Anthrax is a proven risk and of immediate concern. Smallpox is equally urgent because of its capability for person-to-person transmission and the large number of completely susceptible individuals in the United States and around the world. Presenters discussed details of the bioweapons potential and treatments available for each of these threats along with those of three other "high-priority" potential bioterrorist agents: plague, tularemia, and botulinum toxin. However, these are not the only credible bioterrorist agents out there. For example, the former Soviet Union is known to have weaponized at least thirty biological agents, including several vaccine- or drug-resistant strains.

There are many imaginable bioterrorist scenarios, but if the goal is to induce mass casualties, an aerosol attack is probably most likely. Aerosols exhibit wide-area coverage, and their small particle size allows them to deposit very deeply in the lung tissue which is where many agents, including anthrax, induce maximal damage. A large amount of agent disseminated under good meteorological conditions over a substantially sized city could have considerable downwind reach, resulting in large numbers of casualties.

Food-borne bioterrorism, which could encompass a variety of agents, must also be considered an equally likely threat. Agents that cause foodborne illness are easy to obtain from the environment and often have very low-dose requirements. Foodborne pathogens may in fact be the easiest bioterrorism agent to disseminate. In addition to public health risks, there are several important agricultural risks which were mentioned briefly but not discussed in detail during this workshop.

Anthrax

B. anthracis is a very stable organism because of its ability to sporulate. Most naturally occurring anthrax cases are cutaneous and are transmitted from agricultural exposure. The incidence of infection is unknown; most cases occur in underdeveloped countries. Throughout the world, since the late 1930s, attenuated strains have been used as live veterinary spore vaccines and have proven to be highly effective in controlling disease in domesticated animals. Since the 1950s, one of these strains has been used as a live attenuated strain in humans in countries of the former Soviet Union. The molecular pathogenesis of anthrax, including the exact target of its lethal factor, is largely unknown. However, enough is known that we can begin to predict where second-generation vaccines and various antitoxin modalities might work.

Currently, there are three types of preventative or therapeutic countermeasures against anthrax: vaccination, antibiotics, and various adjunctive anti-toxin treatments. In terms of developing new therapeutics, initial immediate efforts should be to evaluate already licensed antibiotics. Longer-term efforts should include identifying protective antigens that are effective against modified strains; developing vaccines that act more quickly and would be more useful in a post-exposure scenario; exploring the combined use of vaccines and antibiotics; and exploring new antitoxin treatments. Critical to all of these efforts is the need for a large-scale central animal testing facility.

Smallpox

Smallpox has several features that make it an attractive bioterrorist agent: it is highly stable; it is infectious by aerosol; it is highly contagious; most clinicians lack experience recognizing the disease; and, because vaccination against smallpox ceased after eradication, most of the world's population is highly susceptible to infection.

Even though a smallpox vaccine exists, there are several unresolved bioterrorism-related issues regarding smallpox vaccination:

- It is not clear which health care providers should be immunized preceding any potential outbreak versus immediately following an outbreak.
- The current supply of vaccinia immune globulin is insufficient for treating all of the expected adverse effects associated with vaccinia immunization, should all 300 million doses of cell-cultured vaccinia that are currently being produced be administered.

- The cell-cultured vaccine that is currently being produced is considered only a stop gap measure since it cannot be used in immunocompromised individuals or children.
- It is unclear whether the available vaccine would protect against aerosol exposure of the type and magnitude that would be expected in a bioterrorist event. Although monkey/monkeypox studies have shown that yes, the vaccine does provide adequate protection, it is unclear how applicable these studies are to smallpox in humans.

Currently, most research and development efforts for smallpox therapeutics are focused on antiviral drugs. Thus far, the leading candidate is cidofovir, which has been approved under an IND for treating disseminated vaccinia but has not been approved for treating smallpox. Despite its promise, however, cidofovir is unsuitable for mass casualty use. More effort needs to be directed toward other therapeutics, such as immunomodulators.

Plague

Plague—a deadly and highly contagious disease—was weaponized in the former Soviet Union for aerosol delivery and engineered for antimicrobial resistance and possibly enhanced virulence. In the WHO modeling scenario that was developed in 1970, a 50-kilogram release over a city of five million would cause about 150,000 cases, or 36,000 deaths, in the first wave. A secondary spread would cause a further 500,000 cases, or 100,000 deaths. Plague requires intensive medical and nursing support and isolation for at least the first forty-eight hours, followed by two to three weeks of slow convalescence. The hospitalization and isolation that would be required for this many people in a single city is nearly unimaginable.

Currently in the U.S., there is no available plague vaccine. The live vaccines that are sometimes used in other countries have unacceptable adverse effects. There are, however, a number of laboratories trying to develop a new generation vaccine, as well as new delivery methods. Several different types of antibiotics that can be used to treat plague are included in the national pharmaceutical stockpile. Antibiotic treatment must be instituted early during the course of infection, otherwise death occurs in three to six days.

Tularemia

Tularemia was weaponized as an aerosol both in the U.S. and the former Soviet Union where it was also engineered for vaccine-resistance. In the WHO modeling scenario of 1970, 50 kg over a city of 5 million would incapacitate 250,000 people and cause 19,000 deaths. Tularemia is highly infectious but not contagious. Treatment is similar to that for plague but more extensive, as is the post-prophylaxis to prevent relapses of disease. The tularemia vaccine is a live attenuated vaccine that was previously available as an investigational drug through DoD and is now being investigated by the Joint Vaccine Acquisition Program. However, it does not offer full protection against inhalational transmission, and it takes about fourteen days for protection to develop. The vaccine has been recommended for use in people who work routinely with the organism in the laboratory, but it is unknown whether it would be useful in first responders at high risk for exposure.

Botulinum Toxin

Botulinum toxin has several features that make it an attractive bioweapon, including its extreme potency and lethality; the ease of its production, transport and misuse; and its profound impact on its victims as well as the health care infrastructure. Like tularemia, it has a very diverse mode of transmission: it can spread through foods, beverages or as an aerosol. Botulinum toxin, of which there are seven serotypes, kills by paralytic ability and is one of the most poisonous substances known.

Although an investigational vaccine exists, immunization is really not a viable option for bioweapons defense: the vaccine is still only investigational even after ten years; its components are aging and losing potency; it only protects against toxins A, B, C, D, and E, not serotypes F and G; it is very painful to receive; it requires a booster at one year; and the use of it would deprive the recipient for life of access to medicinal botulinum toxin.

The army has developed an equine antitoxin that provides coverage against all seven toxin serotypes, but the supply is limited and the drug carries the risk of serious allergic reaction. However, equine antitoxin is inexpensive to produce and could be made in large quantities if a specialized facility were available. A human-derived botulinum antitoxin has been developed as an orphan drug but is very difficult to produce in large quantities and is of limited use because it protects against only five serotypes.

DoD is developing a recombinant vaccine which is not expected to become a licensed product, however, for at least another ten years. Researchers are also developing recombinant human antibodies as an alternative therapeutic option. Antibodies have several distinct advantages as bioweapons defense agents: they induce immediate immunity; they can be produced in unlimited quantities; and they are highly potent. In fact, an unlimited supply of human recombinant antitoxin is probably the best defensive measure against botulinum toxin.

ANTHRAX

Colonel Arthur M. Friedlander*
Senior Military Research Scientist, United States Army Medical Research Institute of Infectious Diseases

> "It has now prevailed and been recognized in this neighborhood about forty years, and notwithstanding all that has been done to prevent it, by ventilation, the use of respirators and other means, it still continues, as severe and frequent as it ever was, overclouding the life of the sorter." (Bell, 1880)

Anthrax has a long history: apocryphal accounts describe it as the fifth and sixth plagues in Exodus, when dust was cast into Pharaoh's eyes; it was the first disease for which a microbial etiology was determined by Robert Koch; and the anthrax vaccine was one of the first live vaccines, developed by Pasteur and one of the first examples of attenuation of a fully virulent organism for use as a vaccine. Physicians in the latter part of the 19th century, particularly in England, were well aware of the clinical and pathological aspects of anthrax. It is now incumbent upon a new generation of physicians to become intimately familiar with this disease.

There are several characteristics of *B. anthracis* that make it a potentially very lethal bioweapon, most importantly its stability and infectivity as an aerosol and its large footprint after aerosol release. An aerosol release of anthrax could potentially affect millions of individuals.

The organism's stability stems from its ability to sporulate. Dormant spores are estimated to have survived in some archaeological sites for hundreds of years. The spores occur in soils worldwide and infect grazing herbivores. After they enter their mammalian host, they germinate into actively replicating vegetative cells. When the mammalian host dies and its carcass is exposed to the air, the bacteria sporulate. It is unknown whether spores go through a germination-replication-sporulation cycle in the soil, or whether amplification of bacterial numbers occurs only within the host.

Under natural circumstances, humans become infected only via contact with infected animals or contaminated animal products. Anthrax is primarily a developing world disease associated with agricultural exposure causing cutaneous or gastrointestinal infection. However it emerged as "woolsorter's disease" in the industrial world in the latter half of the 19th century. With the rise of the industrial revolution, the large quantities of contaminated animal products being processed in enclosed rooms generated aerosols of anthrax spores which caused the first known cases of inhalational anthrax. Today, inhalational anthrax is extraordinarily rare.

---

* This statement reflects the professional view of the author and should not be construed as an official position of the U.S. Army Medical Research Institute of Infectious Diseases.

The incidence of all forms of anthrax is unknown because reporting is unreliable. Large anthrax outbreaks tend to occur during breakdowns of the public health structure, for example during war. Ten thousand human cases occurred in Zimbabwe during the 1970's and early 1980's; only eight of these cases were reported to be inhalational anthrax although no autopsies were performed.

B. anthracis is a large gram-positive, sporeforming, non-hemolytic, non-motile bacillus. Its known virulence factors include a polyglutamic acid capsule, which is antiphagocytic and without which the organism is attenuated; and the well-known lethal and edema toxins. Like most bacterial pathogens, the virulence determinants are encoded on plasmids. The two toxins are encoded on one plasmid, and the genes responsible for synthesis of the capsule are on a smaller plasmid. It is possible, under specific laboratory conditions, to eliminate the smaller plasmid and produce an unencapsulated, attenuated strain of bacteria that still produces both toxins. Since the late 1930's and early 1940's, the Sterne strain and others similar to it have been used throughout the world as live veterinary spore vaccines that have proven to be highly effective in controlling disease in domesticated animals. Since the 1950's, a similar strain has been used as a live attenuated vaccine for humans in the former Soviet Union. It is also possible to delete the toxin plasmid, resulting in an avirulent organism that produces only capsule.

Inhalational anthrax is characterized by lymphadenitis of the tracheobronchial and mediastinal lymph nodes and mediastinitis. On chest x-ray, the lungs are usually clear while there is usually mediastinal widening and often pleural effusions. The incubation period is usually a week or less. The initial symptoms are mild and non-specific and include fatigue, which can be very profound, headache, fever, chills, and sweats. There may or may not be a cough, usually non-productive. Because the disease is centered in the mediastinum, patients sometimes experience a sense of precordial discomfort. Abdominal pain has been prominent in some cases. As the infection progresses, symptoms include the abrupt onset of dyspnea, tachycardia, increased chest pain, and occasionally stridor. Pneumonia may occur but is usually absent. There is a rapid progression to cyanosis, shock, and death.

An early diagnosis is enormously important. The definitive criterion for establishing a diagnosis is isolation of the bacteria from blood, pleural or cerebral spinal fluid, or other tissue. Since there is no other gram-positive bacillus that causes sepsis in healthy individuals, the isolation of a bacillus from the blood should alert every clinical microbiology department to the diagnosis of anthrax. A chest x-ray or CT scan should also show the characteristically enlarged mediastinum. An outbreak would be characterized by large numbers of previously healthy people with these symptoms appearing in emergency rooms and physician offices.

Our knowledge of the gross pathogenesis of this disease is good but, as with many infectious diseases, we know very little about its molecular pathogenesis. The infectious spore enters the body either through a break in the skin, the GI tract, or by the respiratory route. In the skin and GI tract, the spore germinates locally. After entry through the respiratory tract, the spore is transported from the lung via macrophages to the regional lymph nodes. It spreads from node to node producing hemorrhagic necrosis and then extending into the mediastinum. From the lymph it spreads into the systemic circulation. About fifty percent of individuals with inhalational anthrax have evidence of meningitis. Impairment of respiratory function due to interference with lymphatic and vascular outflow associated with mediastinitis and pleural effusions is likely the primary mechanism of death.

The anthrax toxin, which causes edema and cell necrosis, probably contributes to death as well. Although the anthrax toxin is similar to many other bacterial and plant toxins, it is unusual in the sense that the functional domains reside on separate proteins. The individual proteins by themselves are inactive and have no biological function. Only when protective antigen combines with lethal factor does it constitute lethal toxin, and only when it binds with edema factor does it constitute edema toxin.

From cell culture studies, it appears that the anthrax toxins function as outlined in Figure 2-1. Protective antigen binds to a cellular receptor where a cell surface protease cleaves it, releasing a small 20 kD fragment and exposing a cryptic site on the molecule; it then forms a heptamer and subsequently binds to either edema or lethal factor. This whole complex is then internalized by receptor-mediated endocytosis into an acidic vesicle. Under conditions of low pH, the complex inserts into the membrane and, like other toxins, the enzymatic toxin components are delivered to the cytosol. Edema factor raises cyclic AMP levels to pharmacological levels, which is clearly responsible for some of its biological effects. The exact target of lethal factor, a zinc protease, remains, to date, unknown.

As follows from this model there are several potential targets for antitoxin therapeutics (see Collier's discussion of the most promising). The recent identification of a motor protein that controls sensitivity, or resistance to lethal toxin, will hopefully lead to additional antitoxin strategies. The current human vaccine is thought to act predominantly by inducing antibodies that block the binding of protective antigen to the cell surface receptor and block the binding of edema and lethal factor to the cell-bound protective antigen although the antibodies may also act on the organism itself.

It is unclear whether anthrax pathogenesis involves a cytokine cascade. If so, the recent licensure of activated protein C for use in sepsis could have important ramifications for anti-anthrax therapy. The possible role of cytokines in anthrax warrants further evaluation. The multiple studies that have tested adjunctive treatment for sepsis should be used to guide this effort.

There are many unresolved issues with regards to prophylaxis and treatment. For example, which antibiotics should be used? Do we need adjunctive treatments? What if the *B. anthracis strain* is antibiotic- or vaccine-resistant?

About ten years ago, Russian investigators reportedly produced both multidrug-resistant and vaccine-resistant *B. anthracis* strains.

In terms of therapy, there are several points to be emphasized. First, it should be remembered that antibiotics affect only the bacillus, not the spore. Thus it is possible that sufficient numbers of ungerminated spores may persist in an exposed individual after completion of a course of antibiotics, only to cause disease upon subsequent germination when antibiotics are no longer present. The conditions which govern the germination of spores *in vivo* remain obscure. Secondly, the notion that inhalational anthrax is invariably fatal once symptoms occur is likely untrue as evidenced by the survival of some of the current cases. Indeed, there is experimental evidence supporting the efficacy of late-stage intensive treatment in non-human primates showing signs of bacteremia or even mediastinitis. Lastly, there are many other antibiotics that show activity *in vitro* that may extend the therapeutic options for prophylaxis and treatment. These need to be evaluated in animal models before consideration for human use.

The Department of Health and Human Services, with input from the Department of Defense, is currently focusing on three therapeutic issues: testing licensed antibiotics that could be used to treat anthrax; developing human antibodies against the current vaccine, which has been administered to about 500,000 individuals; and assessing combined vaccine and antibiotic use. Other issues that need to be addressed include: identifying near-term, mid-term, and long-term research goals; identifying new protective antigens that are effective against modified strains; producing vaccines that work more quickly, particularly from the perspective of a post-exposure scenario; and, critical to all of these efforts, developing a large-scale central animal testing facility as evaluation of new treatments in humans will likely be extremely difficult.

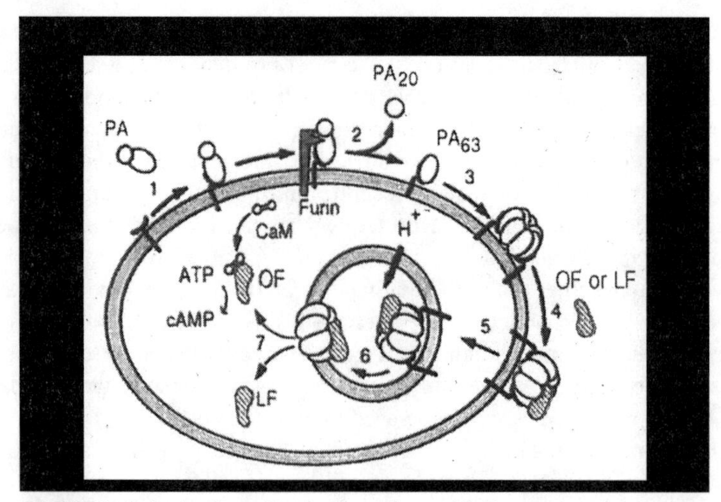

FIGURE 2-1 Anthrax Toxin Function with Protective Antigens

MEDICAL COUNTERMEASURES AGAINST THE RE-EMERGENCE OF SMALLPOX VIRUS

Peter B. Jahrling,* Ph.D.
Senior Research Scientist, United States Army Medical Research Institute of Infectious Diseases

The recent bioterrorist attacks involving anthrax have increased awareness that biological agents are truly weapons of mass destruction. Unlike anthrax, the smallpox virus is a contagious disease with fairly high rates of human to human transmission. As such, the use of smallpox as a bioterrorist agent is considered to pose an even greater threat than anthrax.Following publication of the Institute of Medicine report *Assessment of the Future Scientific Needs for Live Variola Virus* in 1999, collaborative research involving the Department of Defense (DoD) and the Department of Health and Human Services (DHHS) was initiated to address the recommendations suggested by the Report. The subject of today's presentation focuses on the recent research and our development of an animal model for Variola (smallpox) virus infection.

The desirability of animal model development is driven by the proposed Food and Drug Administration Animal Efficacy Rule, which was written to facilitate approval of countermeasures for infectious diseases, such as smallpox, which do not occur naturally in human populations. The Rule requires that pathophysiology of the animal model disease be faithful to the human disease course, and that the efficacy study endpoint must be based on reduced morbidity or mortality. Insight into the "toxemia" of human smallpox might be achieved by application of modern tools of virology and immunology to the model infection. Conventional wisdom holds that variola does not produce smallpox-like disease in any species other than humans; however, cynomolgus monkeys infected parenterally with variola strain were reported to develop non-specific, febrile disease. In studies conducted by the DoD in collaboration with the Centers for Disease Control, we tested four variola strains for virulence in monkeys exposed to high infectious doses delivered by aerosol, intravenous, or a combination of routes. Eventually, we identified a virus strain that produced a lethal disease process resembling rapidly progressive, human smallpox.

Initially, two variola strains (Yamada and Lee) were used to expose monkeys to aerosolized doses of 10^8 plaque-forming units (PFU). Results were disappointing, since the disease courses were mild and nondescript. Subsequent studies used different variola strains (Harper & India 7124), higher doses (10^9 PFU), and included intravenous inoculation, which may be critical. Summary data are shown in Table 2-1.

---

* This statement reflects the professional view of the author and should not be construed as an official position of the U.S. Army Medical Research Institute of Infectious Diseases.

TABLE 2-1 Results of primate exposures to smallpox virus

MK #	Sex	kg	inoc	strain	route	intended dose	Day death
C099	M	7.2	5/31/2001	VAR Harper	Aerosol + IV	10^9	
C625	M	6.5	5/31/2001	VAR Harper	Aerosol + IV	10^9	4
C681	M	8.5	5/31/2001	VAR Harper	Aerosol + IV	10^9	4
C881	M	5.5	5/31/2001	VAR Harper	Aerosol + IV	10^9	6
C171	M	5.5	6/1/2001	VAR 7124	Aerosol + IV	10^9	3
C651	M	6.9	6/1/2001	VAR 7124	Aerosol + IV	10^9	4
C115	M	7	6/1/2001	VAR 7124	Aerosol + IV	10^9	3
C713	M	4.9	6/1/2001	VAR 7124	Aerosol + IV	10^9	13
C373	M	4.6	7/6/2001	VAR 7124	IV	10^9	4
C088	F	4.1	7/6/2001	VAR 7124	IV	10^9	3
C437	M	3.9	7/6/2001	VAR 7124	IV	10^9	6
C956	M	3.9	7/6/2001	VAR 7124	IV	10^9	4
C083	M	8.4	8/24/2001	VAR 7124	IV	10^9	
C003	M	6.2	8/24/2001	VAR 7124	IV	10^9	10
57-394	F	3.2	8/24/2001	VAR 7124	IV	10^8	
C271	M	7.2	8/24/2001	VAR 7124	IV	10^8	
C282	F	6	8/24/2001	VAR 7124	IV	10^8	
C835	M	4.3	8/24/2001	VAR 7124	IV	10^7	
57-245	F	3.5	8/24/2001	VAR 7124	IV	10^7	
C677	M	6	8/24/2001	VAR 7124	IV	10^7	
C382	F	5	8/24/2001	VAR 7124	IV	10^6	
C409	M	4.9	8/24/2001	VAR 7124	IV	10^6	
48-48	F	3.3	8/24/2001	VAR 7124	IV	10^6	

Three of four monkeys exposed to the Harper strain by a combination of aerosol and intravenous routes died rapidly, three to six days after exposure. Likewise, all four monkeys exposed to the India strain died, although one developed a more protracted disease course and died on day 13. Subsequent inoculation of four monkeys with 10^9 PFU India 7124 via the intravenous route alone yielded uniform rapid lethality. In subsequent attempts to obtain a more slowly evolving disease course, lower doses produced systemic infections and more protracted disease courses, but no deaths.

Hematologic evaluation of lethally infected primates revealed profound leukocytosis, with WBC > 50,000/mm^3 (20% monocytes) in acutely ill animals. Platelet counts dropped to an average of 125,000, consistent with coagulation factor perturbations and fibrin deposition associated with evolving disseminated intravascular coagulation (DIC). Serum chemistry evaluations revealed profound increases in serum creatinine and blood urea nitrogen consistent with kidney lesions as well as elevations in AST and ALT consistent with hepatic damage. Infectious virus at concentrations > 10^8 PFU/g were retrieved from all visceral tissues obtained from acutely moribund or terminal monkeys at necropsy. This is consistent with the demonstration of viral antigens by immunohistochemistry in association with pathological lesions in these same tissues. Infectious viral burdens in monkey # C-713 (which died on day 13) were lower. Infectious virus was also retrieved from throat swabs of iv-inoculated animals within 48 hours of exposure, before the evolution of skin lesions or fevers. It is probably that these animals are contagious at this early stage of infection; isolation of virus from throat swabs of human smallpox-infected patients was never systematically attempted. Detection of viral genomes in the blood of inoculated monkeys as early as one day after inoculation was achieved using TaqMan PCR. This assay, which requires less than one hour to run, promises to detect infection during the asymptomatic prodrome, when countermeasures such as antiviral drugs are predicted to be most effective.

Additional insight into the pathogenesis of variola in lethally infected primates is being obtained by evaluation of high-density cDNA microarray data, which measures and classifies gene expression in peripheral blood cells obtained sequentially. RNA was prepared from isolated peripheral blood leukocytes, labeled as fluorescent cDNA for microarray analysis, and hybridized to arrays which include >10,000 uniquely named genes. Using a two- color comparative hybridization format, expression patterns were analyzed according to biological themes. Gene expression analysis identified dramatic response patterns that correlated with lethality and gave insight into pathogenesis. Several relevant biological themes included interference with interferon, IL-18, and TNF-alpha, inhibition of interleukin-1 beta and apoptosis. Activation of coagulation cascade factors, and down regulation of immunoglobulin response and cell mediated immunity-related genes were also related to lethality. Microarray evaluations will be extended to include tissue expression patterns, comparisons with other virus infections (especially monkeypox) in primates, and in vitro systems with primary target cells. Potential benefits of host genome-wide expression profiling include early detection of infected individuals, recognition of variant agents and prognostic markers, identification of virulence and disease, novel therapeutic and prophylactic strategies, and determination of early signatures of a protective immune response to vaccination.

Further refinements of this primate model are necessary before it can be exploited to test antiviral drugs or other countermeasures in accordance with the FDA Animal Efficacy Rule. Ideally, a combination of viral strain, dose, and route can be identified which produces a less accelerated disease course, one that more closely reflects the temporal course of human smallpox. The observation that certain strains of variola can produce fulminent disease in monkeys is a breakthrough in the quest for effective countermeasures for smallpox. This outcome is the exact opposite of what was predicted, since until now it was assumed that the host range of variola was restricted to humans.

Evaluation of the model disease course has yielded considerable insight into the nature of the "toxemia" associated with human smallpox. In addition, clinically relevant samples of blood and tissues have been obtained and tested, to validate modern diagnostic strategies such as TaqMan PCR. Using this technique, we have demonstrated the feasibility of obtaining a definitive diagnosis of exposure to smallpox virus during the prodrome. This capability should improve the likelihood of successful intervention using antiviral drugs. The production of these clinical samples is a significant byproduct of animal model development. The samples constitute a national resource for validation of diagnostic strategies based on detection of smallpox viral genomes and antigens now, and in the future.

Further refinement of the primate model for smallpox will include evaluation of the sub lethal, yet severe, infections associated with lower doses of inoculum virus. Objective, quantifiable correlates of disease severity may eventually be substituted for lethality in efficacy evaluations of candidate antiviral drugs and other countermeasures. As model refinement continues, concurrent advances in the identification of useful antiviral drugs are anticipated. These advancements, combined with further enhancements in diagnostic strategies, are reasonable milestones projected for this collaborative DoD/DHHS research program. Effective mitigation of an adversary's most potent biological weapon (smallpox) must be a national priority.

TULAREMIA AND PLAGUE: ASSESSING OUR UNDERSTANDING OF THE THREAT

David T. Dennis,* **M.D., M.P.H.**
Division of Vector-Borne Infectious Diseases
Centers for Disease Control and Prevention

Yersinia pestis and *Francisella tularensis* are category A critical biological agents that pose a risk to national security because they could be easily disseminated (both agents) or transmitted person-to-person (*Y. pestis*), could cause a high mortality, and require special action for public health preparedness (CDC, 2000). Both agents have been weaponized as aerosols, the expected mode of delivery in a bioterrorist attack (Inglesby et al., 2000; Dennis et al., 2001). Plague holds special concern because of its potential to cause panic, its contagiousness in the pulmonary form, its fulminating clinical course and high fatality, and the possibility that it could be engineered for plasmid-mediated resistance to multiple antimicrobial agents (Galimand et al., 1997). Sepsis with either agent can result in catastrophic physiological consequences of compliment and cytokine cascade (systemic inflammatory response syndrome [SIRS]). Severity of illness is expected to require intensive medical care, including respiratory and other organ support that might readily overwhelm hospital response capacity (Inglesby et al., 2001). Pneumonic plague's contagiousness would require isolation and possible quarantine, which would complicate medical and public health management. A WHO model of the release of 50 kg of *Y. pestis* over a city of 5 million predicts 500,000 cases with 100,000 deaths when both primary and secondary transmission are considered; a similar model for release of *F. tularensis* predicts 250,000 persons incapacitated and 19,000 deaths (WHO, 1997).

BOX 2-1 Bioterrorism Aphorisms

- Do not assume anything
- Expect the unexpected
- What we know is not much
- Bioterrorism is, among other things, unnatural
- We do not know what we do not know
- We are not ready

* The information provided in this paper reflects the professional view of the author and not an official position of the U.S. Department of Health and Human Services or the Centers for Disease Control and Prevention.

Standard, classical microbiological diagnostic tests would be of limited value in a major bioterrorism event, since they are time consuming and labor intensive; unfortunately, newer, rapid testing methods for these agents, such as antigen detection and DNA amplification, are neither standardized nor widely available.

Recommendations for antimicrobial treatment of plague or tularemia patients in a bioterrorist attack have been developed for both the mass- and contained-casualty situations (Inglesby et al., 2000; Dennis et al., 2001). The principal recommended antimicrobials are available through the National Pharmaceutical Stockpile. Two of these, gentamicin and ciprofloxacin, are not FDA-approved for treating plague or tularemia. Post-exposure antimicrobial prophylaxis is recommended to protect certain populations in the event of bioterrorist use of either plague or tularemia agents, but it would be difficult to identify populations at risk and administer drugs to them in a timely fashion. In the case of plague, isolation of cases and their close contacts, and quarantine of exposed populations could be difficult to enforce and would likely create fear and chaos (Inglesby et al., 2001).

Historical vaccines for plague and tularemia, based on killed and live-attenuated preparations respectively, are currently unavailable in the United States, and a newer generation of vaccines and immunotherapeutics is greatly needed to address pre- and post-exposure prevention of disease. Recombinant protein subunit vaccines (against F1 and V antigens singly, in combination, and as fusion products) have been developed for plague (Titball and Williamson, 2001). One subunit combination product has recently undergone Phase I clinical testing in the U.K. A microencapsulated formulation shows promise for respiratory tract delivery (Eyles et al., 1998). Further, oral administration of a *Salmonella typhimurium* mutant expressing the F1 antigen of *Y. pestis* protects mice against subcutaneous inoculation of a virulent plague strain (Titball et al., 1997). Intraperitoneal administration of monoclonal antibodies to the F1 antigen protects mice against both parenteral and aerosol challenge with human pathogenic *Y. pestis* strains (Anderson et al., 1998). Similar vaccines and recombinants have not yet been developed for use against tularemia, but recent progress in sequencing of the *F. tularensis* genome may lead to identification of candidate immunoprotective proteins (Karlsson et al., 2000).

Recent advances in *Y. pestis* and *F. tularensis* strain typing, using multiple-locus, variable-number tandem repeat analyses are expected to provide rapid tracking of outbreak strains as well as providing a foundation for deciphering global genetic relationships of these organisms that could be useful in the event of a BT attack (Johansson et al., 2000; Klevytska et al., 2001; Farlow et al., 2001).

Personal experience gained by participating in CDC responses to bioterrorist use of *Bacillus anthracis* reinforces the need to think freely about potential misuse of *Y. pestis* or *F. tularensis* (Box 2-1), to heed lessons learned (Box 2-2), and to ensure that preparedness and response needs are met before a critical event occurs (Box 2-3).

BOX 2-2 Lessons Learned

- Primary response is local
- Federal support must be immediately available
- Better agency integration needed (DHHS, Justice, Defense, State and Local)
- Surge capacity vital
- We don't know what we don't know

BOX 2-3 Critical Public Health Preparedness and Response Needs

- Pre-approved, integrated, organizational plans, policies, protocols
- Systems for surveillance, case id, contact tracing (plague), rapid epidemic/environment assessment
- Surge capacity for outbreak and consequence management, including epidemiology/survey sampling, lab, Rx, Px, logistics
- New/improved diagnostics and molecular id
- New/improved vaccines and therapeutics

BOTULINUM TOXIN AS A BIOWEAPON

Stephen S. Arnon, M.D.
Founder and Chief
Infant Botulism Treatment and Prevention Program
California Department of Health Services

Botulinum Toxin and Human Botulism

Botulinum toxin is the most poisonous substance known. One gram, evenly aerosolized and inhaled, could kill over one million people; one hundred grams, evenly distributed in a food or beverage and ingested, could also kill over one million people (Arnon et al., 2001). Botulinum toxin is considered a plausible prime bioweapon threat because of its extreme potency and lethality, its ease of transport and misuse, and its profound impact on victims and the health care infrastructure. Botulinum toxin is the only non-replicating member of the six Centers for Disease Control and Prevention (CDC)-designated Class A (highest threat) agents. Other aspects of the toxin may also make it attractive as a bioweapon (Table 2-2).

Botulinum toxin is a simple di-chain protein whose "heavy chain" (100 kD) contains the binding and internalization domains and whose "light chain" (50 kD) contains the catalytic (Zn^{++}-proteinase) domain. The toxin binds at peripheral nerve cholinergic synapses, the most important of which clinically is the

TABLE 2-2 Features making botulinum toxin attractive as a bioweapon

Attribute	Consequence
• Extreme potency and lethality	• Can cause clusters of 10–500 fatalities in any city at any time
• Ease of production, transport and misuse	• Convenience and accessibility
• Profound impact on victims and health care infrastructure	• Can target key small groups (e.g. Congress, Supreme Court); can overwhelm urban hospital ICU capacity
• Aerosols degrade quickly	• Enemy's infrastructure or arms can be captured intact; no decontamination needed
• Not person-to-person transmissible	• Use does not place user at risk
• Versatility in use (foods, beverages, aerosols)	• Can terrorize almost at will
• Rapid turnover of publicly served foods and beverages	• Evidence may be discarded before illness presents; easy to escape detection and capture

neuromuscular junction. After uptake into the neuronal cytoplasm, the toxin cleaves one or more of the "SNARE-complex" proteins, thereby preventing the release of the acetylcholine-containing vesicles that normally cause muscle contraction. The net result of the toxin's action is a flaccid muscle paralysis. Botulinum toxin is produced by the spore-forming anaerobic bacterium *Clostridium botulinum*, whose natural home worldwide is the soil and dust, from which it can be isolated with undue difficulty.

Botulinum toxin exists in seven different serotypes arbitrarily assigned the letters A-G; antibody that neutralizes one toxin type does not neutralize any other serotypes (e.g., anti-A antitoxin does not neutralize toxins B-G, etc.). The toxin is stable in foods and unchlorinated beverages for extended periods of time, but is easily inactivated by heating (e.g., 85°C x 5 minutes). Consequently, foodborne botulism (whether natural or bioterrorist) can result only from eating foods that are not heated, or not heated thoroughly. Waterborne (or beverage-borne) botulism has never been reported but is scientifically possible. Botulinum toxin is colorless, odorless, and as far as is known, tasteless.

Botulinum toxin is the only Class A agent that is also a licensed medicine, a fact that complicates the design of defenses against possible bioweapon use of the toxin. In the United States type A and type B botulinum toxins are licensed for the treatment of blepharospasm, strabismus and cervical dystonia. However, both toxins are extensively used "off-label" to treat a range of more widely prevalent disorders that include spasticity (from stroke, head trauma, cerebral

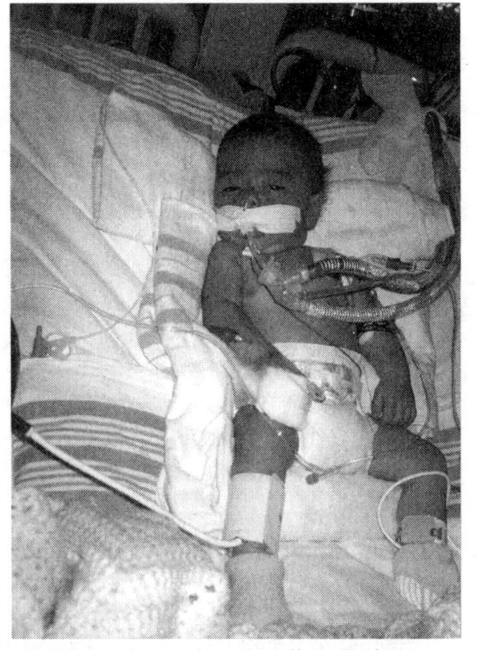

FIGURE 2-2 Infant Botulism Patient

palsy, multiple sclerosis, etc.), headache (both migraine and tension), low back pain, benign prostatic hypertrophy, and even facial wrinkles and other cosmetic concerns. Many other therapeutic uses of the toxin are under investigation.

Human botulism has several forms: foodborne, waterborne (potentially), inhalational, infant and wound. Only the first three varieties represent bioterrorist possibilities. Recognition of a botulism outbreak depends on the astute clinician who promptly notifies public health authorities. Clinically, botulism always begins in the muscles of the head, eyes, face and throat, most probably because of their relatively greater blood flow and density of innervation per unit muscle mass. The illness then progresses as a symmetric, descending flaccid paralysis. Severe cases are resource-intensive because they require antitoxin treatment, mechanical ventilation, gastrointestinal tube or intravenous feeding, and 24-hour intensive nursing care (Figure 2-2). Death results either from obstruction of the upper airway by unswallowable secretions and flaccid pharyngeal muscles or from complications of mechanical ventilation and intensive care.

Production and Delivery of Botulinum Toxin as a Bioweapon

Botulinum toxin is readily available as a bioweapon because of the relative ease with which its source, *C. botulinum*, may be isolated from nature or otherwise obtained. A minimal amount of laboratory equipment and microbiological expertise is needed to cultivate *C. botulinum* and concentrate its toxin to weapon-utilizable material for oral use. With access to an autoclave, 500–1000 human oral lethal doses could be produced for a few hundred dollars. Making effective toxin aerosols would require greater scientific sophistication and resources.

The natural occurrence of foodborne botulism brought the existence of botulinum toxin to medical attention over 200 years ago; this route of toxin dispersal remains available to terrorists today. No instances of waterborne botulism have been reported, and standard potable water treatments rapidly inactivate the toxin. One instance of accidental inhalational botulism occurred approximately 30 years ago among veterinary technicians performing animal autopsies, thereby confirming the feasibility of the aerosol delivery route for humans (Arnon et al., 2001).

Present Methods of Controlling the Botulinum Toxin Threat

The botulinum toxin threat may be analyzed by means of a three-part schema consisting of *sources* (the "reservoir"), *terrorists* (the "vectors") and *potential victims* (the "population"). The *sources* of *C. botulinum* are many and uncontrollable. They include rogue nations, unemployed former bioweapons scientists, open-access microbiological culture collections in various countries, the black market, and most fundamentally, the ubiquitous presence of *C. botulinum* in soils worldwide. The *terrorists* are by definition uncontrollable but may be thwarted in whole or in part by good *a priori* intelligence. Hence, the *potential victims* represent the only point of intervention in defending against bioweapon use of botulinum toxin. The interventions currently available to protect potential victims consist of 1) surveillance and early detection of toxin release, 2) immunization with botulinum toxoid, and 3) provision of antitoxin and supportive care.

The United States has a well-established botulism surveillance and detection system directed by CDC that was recently enhanced by daily reporting of all suspected botulism cases by the states. Availability of biosensors in key locations (e.g., airports) to detect aerosol release of botulinum toxin would enhance present surveillance capabilities.

Immunization of either civilian or military populations with botulinum toxoid is not a practical defense against weaponized botulinum toxin for several reasons. The existing pentavalent botulinum toxoid contains only A-E toxoids (i.e., it lacks F and G), and it remains an Investigational New Drug that requires informed consent before administration. The pentavalent toxoid is painful and highly reactogenic, and the full series of immunizations takes a year to complete (0-2-12-52 weeks). Finally, administration of toxoid deprives the recipient of access to any of the therapeutic benefits of medicinal botulinum toxin because the toxoid induces formation of toxin-neutralizing antibodies and the immunologic memory cells that produce them.

Antitoxin and supportive hospital care constitute the mainstays of the treatment of patients with botulism. Licensed and Investigational New Drug (IND) equine antitoxins exist in limited supply, as does an Investigational (IND) human-derived antitoxin developed for the treatment of infant botulism (Arnon, 1993). The licensed equine antitoxin is only a bivalent (anti-AB toxins) product, while the Investigational (IND) equine antitoxin is a heptavalent (anti-ABCDEFG toxins) product. The human-derived antitoxin (Botulism Immune Globulin Intravenous; BIG-IV) is a pentavalent product (anti-ABCDE toxins) whose license application, recently filed with the U.S. Food and Drug Administration, requested labeling only for its anti-A and B toxins activity because virtually all U.S. infant botulism cases are caused by these two toxin types.

Equine botulinum antitoxin is relatively inexpensive to make and can be produced in relatively large amounts. Its production requires a horse farm and staff, full-time veterinarians and a plasma-fractionation facility. At present a substantial reserve of frozen equine heptavalent plasma exists under CDC management, and efforts are underway to fractionate it into an available vialed product. However, equine antitoxin has a short (ca. 1-week) half-life, and its use carries the risk of serious allergic reactions to equine proteins.

Human-derived botulinum antitoxin (BIG-IV) is made from volunteers initially immunized with pentavalent botulinum toxoid for occupational safety reasons and then boosted with toxoid to obtain hyperimmune source plasma. BIG-IV has a long (ca. 1 month) half-life and a negligible allergic risk. The pool of potential plasma donors is small and insufficient to meet national needs. Treatment with human botulism antitoxin is effective as well as safe. In a randomized, placebo-controlled, double-blinded clinical trial, use of BIG-IV to treat infant botulism patients shortened their mean hospital stay by over 50%, from 5.7 weeks to 2.6 weeks, and reduced their mean hospital charges from approximately \$130,000 to approximately \$60,000 per case (California Department of Health Services, presently unpublished data).

Novel Methods for Containing the Botulinum Toxin Threat

The development of recombinant botulinum vaccines by the Department of Defense (DoD) for toxin types A, B, C, E and F began several years ago by expressing fragments of toxin in yeast; the vaccine against type A toxin is highly immunogenic and protective in mice (Byrne and Smith, 2000). DoD currently projects that a licensed recombinant vaccine product may be available in about 10 years. Its use would require special consideration because recipients would be deprived thereafter of the therapeutic benefits of medicinal botulinum toxin, one of which, post-head-trauma spasticity, is a likely consequence of combat situations. In addition, an adversary who knew the toxin types contained in the vaccine might choose to weaponize toxin types D or G in disregard of the current view that technical difficulties preclude doing so.

Recombinant human antitoxin antibody represents another way to control the bioweapon threat posed by botulinum toxin. A highly potent preparation of recombinant human (*biotech*) antibodies that neutralizes botulinum type A was recently reported (Marks, 2001), thus establishing "proof-of-concept" for these products. The *phage-display* technology that underlies the development of these recombinant human antitoxin antibodies also enables neutralizing human antibodies to the remaining six (B-G) botulinum toxin types to be rapidly created. The phage-display technology is versatile as well as fast, and it could be used to make human anti-anthrax antitoxin as well as the human Vaccinia Immune Globulin (VIG) needed to support the widespread administration of some or all of the 300 million doses of smallpox vaccine that the United States government recently decided to purchase.

Recombinant human antitoxin has several important advantages over vaccines, toxoids and existing equine antitoxins: 1) recombinant human antitoxin (RHA) provides immediate immunity when given, and there is no delay in waiting for the recipient's immune system to make its own antibody in response to the vaccine or toxoid administration; 2) RHA can be made in the substantial quantity that is needed for the United States' and its allies' stockpiles; 3) RHA is very potent and has a long (ca. one-month) half-life; 4) RHA is safe and nonallergenic and so can be given multiple times; 5) RHA is practical for widespread use in civilian and military populations; 6) the licensure pathway for RHA is understood, because FDA has already licensed 10 monoclonal antibody products; and perhaps most importantly, 7) RHA can be used either prophylactically or therapeutically. The ability to provide RHA prophylactically to selected populations potentially at high risk (e.g., an overseas military force, Congress, etc.) and thereby provide with a single injection an immediate and long-lasting (6–12 months) immunity to all 7 botulinum toxin serotypes might effectively remove weaponized botulinum toxin from an adversary's arsenal, simply because the opponent would know in advance that the toxin bioweapon would not work.

Second-Generation Botulinum Toxin Bioweapons

Modern molecular biology techniques enable the gene for the enzymatic "light chain" of botulinum toxin to be fused with the gene of a targeting molecule that will take the toxic combination to various non-neuronal cell types. Secretory and other cell types that contain the substrates for botulinum toxin "light chain" may be found in the pancreas, liver, thyroid, adrenal and heart (Rosetto et al., 1996). Thus, botulinum toxin "light chain," if redirected to the pancreas, could block the secretion of insulin and thereby cause epidemic diabetes. Such recombinant, "second-generation" toxic molecules have been made, usually with the goal of finding an improved anti-cancer or equivalent therapeutic agent.

Also, the genes for the both the light chain and heavy chain of botulinum toxin have been engineered for high efficiency expression in *Escherischia coli* for the stated purpose of enhancing the production of medicinal botulinum toxin, which presently commands a market of several hundred million dollars per year. However, virtually all humans carry *E. coli* as part of their normal intestinal microflora. Strains of *E. coli* that commonly cause diarrhea are often spread through contaminated foods and may be further disseminated by the fecal-oral route. An enteropathogenic strain of *E. coli* engineered to express the genes for both chains of botulinum toxin would have the potential to eliminate much of the human race, a potential that underscores the need for prompt development of effective countermeasures to botulinum toxin.

Priorities for Preparedness

The major preparedness needs for the United States can be arranged into public health, clinical medicine and research categories (Table 2-3). The three most important immediate needs are 1) fractionating and vialing the 7,000 liters of frozen heptavalent equine plasma for the National Pharmaceutical Stockpile, 2) ensuring an adequate surge capacity for ventilators and mobile intensive care units, and 3) rapidly developing a heptavalent human recombinant antitoxin to provide an unlimited supply of the key defensive commodity.

TABLE 2-3 Priorities for preparedness

Public Health
- Fractionate, vial and IND the 7000l of frozen Army-CDC equine heptavalent plasma
- Develop more surge capacity at all levels (federal, state, local); both laboratory and epidemiology
- Develop rapid in vitro toxin detection methods
- Produce human recombinant antitoxin for stockpiling as well as current use

Clinical Medicine
- Improve communications with public health colleagues for early detection
- Ensure adequate surge capacity for ventilators and mobile intensive care units and their staffing
- Produce human recombinant antitoxin for stockpiling as well as current use

Research
- Begin development of human recombinant antibodies against toxin serotypes B-F
- Begin scale-up production of existing human recombinant anti-A antibodies to establish pathways and capabilities
- Obtain intelligence on toxin-derived bioweapons research in other countries to enable defensive recombinant human antitoxin development

Summary and Conclusion

Existing in vitro technologies could produce the stockpiles...needed both to deter terrorist attacks and to avoid the rationing of antitoxin that would be required in a large outbreak of botulism. A single small injection on oligoclonal human antibodies could, in theory, provide protection against toxins A-G for many months. "*Until such a product becomes available, the possibilities for reducing the population's vulnerability to the intentional misuse of botulinum toxin remain limited.*" (Working Group on Civilian Biodefense, 2001)

RESEARCH CONSIDERATIONS FOR BETTER UNDERSTANDING OF BIOLOGICAL THREATS

Kenneth Alibek, M.D., Ph.D., Sc.D.
Distinguished Professor of Medical Microbiology and
Executive Director, Center for Biodefense
George Mason University
President
Advanced Biosystems, Inc., a subsidiary of Hadron, Inc.

Because they are our best protection against infectious disease, it is necessary that we continue to develop, approve, and introduce new vaccines against many naturally occurring infectious diseases and against some biological weapons threat agents. Yet the U.S. tends to focus its discussion of bioweapons vaccines and therapeutics on only a handful of potential agents, even though the former Soviet Union is known to have developed at least thirty biological agents for use as bioweapons. Alarmingly, it takes only two to three years to develop a biological weapon but, in the best-case scenario, eight to ten years to develop a new vaccine.

It has been suggested that live vaccines are too reactogenic for general use. But, for the purpose of boosting our biodefense arsenal, perhaps this issue requires reevaluation. In Russia, for example, all major vaccines—including anthrax, plague, and tularemia—are live vaccines. The United States had a good live plague vaccine and has a very strong live tularemia vaccine. The latter may not be approved for human use, but its protective efficiency is very high. Bioengineered vaccines are another possibility.

The use of alternate methods of vaccine administration must also be addressed. Biodefense vaccine administration techniques should not only be safe but must also provide for the vaccination of large numbers of people in a very short amount of time. Currently the U.S. focuses on injection vaccines, but there have been many studies on aerosol, inhalational, and oral vaccines. An aerosol plague vaccine, for example, can be used to immunize more than 1,200 people per hour, while a single operator can administer injection vaccines to only 20 to 30 people per hour. Inhalable plague, anthrax, and tularemia vaccines have all been extensively studied in Russia and have not shown any significant side effects. This type of vaccine could work both systemically and locally, for example to induce mucosal immunity in the respiratory tract.

However, though vaccines have proven extremely effective against infectious diseases in general, they are of limited utility in the defense against infections caused by biological weapons. Vaccination is a successful defense only when the target population is well-defined and can be identified well in advance of an attack; when the biological threat agents in the enemy's biological weapons arsenal are known; when vaccines for those agents have already been developed; and when the biological agents used are not genetically altered strains capable of circumventing a vaccine. Most military and nearly all terrorist scenarios will not meet all of these criteria. Therefore, vaccination of the general population against biological weapon agents is neither feasible nor advisable. In the context of biological weapons, the best use of vaccination is for troop protection, where both the target population and potential threat are more defined.

Other, non-vaccine biodefense products need to be more seriously considered. In particular, over the past twenty years there has been extensive research on immunomodulators and their role in protecting against viral and bacterial pathogens. Although several such products have been developed that could potentially resolve many of our biodefense issues, none of them have been introduced yet into the field of biological weapons defense. In a biological attack, the target population would likely be large and poorly defined, the scale or even the fact of the attack may not be immediately apparent, and the biological agent used in the attack may not be immediately identified. For either military or terrorist use of biological weapons, creation of an aerosol cloud—usually accomplished by explosion or spraying—is by far the most effective mode for deploying biological weapons. This method can be used against large target areas and with practically any biological threat agent. Therefore, effective medical defense against biological weapons must incorporate protection against aerosol deployment. The nature of a biological weapons attack also dictates that the most successful medical defenses will be prophylactic, rapid-acting, long-lasting, effective against a broad spectrum of threat agents, and relatively easy to deliver to a large population.

Finally, we need to re-examine our knowledge about the pathogenesis of bacterial and viral infections. For example, we still refer to the three major virulence factors in any discussion of anthrax. However, many other virulence factors may exist. My laboratory has conducted hundreds of experiments on anthrax lethal toxin, and thus far we have found no evidence whatsoever that it is capable of inducing normal healthy donor immune cells to produce any cytokines involved in the development of septic shock. Human endothelial cells are, however, relatively susceptible to the effect and the action of lethal toxin. It is becoming apparent that current theories on the role of anthrax toxin must be reexamined and revised. Our experiments have indicated that other overlooked anthrax virulence factors exist, including cell wall components and (possibly) hemolysins. Our work has indicated that the only factor of the anthrax bacterium capable of consistently inducing the mediators of septic shock seen in late-stage anthrax infection is a component of the cell wall skeleton, not lethal toxin, and that there exist many other overlooked exogenous and endogenous mediators which contribute in the development of anthrax sepsis and septic shock. Although these are preliminary data with which we are not able to make any final conclusions, they do highlight the need to re-examine the pathogenesis of anthrax. They may also explain why people die when antibiotics are administered in the late stages of infection: these cell wall skeleton components are very powerful inducers of septic shock mediators, and their concentration remains very high in the bloodstream after bacterial death. The preliminary work we have conducted to this point has led us to the belief that the most successful strategy for anthrax treatment would be dependant on the stage of infection (post-exposure, lymphatic, systemic, or late-stage).

It is our recommendation that a task force be established to more carefully analyze the events that occur during anthrax infection. It is very important that we re-evaluate our knowledge of pathogenesis and identify what we have missed in the field of protection and treatment of infectious diseases caused by biological weapons.

AEROSOL TECHNOLOGY AND BIOLOGICAL WEAPONS

C.J. Peters, M.D.
Professor, Departments of Microbiology and Immunology and Pathology
University of Texas Medical Branch at Galveston

Richard Spertzel, VMD, Ph.D.
Former head of the biological weapons inspection team for the UN Special Commission on Iraq

William Patrick III
President
Biothreats Assessment Co.

The threat of international terrorism to this country and others has never been as serious as it is today. The U.S. abandoned its program for offensive biological warfare in 1968, but the successful effort to weaponized infectious organisms and toxins should have educated the U.S. government to their dangers. The revelations that the Soviet state had a more extensive and similarly successful undertaking (Alibek, 1999) only increases the likelihood that the development of weapons of mass destruction lies within the reach of others. The extensive programs used by these two states are often cited as hurdles too high for other countries to surmount, but it must be remembered that they worked with multiple agents and manufactured literally tons of product in a time before many of today's advances in biology and fermentation technology. The Gulf war brought the potential of other states to produce biological weapons home, but the weapons were not used in spite of the argument that they are intensely destabilizing (Haselkorn, 1999). The spate of anthrax laced letters in 2001 has led to a re-awakening of concern.

Numerous commissions have reviewed the threat of bioterrorism (Countering the Changing ... , 2000; Second Annual Report, 2000) in recent years and uniformly concluded that the US was vulnerable and that the likelihood of such an event was high. Nations suspected of having offensive biological warfare programs have been named by the Office of Technology Assessment (1993) and these same states are often also identified as terrorist sponsors. In light of these agreed-upon threats, why has there been so little concern about this possibility in many quarters and why has so much surprise been expressed over the outcome of a handful of letters containing anthrax spores dispatched through the mail? One important component is the lack of familiarity with the concepts which were the pillar of the old US biowarfare program and also the Soviet program, particularly as refers to the danger of aerosols. Another reason may be that no one expected that the manufacturer of such a deadly powder as was placed in Congressional letters in autumn of 2001 would employ an envelope as a delivery route rather than using a more stealthy and lethal dissemination system. This complacency was reinforced by the large number of "anthrax" powder hoaxes that have occurred over the last few years, a social phenomenon worthy of study in itself.

This short essay attempts to outline why we should be concerned about use of biological weapons in terrorism, why some scenarios are more dangerous than others, and some general observations concerning what we could do to combat bioterrorism.

Nature of the Threat

When one considers the ways in which BT might be carried out, the first rule should be that we do not know who the possible terrorist will be, his or her motivation, or the wherewithal that may be available for the attack. Thus, attacks to incapacitate selected persons to gain attention, or to cause serious illness for revenge might have a very different approach than attacks designed to cause mass casualties. An effort by a disgruntled clinical laboratory worker could have a very different scope than one by a well-funded non-state organization or a state-sponsored group. Parenthetically, the failure of the Japanese Aum Shinriko cult to succeed with biological terrorism should not provide much comfort concerning the need for state sponsorship given the manifest ineptitude of the perpetrators (Smithson, 2001).

The second issue is the dissemination of factual information concerning the real dangers of such an attack. Some would argue that the less said the better. Unfortunately for this approach, US society does not respond without facts and public opinion supporting programs. This means that the actual dangers must be explained to the public and responsible political leaders without inflammatory rhetoric or divulging detailed methods for the assaults, and we must take the chance that plain-speaking might motivate some to undertake the very actions we are trying to prevent. Furthermore, discussion of the facts may help the public, media, and health authorities respond in a calmer and more rational fashion than was observed with the aerosol anthrax attacks of 2001.

If we consider the methods by which microbes might be delivered to a target population there are multiple routes. Direct inoculation, infection of natural vectors or reservoirs and loosing them on the target population, or infection of a few persons and counting on their spreading the infection even further are some possibilities. If we focus on terrorist strategies that can inflict mass casualties none of these possibilities is highly feasible today with the exception of the use of smallpox, a virus that is well-known to spread from human to human after a long and successful career in that evolutionary niche. If we conclude that other organisms must be delivered directly to the target host, we should also consider water, food, and aerosols as potential vehicles of infection. Contaminated water from wells and storage containers has been associated with outbreaks of disease, but the odds are against this approach for causing mass casualties because of the dilution factor, chlorination, and the usual treatment of water before consumption in this country. Food-borne pathogens have caused many outbreaks in the US and are a major cause of morbidity and mortality. Even though our distribution system is highly centralized, food items are usually not consumed synchronously except at special events. Improved surveillance of food-borne disease and newer methods of molecular typing of offending organisms should provide a counterweight to the wide dissemination of contaminated food. If a few cases are recognized and traced to a food source, warnings and recalls may well serve to protect us.

The over-all societal impact of any one of these dissemination methods could be considerable, regardless of the actual health damage. We have case studies already, including non-lethal *Salmonella* infection of a few hundred citizens (Torok et al., 1999), Sarin gas attacks with only 20 deaths (Smithson, 2001), food tampering, and most recently anthrax delivered by letter or even anthrax hoaxes. Aerosols, however, are an important route of attack because of their ability to cause really large numbers of casualties. The deficiencies of aerosols such as dependence on metrological conditions, the unsuitability of most organisms for air-borne spread, and the technical demands may be counterbalanced in the hands of skillful perpetrators by the advantages of stand-off attack, silent spread of incapacitating or lethal disease, and wide-area coverage.

Aerosol Infections

Aerosol infection has long been recognized as a route of microbial transmission. Measles, influenza, smallpox, and tuberculosis are all known to be transmissible between patients by aerosols and additionally in the laboratory tularemia, rickettsiae, viral hemorrhagic fevers, and many other agents are threats to the microbiologist (DHHS, 1999). Decreases in human tuberculosis and virtual elimination of diseases such as measles and smallpox from common medical experience as well as development of enhanced methods of protecting laboratory workers (ironically using technology developed during the US biowarfare program) have resulted in a loss of appreciation for this route of infection, but the US and Soviet BW programs were largely based on the properties of selected agents for causing large scale infection of human populations under the proper metrological conditions and with carefully developed methods of dissemination. Moreover, the terrorist could attack enclosed environments such as stadiums or large buildings in order to eliminate the meteorological factors that degrade a small particle aerosol.

Biological agents have not seen widespread use in warfare so it is not surprising that there is skepticism as to their efficacy. It is not appreciated that the US program in offensive biological warfare (terminated on November 28, 1969) rigorously tested each step in the link between a microorganism selected by several criteria and the delivery of a credible biological attack (US Congress, OTA, 1993a, b; Rosebury, 1947; Hersh, 1968; McDermott, 1987; Cole, 1997; Sidwell et al., 1997; Patrick, 2001). Tularemia would be an excellent example because extensive information is available in the published literature, congressional hearings, and popular press. From the initial isolation of the organism by Francis and coworkers, it was notorious for causing infections in the laboratory, a frequent hallmark of aerosol infectivity. The aerosol properties were intensively studied and methods were found to enhance its stability in storage and in aerosols. Animals and later humans were challenged with graded doses of the bacterium delivered in different particle sizes to establish the quantitative properties of these aerosols. Open air dissemination was mimicked using a surrogate organism, *Serratia marscens*, and this confirmed that an organism with the aerosol stability and infectivity of *Francisella tularensis* could cause mass casualties over large geographic areas provided attention was given to metrological conditions. The areas affected could reach thousands of square kilometers. The resulting environmental transmission from the large number of different nonhuman mammalian and arthropod species that would be infected in a tularemia attack cannot be evaluated. Thus, there is little doubt that large numbers of human casualties could be caused by efficiently weaponized organisms readily available from nature.

A relatively small number of agents are suitable for causing literally thousands or hundreds of thousands of casualties, and this may provide a basis for prioritization of medical and other measures to deny the intent of terrorists. The agents in Table 2-4 and Table 2-5 contain a substantial proportion of the organisms capable of achieving mass casualties. Biological toxins, even the highly potent botulinum toxin, and chemical agents are absent because of their relatively low potential to cause casualties whether evaluated on the basis of purified agent or the basis of the likely highest concentration practically achievable. It is impossible to list every possible agent and there are always discussions among experts as to whether some should be added or omitted. However, there is general agreement that those in Table 2-4 comprise the most deadly and the most likely to seriously destabilize governmental function and civil society. All these agents were weaponized by the USSR or the U.S. (pre-1969) or both. They grow to excellent titer for more efficient manufacture and they are highly stable and infectious in aerosols when properly prepared. Agents in Table 2-5 are of lesser threat, particularly those at the bottom of the list. Some, such as typhus or glanders may well belong in Table 2-4. Toxins are inherently of lesser efficiency because they cannot match the killing or incapacitating power of the highly infectious organisms; the toxins have to produce their effects as delivered, but the infectious agents grow and produce toxins or other effects in the recipient's body.

According to a WHO scenario several infectious agents would be expected to produce 35,000 to >100,000 casualties if 50 kg were delivered in a line source and carried down-wind over a populated area. In the case of some of the more stable agents, down-wind reach would exceed 20 km. The Office of Technology Assessment (1993) has published similar figures. It must be borne in mind that the US and Soviet programs prepared literally metric tons, not kilograms, of agent and that appropriate devices for delivering line sources or multiple overlapping point sources were available (Alibek, 2001; Sidell et al., 1997). Impact on the infected members of the population would depend on the agent used and the nature of the response (for example, alacrity of recognition of initial patients, public health and medical infrastructure, vaccine and antibiotic stockpiles).

The additional impact that might be possible through modification of naturally occurring organisms by methods well within the reach of simple biotechnology including induction of antimicrobial resistance, enhancement of viru-

TABLE 2-4 Some diseases and their causative agents considered to be aerosol biological warfare threats capable of causing mass casualties (CDC Category A Biological Agents/Diseases)

Variola major (Smallpox)
Bacillus anthracis (Anthrax)
Yersinia pestis (Plague)
Francisella tularensis (Tularemia)
Viral hemorrhagic fevers (Filoviruses, Arenaviruses, and Rift Valley virus)

lence by addition of toxin genes, or selection of more stable or virulent organisms is formidable. Issues surrounding the more extensive engineering of threat agents are beyond the scope of this discussion, however, it is important to note that the potential exists, but that there would probably be a need for human testing of any resulting candidate.

Properties of Small Particle Aerosols

The basic properties of aerosols must be understood to appreciate the way in which such weapons could be used. The optimum aerosol particle size is thought to be 1–5 microns. This size provides two critical properties: the particles do not settle out over a several hour time period but rather are truly airborne and are carried on wind currents or through heating, ventilation, and air conditioning systems (HVAC) and they are of an optimum size to reach the terminal respiratory bronchiole or alveolus of humans and deposit in those critical areas to set up infection. Generation of small particle aerosols requires energy as evidenced by classical examples such as laryngeal tuberculosis, the wide dissemination of rubella by a disco singer (Marks et al., 1981), or the persistent cough of the index case of a nosocomial Lassa fever outbreak (Carey et al., 1972). Such particle sizes can be achieved intentionally by generating a liquid aerosol with a spray device or by using an appropriately manufactured powder. The dissemination system for a liquid must have the correct relationship between viscosity and solids content of the liquid, air pressure, orifice diameter, and other variables to attain the critical particle size, as well as have the proper stabilizers in the solution to assure that infectivity is not lost. The dry powders are difficult to manufacture, but they are extremely dangerous because they can be prepared so as to aerosolize with minimal energy input and can be manufactured in very fine particle sizes. If the skills to prepare these particles are avail-

TABLE 2-5 Some agents often mentioned as potential aerosol biological warfare or biological terrorist threats

Tick-borne flaviviruses
Typhus and other Rickettsiae
Glanders
Alphaviral encephalitidies
Brucellosis
Q fever
Melioidosis
Nipah virus
Staph Enterotoxin B
Ricin
Tricothecene mycotoxins

able, it may also be possible to formulate encapsulated weapons of greater biological stability.

It is obvious that the preparation and testing of such weapons requires microbiological skills to attain the high concentrations of organisms needed. Equally as important is the expertise for dissemination of liquid or powder aerosols in a stable form in the correct particle size. Such capabilities may be available from many people in different walks of life. Nonetheless, assembling the needed skill sets implies an organization with some resources, particularly as one moves into the powders and into greater quantities of agent. Practice or trials would be important to assure success, although, as the saying goes, "the proof of the pudding … ". Persons with previous experience in offensive biowarfare programs could be extremely valuable resources to such an endeavor.

When these aerosol clouds are generated, there is a period of instability and larger particles or agglomerates fall out near the dissemination device with considerable surface contamination possible. Once the small particle aerosols are formed, they will move with wind currents and traverse the landscape, being gradually diluted by mixing and decay. Because of their dependence on wind currents, aerosols may be diverted from their planned target; warming of the earth's surface after sunrise will result in their being carried to higher levels of the atmosphere unless inversion conditions are present. Loss of infectivity by biological decay will also occur, depending on stabilizers in the suspension medium, ultraviolet intensity, humidity, and temperature. They will enter buildings through HVAC systems, but the urban landscape profile may result in extensive disturbances of air flow. The ultraviolet light sensitivity of most of the organisms, combined with the meteorological needs, will favor the use of an evening, night time, or early morning, attack if done outdoors. Terrorist attacks may also be directed toward buildings or enclosed stadiums, making the air conditioning systems the obvious route of the delivery of the aerosol.

The extreme "fluffiness" or ease of aerosolization of the most dangerous powders is difficult to imagine. The material in the Hart Senate office building seems to offer an excellent example: all present in the room when the offending letter was simply opened had spores in their noses when tested 4 hours later; anthrax spores traveled to adjacent rooms through the HVAC; extensive surface contamination was present. Subsequent examination of the powder showed it to have a very high concentration of anthrax spores (reportedly 10^{12}/g), finely dispersed particles, and it was readily aerosolized with the slightest disturbance. Undoubtedly, without antibiotic prophylaxis most of those in the room would have suffered inhalation anthrax. Had this material been introduced clandestinely into the air conditioning intake, there would have been no warning until the first cases were recognized, perhaps too late for therapy. The magnitude of the exposure can be seen from studies done by the Canadian Defense Forces (Kournikakis et al., 2001) in a simulation using only $1/10^{th}$ the number of spores and $1/20^{th}$ the estimated quantity of a readily aerosolized powder containing

Bacillus globigius as a surrogate for *Bacillus anthracis*. In less than one minute, spores spread throughout the room in similar concentrations as observed at the site of the envelope opening. Testing of the filter from the respirator worn by the subject opening the letter yielded 80,000 infectious units, equivalent to an estimated 15-320 human LD50 for anthrax spores.

These principles governing airborne particles in the 1–5 micron range are essential in our response to the possibility of an aerosol attack. First of all, buildings with their HVAC may provide little protection against small-particle aerosols if currents of air bring a widely disseminated agent into the zone of their air intake. Indeed, the HVAC systems may provide a particular vulnerability to bioterrorism. Secondly, environmental sampling of surfaces or clothing in areas exposed to such aerosols or nasal swabs of potentially exposed persons will not predict if a person was infected with any certainty; finding agent in these situations demonstrates that an attack has taken place. Thirdly, surgical masks and similar defenses are not effective against such small particles and only give a false sense of protection; highly efficient masks that can protect against small particles (e.g., N100 masks designed for medical staff working with tuberculosis patients) and which are properly fitted on trained personnel are needed (Lowe et al., 1999). Obviously, protective gear must be worn during the risk period, but these small-particle aerosols are odorless and invisible to the eye so the general utility of personal protective gear is limited.

Secondary Aerosols

The issue of secondary aerosols is an important one. Infectious agents on solid surfaces such as soil, counter tops, machinery are thought not to be subject to aerosolization unless considerable energy is applied. Studies of *Bacillus* spores have shown aerosols intentionally deposited on the ground are very difficult to re-suspend with ordinary traffic or even intentional beating of the surface (Patrick, 2001). Soil subjected to high air flow yields few particles in the dangerous 1–5 micron range (Chinn et al., 1990). These field evaluations show that even when the contamination of a surface reaches 10^7/meter^2 the concentration above the surface, even with considerable disturbance, will be extremely low. Thus, in a field biowarfare situation there is little danger from secondary aerosolization.

There is little experience with transfer of infectious powders from one solid surface to another or with the deposit of larger clumps of highly aerosolizable particles on hard solid surfaces. Surrogate infectious agents, fluorescent tracers or radioactive particles predict that highly concentrated biological agents (titers $>10^9$/g and perhaps as much as 10^{12}/g) will extensively contaminate surfaces they impact and, if viable, pose a contact risk. Large quantities of organisms from dangerous powders can result in short-term presence of organisms in the external nares of exposed persons and the fall-out of larger particles can lead to environmental contamination near the site of dissemination. The possibility of secondary aerosols in this situation is thought to be small, but application of high energy sources or the presence of physical clumps of particles could be problematic.

The risk of aerosol infection to an exposed human would depend on the amount of material aerosolized and the infectious dose for humans. The actual amount of tularemia or Q fever required to infect 50% of exposed humans is known and is on the order of 1–10 organisms. For other organisms such as anthrax it is necessary to extrapolate from cynomolgus monkeys or other experimental animals. In this case the lethal dose for 50% of animals is 8,000 spores by aerosol. The LD50 is determined by exposing animals to graded doses of the infectious agent and calculating the linear relationship between the logarithm of the dose increment and the increase in response of the target animals (Finney, 1964). This is usually done between 20–80% lethality and the LD50 calculated. In fact the linearity can probably be extrapolated further to furnish at least an approximation of the risk from lower doses; in the case of anthrax, published values for the slope (Glassman et al., 1965; Chinn et al., 1990) suggest that inhaling a dozen spores could be risky in a small percentage of the population.

These concepts are important to the practical management of situations in which a suspicious powder is involved. The physical properties of a readily aerosolized powder will be recognized by an experienced observer or by laboratory analysis. An ordinary dried culture of, for example, anthrax will not pose a great hazard beyond the readily recognized and treated cutaneous anthrax. Decontamination of a building needs to address the dangerous states of the contaminating organism. Safety is the goal, not "sterility". In the case of anthrax spores, significant quantities of aerosolizable particles is the criterion. Sterility is less important than being certain that any residual infectivity is earth bound.

Relative Importance of Different Agents

Consideration of the different bioterrorism agents and some of their properties is the first step to prioritize defenses against them. Each has different properties as we see them today and thus each presents different threats and different opportunities for control. This discussion has been cast in terms of the worst case scenarios (effective broad-scale aerosol dissemination) but we must recognize that, although protection against this situation is important, the most likely eventuality is a less extensive or less successful attack. Fortunately, attention to the worst case is a step toward the more general solution, although the lesser eventuality should also be in the mind of planners.

It is also important to note that biodefense efforts meld with the general struggle against infectious diseases. For example, strengthening the public health system will provide benefits regardless of whether a bioterrorist attack occurs. Money spent on communications within the public health system is long overdue. Planning will help in disaster response, regardless of the nature of the event. Perhaps much of the money spent on increasing smallpox vaccine stocks will eventually be "sunk costs" but we should not regard research and vaccine development on other agents as anything other than an benefit for human-kind.

Smallpox provides a threat whose consequences are simply unacceptable, regardless of the probability of its use. Therefore we must develop clinician awareness, diagnostic systems, and stockpiles of existing vaccine that give us a validated countermeasure to deploy in case of attack. Whether additional antiviral drug and vaccine development is justified is a matter of prioritization against other threats.

Anthrax also is a special case. It is widely distributed in nature and thus readily available to terrorists in virulent form. The spores are extraordinarily stable on storage and in aerosols obviating many of the terrorist's research needs to develop an effective weapon. Inhalation anthrax is a fearsome disease if not treated early with effective antibiotics, and production of antibiotic-resistant anthrax is readily achieved.

Plague and tularemia are both severe diseases but they require another level of sophistication in weaponization. Their cultivation in virulent form and their dissemination in stable aerosols is more difficult than for anthrax. The viral hemorrhagic fevers are essentially without therapy, have severe psychological impact, and carry a high mortality. Their production is still more difficult, but the technology is readily accessible to an experienced microbiologist (Peters, 2000).

When considering the impact of limited or massive dissemination of the agents in tables 2-4 and 2-5, one must factor in the disruption of the health care system, the role of antibiotic resistance, the fear-factor in the population and medical staff, as well as the state of defensive preparations. One element that is often neglected is the influence of a communicable disease on travel and commerce. Any of these diseases could lead to severe disruptions in the free travel of U.S. citizens and others within the US and in international air transport systems. If the agent is also an agricultural pathogen, then internal movement of animals would be frozen and exports would be stopped, resulting in even more severe economic consequences.

Strategies to Confront the Problem

Any attempt to deal with BT should consider the entire spectrum of responses, including state and local organization supported by a comprehensive federal plan. The public health system will be the back-bone, but there will have to be participation of the entire society. Recognition by the clinician, laboratory diagnosis, and mobilization of countermeasures will all play a part. As noted above, the strategy should be tailored to each agent or group of agents.

It must be emphasized that environmental detection and patient diagnosis of the specific agent employed are keystones of an improved response to the threat. Detection suffers from the need to be active at the time and site of an attack, so economics will probably limit its future usefulness to selected high risk venues.

Diagnostics are, in principle, more focused and also require the suspicions of informed clinicians; widely deployed they are a very significant expense. An additional demand on detection and diagnostics is the recognition of subversion of our defenses by inducing resistance to anti-infectives or other protective modalities.

Vaccine approaches are suitable for selected at-risk groups, particularly for specific high-priority BT agents. However, specific vaccines are not general remedies for the threat to the civilian population. The expense and difficulty of administration and the inevitable side effects will limit their widespread use. They remain important elements of our response in selected populations and specific circumstances.

Anti-infective drugs could be a very effective response if problems of drug-development, drug-resistance, safety and efficacy testing, stockpiling, and distribution can be solved. Other supportive measures directed to bacterial toxins or the over-exuberant inflammatory responses induced by some viruses could be useful, as well. Further definition of the Toll-like receptor family could open the way to broadly protective remedies that could be used in the event of BT attacks.

Some agents pose sufficient problems to demand immediate and thorough attention. Smallpox, because of its track record of interhuman transmissibility and high case fatality, is clearly a first-echelon target. Anthrax, because of its ease of weaponization, deserves attention to the development of more effective therapy beyond antibiotics. Antitoxic strategies at the level of the toxin molecules as well as their down-stream effects should be developed in a very short time frame. Furthermore, the terrorist use of antibiotic-resistant strains should be anticipated.

Plague and tularemia might seem to be resolved in principle because of the existence of effective antibiotics, but their relatively short incubation periods place high demands on availability of effective antimicrobials and the facility with which antibiotic resistance can be induced has important implications for defensive strategies. This is complicated because the log-normal distribution of incubation periods is "front-loaded" (Sartwell, 1950) and because late treatment can fail even though the bacteria are eradicated.

The viral hemorrhagic fever agents would induce widespread fear and even panic among the both general population and health-care providers. The arenavirus drug ribavirin should be stockpiled in modest amounts in the mean while, but more general strategies against the arenaviruses and other viral threat pathogens should be pursued.

Of course intelligence information and any dissuasion afforded by international agreements would be most welcome. We clearly cannot depend on these modalities to protect us completely. Many of the agents are widely available and so measures designed to limit their access are illusory in their effectiveness; anthrax is a case in point. However, limiting access to certain agents such as Ebola, Marburg, and smallpox viruses should be pursued. The equipment needed to produce limited amounts of biological agents is readily available and we cannot control or monitor access, but perhaps we can develop measures to track high output equipment and the movement of particularly sensitive expertise and genetic material.

A strong research program and the industrial base to develop promising research leads into practical human countermeasures will be the best defense. One of the impediments, in addition to the perennial need for funding, is the lack of suitable containment laboratories. Furthermore, the diminution of expertise and suitable laboratories to study infectious aerosols is alarming. Another variable in play is the concern for limiting dissemination of research results; we have to be very careful not to suffocate our defensive effort with excessive secrecy unless the controls can be shown to add to our safety.

REDUCING THE RISK: FOODBORNE PATHOGEN AND TOXIN DIAGNOSTICS

Susan E. Maslanka,* Jeremy Sobel,* and Bala Swaminathan*

Foodborne and Diarrheal Diseases Branch
Division of Bacterial and Mycotic Diseases
National Center for Infectious Diseases
Centers for Disease Control and Prevention

Estimates of Foodborne Illness in the United States

The spectrum of illnesses caused by consumption of contaminated foods may range from self-limiting mild gastroenteritis to life-threatening neurologic, hepatic and renal syndromes (Mead et al., 1999). recently estimated the number of illnesses, hospitalizations and deaths in the United States using data from various national surveillance systems. Their estimates indicate that contaminated foods cause approximately 76 million illnesses, 325,000 hospitalizations and 5,000 deaths in the United States each year. The economic burden is estimated to be 9 to 32 billion U.S. dollars. More than 200 known diseases are transmitted through foods; the agents of foodborne illnesses include viruses, bacteria and their toxins, fungi and their toxins, parasites, poisonous plant components, marine biotoxins, heavy metals and possibly, prions. However in 82% of foodborne illnesses the identity of the pathogen is unknown. Of 1,500 deaths each year due to known pathogens, 75% are caused by *Salmonella, Listeria monocytogenes* and *Toxoplasma*.

---

* The information provided in this paper reflects the professional view of the authors and should not be construed as an official position of the U.S. Department of Health and Human Services or the Centers for Disease Control and Prevention.

Changes in the Foodborne Disease Outbreak Scenario

The epidemiology of foodborne diseases has undergone profound changes in the last 2 decades. Some factors influencing this change are the global distribution of food supplies to meet increasing consumer demands for greater diversity of foods, centralization of food production, processing and distribution to improve efficiencies and reduce costs, demographic changes occurring in industrialized nations that have resulted in increases in the proportion of the population with heightened susceptibility to severe foodborne infections, changes in food-related behavior of consumers and dramatic increases in world travel (Kaferstein et al., 1997; Swerdlow and Altekruse, 1998). One negative effect of high-degree consolidation of food production, processing and distribution is that food safety-related failures may affect large numbers of people over large geographic areas and may have disastrous public health consequences. Because of the explosive increases in international travel, new and emerging pathogens from one corner of the world are able to arrive at a location thousands of miles away in a matter of hours. The transcontinental flights themselves offer ample opportunities for transmission of foodborne disease during travel (Tauxe et al., 1987). In addition, the manufacturers and/or the distributors of the contaminated food are likely to encounter dire financial and public relations consequences following the implication of their products as a source of widespread illness.

These changes in food diversity and consumer demands have changed the way outbreaks are investigated. In the past, the majority of foodborne outbreaks occurred locally and could be readily detected by epidemiologic surveillance methods. An outbreak could be detected by an acute increase in foodborne illness and local food handling mistakes could be identified and controlled following epidemiology investigations. The "New Scenario" foodborne outbreak may involve a complex multistate investigation that may also be separated by time of onset of illness. While epidemiology investigations still provide needed information; laboratory data, particularly subtyping data, is now critical to implicate a food source and to link cases which may be geographically unrelated. A new level of quality (validation and standardization) of laboratory methods is required because of the potential adverse effects on a manufacturer of an implicated product. A once local problem, managed locally, now requires extensive resources to investigate and control.

Large Foodborne Disease Outbreaks

Examples illustrating large-scale (several thousands of cases) foodborne outbreaks are listed in Table 2-6.

The 1985 outbreak of *Salmonella* ser. Typhimurium infections was most likely caused by improper switching of the stainless-steel pipes in the milk processing facility, which resulted in raw milk coming in contact with pasteurized

TABLE 2-6 Foodborne outbreaks

Year	Location	Etiologic agent	Food vehicle	Number of persons affected
1985	Midwestern U.S.A.	*Salmonella* serotype Typhimurium	2% pasteurized milk produced by a large dairy	250,000
1994	Nationwide, U.S.A.	*Salmonella* ser. Enteritidis	Ice cream	224,000
1997	Sakai city, Japan	*E. coli* O157:H7	School lunch, radish sprouts	10,000

milk (Ryan et al., 1987). Interestingly, the outbreak was first recognized as a potentially large one when clinical laboratories in the region ran out of laboratory supplies for culturing *Salmonella* from ill persons. The ice cream-associated outbreak of *Salmonella enteritidis* infections in 1994 was caused by improper cleaning and sanitation of the ice cream premix tanker that was used previously to transport raw liquid eggs (Hennessy et al., 1996). The Japanese outbreak of *E. coli* O157:H7 infections was most likely caused by contamination of seeds used for sprouting or contamination of water used in the sprouting process (Michino et al., 1999).

Foodborne Pathogen/Toxins as Agents for Bioterrorism

Intentional contamination of our food and water supply is a real threat. Before this year, the only acts of bioterrorism in the U.S. involved foodborne agents. In 1984, members of a religious commune in Oregon attempted to influence the outcome of a local election by intentionally contaminating salad bars in several restaurants with *Salmonella* ser. Typhimurium. The outbreak affected at least 750 persons and *S.* Typhimurium was cultured from stool specimens of 388 persons (Török et al., 1997). In 1996, 12 of 45 laboratory workers at a large medical center in Texas became infected with *Shigella dysenteriae* type 2; the outbreak was associated with eating pastries or doughnuts that had been placed in the staff break room on a specific day. Epidemiologic and laboratory investigations strongly suggested intentional contamination of pastries by someone who had access to the bacterial stock cultures in the medical center's laboratory and who was familiar with the methods of culturing the bacteria (Kolavic et al., 1997).

Unlike some potential threat agents (i.e., smallpox) for which the sources are limited, many foodborne agents such as *Salmonella*, *E. coli* O157, and even botulinum toxin are relatively easy to obtain or produce. Many of the agents are stable under a variety of conditions and so could easily be added to food and water supplies before consumption. Although there was no reason to suspect foul play in any of the three foodborne outbreaks listed above, each could have easily been caused by intentional contamination by one or more persons involved in some way in food processing, preparation or transport.

Some foodborne disease agents require only a small inoculum to cause disease. Shigellosis can be caused by as few as a few hundred organisms; the infective dose of E. coli O157:H7 is thought to be even less (Hornick, 1998). Botulinum toxin is one of the most potent toxins known; it has been estimated that 1 gram of botulinum toxin is enough to kill 1.5 million people. Introduction of botulinum toxin into a food source would severely strain the resources of the health care system (e.g. antitoxin, hospital support, mechanical ventilators, etc) to adequately respond.

Although perhaps less deadly, other pathogens intentionally introduced into food and/or water supplies could also negatively affect the ability of a community to respond to the disease. Widespread disease could easily overburden the health-care system (hospitals, doctors, medical supplies), the public health system (epidemiologists, diagnostic testing laboratories), and emergency response teams (police, paramedics, decontamination crews). In addition, lack of consumer confidence in the quality of the food and water supply would be an additional burden on community governments. Capacity for early detection of intentional contamination of the nation's food and water supply is vital to minimize the impact on community health.

Challenges to Rapid Response

There are a number of challenges to providing a rapid response to intentional or unintentional widespread foodborne outbreaks (Mead et al., 1999).

Specimen collection. Although a mundane and easily overlooked aspect of response, a standard protocol for specimen collection is needed. Different foodborne agents (bacterial, viral, parasitic, etc) have different requirements for preservation to ensure efficient recovery for laboratory detection (Kaferstein et al., 1997).

Cost-reduction initiatives in healthcare. There is a move toward non-culture diagnostic and anti-microbial susceptibility tests to reduce healthcare costs. In some cases, tests for certain agents are not performed unless specifically requested by the physician (Swerdlow and Altekruse, 1998).

Need to differentiate between sporadic and outbreak cases. Foodborne illness occurs daily in the United States. Subtyping methods are needed which can rapidly and accurately separate sporadic cases from outbreak cases (Tauxe et al., 1987).

Lack of monetary incentives for commercial companies. The development, validation, and standardization requirements needed to produce a test kit that can be used for clinical specimens are time consuming and expensive. The lack of potential for profit prevents commercial companies from developing the required subtyping methods (Ryan et al., 1987).

Demand for real-time data. Currently there is no standardized computer system that will allow real-time data exchange between laboratories. The lack of real-time data exchange increases the time for establishing interventions to disease.

PulseNet as a Model

The National Molecular Subtyping Network for Foodborne Disease Surveillance (PulseNet) is CDC's network of public health and food regulatory agency laboratories. As a model, this network fulfills some of the needs for rapid response to outbreaks. Each state health department has the capacity to perform DNA "fingerprinting" of foodborne pathogens using CDC's standardized pulsed-field gel electrophoresis protocols. DNA patterns are analyzed using a standard software package using parameters set by CDC. Testing laboratories are able to communicate electronically via the Internet. CDC maintains a national database of DNA "fingerprint" patterns which is updated as new patterns are confirmed. This database has allowed state health departments to have early recognition of case clusters and helps to identify or confirm potential sources of disease. PulseNet is a rapid, effective means of communication between public health laboratories.

Integrated Approach to Foodborne Diagnostics

An integrated approach is needed to respond to intentional and nonintentional outbreaks of foodborne disease. (1) Sample collection, (2) improved diagnostics (including pathogen identification without isolation, rapid characterization and subtyping without isolation, and preservation of samples for subsequent pathogen recovery if needed), and (3) implementation of a real-time communication network must be seamlessly interconnected in order to effectively apply intervention measures during widespread outbreaks.

REFERENCES

A. Friedlander:

Bell JH. 1880. On Woolsorter's Disease. *Lancet* June 5, 871.

Friedlander AM. 2000. Anthrax-Clinical Features, Pathogenesis, and Potential Biological Warfare Threat. In: Remington JS and Swartz MM, eds. *Current Clinical Topics in Infectious Diseases*, Malden, MA: Blackwell Scientific Publications 20:335.

Friedlander AM. 2001. Tackling anthrax. *Nature* 414(6860):160–161.

Inglesby TV, Henderson DA, Bartlett JG, Ascher MS, Eitzen E, Friedlander AM, Hauer J, McDade J, Osterholm MT, O'Toole T, Parker G, Perl TM, Russell PK, Tonat K. 1999. Anthrax as a biological weapon: medical and public health management. *Journal of the American Medical Association*. 281:1735–1745.

Jernigan JA, Stephens DS, Ashford DA, Omenaca C, Topiel MS, Galbraith M, Tapper M, Fisk TL, Zaki S, Popovic T, Meyer RF, Quinn CP, Harper SA, Fridkin SK, Sejvar JJ, Shepard CW, McConnell M, Guarner J, Shieh WJ, Malecki JM, Gerberding JL, Hughes JM, Perkins BA. 2001. Bioterrorism-related inhalational anthrax: the first 10 cases reported in the United States. *Emerging Infectious Diseases* 7(6):933–944.

Leppla SH. 2000. Anthrax toxin. In: Aktories K and Just I, eds. Bacterial Protein Toxins. *Handbook of Experimental Pharmacology*, Berlin: Springer Verlag, Berlin. 145:445.

D. Dennis:

Anderson GW Jr, Worsham PL, Bolt CR, Andrews GP, Welkos SL, Friedlander AM, Burans JP. 1997. Protection of mice from fatal bubonic and pneumonic plague by passive immunization with monoclonal antibodies against the F1 protein of *Yersinia pestis*. *American Journal of Tropics and Medical Hygiene* 56:471–473.

Centers for Disease Control and Prevention. 2000. Biological and chemical terrorism: strategic plan for preparedness and response: recommendations of the CDC Strategic Planning Workgroup. *Morbidity and Mortality Weekly Report* 49(RR–4):1–14.

Dennis DT, Inglesby TV, Henderson DA, Bartlett JG, Ascher MS, Eitzen E, Fine AD, Friedlander AM, Hauer J, Layton M, Lillibridge SR, McDade JE, Osterholm MT, O'Toole T, Parker G, Perl TM, Russell PK, Tonat K; Working Group on Civilian, Biodefense. 2001. Tularemia as a biological weapon: medical and public health management. *Journal of the American Medical Association* 285:2763–2773.

Eyles JE, Sharp GJ, Williamson ED, Spiers ID, Alpar HO. 1998. Intra nasal administration of poly-lactic acid microsphere co-encapsulated *Yersinia pestis* subunits confers protection from pneumonic plague in the mouse. *Vaccine*. 16:698–707.

Farlow J, Smith KL, Wong J, Abrams M, Lytle M, Keim P. 2001. *Francisella tularensis* strain typing using multiple-locus, variable-number tandem repeat analysis. *Journal of Clinical Microbiology* 39:3186–3192.

Galimand M, Guiyoule A, Gerbaud G, Rasoamanana B, Chanteau S, Carniel E, Courvalin P. 1997. Multidrug resistance in Yersinia pestia mediated by a transferable plasmid. *New England Journal of Medicine* 337:677–680.

Health Aspects of Chemical and Biological Weapons. 1970. Geneva, Switzerland: World Health Organization: 98–109, 105–107.

Inglesby TV, Dennis DT, Henderson DA, Bartlett JG, Ascher MS, Eitzen E, Fine AD, Friedlander AM, Hauer J, Koerner JF, Layton M, McDade J, Osterholm MT, O'Toole, T, Parker G, Perl TM, Russell PK, Schoch-Spana M, Tonat K. 2000. Plague as a biological weapon: medical and public health management. *Journal of the American Medical Association* 283:2281–2290.

Inglesby TV, Grossman R, O'Toole T. 2001. A plague on your city: observations from TOPOFF. *CID*. 32:436–445.

Johansson A, Goransson I, Larsson P, Sjostedt A. 2001. Extensive allelic variation among Francisella tularensis strains in a short-sequence tandem repeat region. *Journal of Clinical Microbiology* 39:3140–3146.

Karlsson J, Prior RG, Williams K, Lindler L, Brown KA, Chatwell N, Hjalmarsson K, Loman N, Mack KA, Pallen M, Popek M, Sandstrom G, Sjostedt A, Svensson T, Tamas I, Andersson SG, Wren BW, Oyston PC, Titball RW. 2000. Sequencing of the *Francisella tularensis* strains SCHU4 genome reveals the shikimate and purine metabolic pathways, targets for the construction of a rationally attenuated auxotrophic vaccine. *Microbial Comparative Genomics*. 5:25–39.

Klevytska AM, Price LB, Schupp JM, Worsham PL, Wong J, Keim P. 2001. Identification and characterization of variable-number tandem repeats in the *Yersinia pestis* genome. *Journal of Clinical Microbiology* 39:3179–3185.

Titball RW, Howells AM, Oyston PCF, Williamson ED. 1997. Expression of the *Yersinia pestis* capsular antigen (F1 antigen) on the surface of an *aroA* mutant of *Salmonella typhimurium* induces high levels of protection against plague. *Infection and Immunity* 65:1926–1930.

Titball RW, Williamson ED. 2001. Vaccination against bubonic and pneumonic plague. *Vaccine.* 19:4175–4184.

S. Arnon:

Arnon SS, Schechter R, Inglesby TV, Henderson DA, Bartlett JG, Ascher MS, Eitzen E, Fine AD, Hauer J, Layton M, Lillibridge S, Osterholm MT, O'Toole T, Parker G, Perl TM, Russell PK, Swerdlow DL, Tonat K; Working Group on Civilian Biodefense. 2001. Botulinum toxin as biological weapon: medical and public health management. Consensus Statement of the Johns Hopkins Working Group on Civilian Biodefense. *Journal of the American Medical Association* 285:1059–1070.

Arnon SS. 1993. Clinical trial of human botulism immune globulin. In: DasGupta BR ed. *Botulinum and Tetanus Neurotoxins: Neurotransmission and Biomedical Aspects.* Plenum Press, New York, pp. 477–482.

Byrne MP, Smith LA. 2000. Development of vaccines for prevention of botulism. *Biochemie* 82:955–966.

Concluding paragraph, Consensus Statement of the Working Group on Civilian Biodefense 2001. Journal of the American Medical Association 285:1059–1070.

Marks JD. 2001. Proceedings of Biological Threats and Terrorism: Assessing the Science and Response Capabilities Workshop, November 27–29, 2001.

Rossetto O, Gorza L, Schiavo G, Schiavo N, Scheller R, Montecucco M. 1996. VAMP/synaptobrevin isoforms 1 and 2 are widely and differentially expressed in non-neuronal tissues. *Journal of Cell Biology* 132:167–179.

C.J. Peters:

Alibek K, Handelsman S. 1999. Biohazard. New York: Random House.

Carey DE, Kemp GE, White HA, Pinneo L, Addy RF, Fom AL, Stroh G, Casals J, Henederson BE. 1972. Lassa fever: epidemiological aspects of the 1970 epidemic, Jos, Nigeria. *Transactions of the Royal Society of Tropical Medicine and Hygiene*, 66:402–408.

Chinn KSK, Adams DJ, Carlon GR. 1990. Hazard Assessment for Suspension of Agent-Contaminated Soil. Joint Operational Test & Information Directorate, Dugway Proving Ground, Technical Report.

Cole LA. 1997. The Eleventh Plague: the politics of biological and chemical warfare. New York:WH Freeman and Co.

Countering the Changing Threat of International Terrorism, Report of the National Commission on Terrorism, Pursuant to Public Law 277, 105th Congress, Ambassador L. Paul Bremer III, Chairman, June, 2000.

Finney, D.J. 1964. Probit Analysis. Cambridge:Cambridge University Press.

Glassman HN. 1965. Industrial Inhalation Anthrax, Discussion. *Bacteriology Review* 30:657–659.

Haselkorn A. 1999. The Continuing Storm. Iraq, Poisonous Weapons, and Deterrence. New Haven: Yale University Press.

Hersh SM. 1968. Chemical and Biological Warfare. America's Hidden Arsenal. New York: Bobbs-Merrill.

Kournikakis B, Armour SJ, Boulet CA, Spence M, Parsons B. 2001. Risk Assessment of Anthrax Threat Letters. Technical Report DRES TR 2001–2048. Canada: Defense R&D Defense.

Lowe K, Pearson GS, Utgoff V. 1999. Potential values of a simple biological warfare protective mask. 263–281 In: Lederberg J, ed. Biological Weapons. The Limiting Threat, Cambridge: The MIT Press.

Marks, J.S., M.K. Serdula, N.A. Halsey, M.V. Gunaratne, R.B. Craven, K.A. Murphy, G.Y. Kobayashi, and N.H. Wiebenga. 1981. Saturday night fever: a common-source outbreak of rubella among adults in Hawaii. *American Journal of Epidemiology* 114:574–583.

McDermott J. 1987. The Killing Winds. The Menace of Biological Warfare. New York: Arbor House.

Patrick W, III. 2001. Biological warfare scenarios. 215223. In: Layne SP, Beugelsdijk TJ, Patel CKN, eds. Firepower in the Lab. Automation in the Fight Against Infectious Diseases and Bioterrorism. Washington, D.C.: Joseph Henry Press.

Peters CJ. 2000. Are hemorrhagic fever viruses practical agents for biological terrorism? In: Scheld WM, Craig WA, Hughes JM, eds. Emerging Infections 4:203–211. Washington, D.C.: ASM Press.

Rosebury T. 1947. Experimental Airborne Infection. Baltimore: Williams and Wilkins Co.

Sartwell PE. 1950. The distribution of incubation periods of infectious diseases. *American Journal of Hygiene* 51:310–318.

Second Annual Report, "Toward a National Strategy for Combating Terrorism," Report of a Commission, Chairman, Gov. James S. Gilmore III, December 14, 2000.

Sidell FR, Takafuji ET, Franz DR. 1997. Medical Aspects of Chemical and Biological Warfare. Washington, D.C.: Office of the Surgeon General.

Smithson AE. 2001. Rethinking the Lessons of Tokyo, Chapter 3 in Stimson Center Report No. 35, Ataxia: The Chemical and Biological Terrorism Threat and the US Response authored by Smithson, A., and Levy, L-A.

Torok TJ, Tauxe RV, Wise RP, Livengood JR, Sokolow R, Mauvais S, Birkness KA, Skeels MR, Horan JM, Foster LR. 1999. A large community outbreak of Salmonellosis caused by intentional contamination of restaurant salad bars. Pp. 167–184 In: Lederberg J, ed. Biological Weapons. Limiting the Threat. Cambridge: The MIT Press.

U.S. Congress, Office of Technology Assessment, 1993. Proliferation of weapons of mass destruction: Assessing the risks. OTA-ISC-559 Washington, DC: US Government Printing Office.

U.S. Congress, Office of Technology Assessment, 1993. Technologies Underlying Weapons of Mass Destruction, OTA-BPISC-115 Washington, DC: US Government Printing Office.

U.S. Department of Health and Human Services, Centers for Disease Control and Prevention, and National Institutes of Health. 1999. Biosafety in Microbiological and Biomedical Laboratories. Washington: U.S. Government Printing Office.

S. Maslanka:

Hennessy TW, Hedberg CW, Slutsker L, White KE, Besser-Wiek JM, Moen ME, Feldman J, Coleman WW, Edmonson LM, MacDonald KL, Osterholm MT. 1996. *New England Journal of Medicine* 334, 1281–1286

Hornick, RB. 1998 In: Infectious Diseases, eds. Hoeprich, P.D., Jordan, M.C. & Ronald, R.C. (J.B. Lippincott, Philadelphia), pp 736–741.

Kaferstein FK, Motarjemi Y, Bettcher DW. 1997. *Emerging Infectious Diseases*. 3, 503–510.

Kolavic SA, Kimura A, Simons SL, Slutsker L, Barth S, Haley CE. 1997. *Journal of the American Medical Association*. 278, 396–398.

Mead PS, Slutsker L, Dietz V, McCaig LF, Bresee JS, Shapiro C, Griffin PM, Tauxe RV. 1999. *Emerging Infectious Diseases* 5, 607–625.

Michino H, Araki K, Minami S, Takaya S, Sakai N, Miyazaki M, Ono A, Yanagawa H. 1999. *American Journal of Epidemiology*. 150, 787–796.

Ryan CA, Nickels MK, Hargrett-Bean NT, Potter ME, Endo T, Mayer L, Langkop CW, Gibson C, McDonald RC, Kenney RT. 1987. *Journal of the American Medical Association* 258, 3629–3274.

Swerdlow DL, Altekruse SF. 1998. In: Emerging Infections 2, ed. Scheld, W.M., Craig, W.A. & Hughes, J.M. (ASM Press, Washington, D.C.) Pp 273–294.

Tauxe RV, Tormey M.P., Mascola L Hargrett-Bean N, Blake, PA. 1987. *American Journal of Epidemiology*. 125, 150–157.

Török TJ, Tauxe RV, Wise RP, Livengood JR, Sokolow R, Mauvais S, Birkness KA, Skeels MR, Horan, J.M. & Foster, L.R. 1997. *Journal of the American Medical Association*. 278, 389–395.

3
Vaccines: Research, Development, Production, and Procurement Issues

OVERVIEW

Vaccines not only afford the best protection against infectious disease but can serve as strong deterrence factors as well. From a bioterrorist perspective, vaccine-resistant agents are more difficult to engineer than drug-resistant agents. But the potential market has been too small and uncertain to encourage the vaccine industry to make large investments in research, development, and manufacturing of new products. This is alarming considering the eight to ten years often needed to develop a new vaccine, compared to only two to three years to develop a new bioweapon.

Even among the four major vaccine manufacturers, there is insufficient production capacity. It was suggested during this session that in order to move animal and clinical testing forward, incentives need to be established to reduce the current challenges of vaccine development; vaccine production priorities need to be set and a central office or leader authorized to declare top priorities; and the role of the major vaccine manufacturers needs to be facilitated by clear directions and active collaboration between industry and government.

The use of vaccines as a civilian biodefense measure presents multiple challenges that are quite different from those of vaccine use by the military. Much of the challenge is due to the fact that the threats are uncertain and risk-benefit information difficult to assess. The very nature of terrorism produces a high level of uncertainty about what to expect and how to prepare. Additionally, DoD has developed vaccines to be used in normal healthy adults between the ages of 18 and 65, not pediatric, geriatric, immunocompromised or other subsets of the civilian population. Currently, there is no policy in place for immunizing the civilian population as a bioweapons defense measure, however several government agencies are working at unprecedented speed to put the correct policies into place.

The threat of a global pandemic makes smallpox one of the top vaccine priorities. An aggressive clinical development plan is currently in place; its goal is to build the stockpile with enough vaccine to protect the entire country within the year. The vaccine immune globulin (VIG) supply also needs to be expanded. Long-term goals include developing a safer vaccine that can be used in immunocompromised or other at-risk individuals.

Anthrax vaccine is another top priority. As of May 2001, over two million doses of the current anthrax vaccine have been administered to over 500,000 individuals, mostly military personnel. But there is an urgent need for more anthrax vaccine for the immunization of high risk civilian populations, as well as for use in medical management of exposed individuals in conjunction with antibiotics. Currently, there is only one manufacturer of licensed anthrax vaccine, but production is limited because of regulatory problems. Several commercial firms have offered to aid in scaled-up production, but the inherent variability of the manufacturing process and the risk of failure when scaling up so rapidly to such a high volume could create problems. Other mid to long-term anthrax vaccine needs include the development of a second-generation vaccine (e.g., a recombinant protective antigen vaccine) as well as better delivery technologies (e.g., plasmid DNA).

Of lesser importance than vaccines against smallpox and anthrax are vaccines against bacterial infections for which antibiotics can be used and other viral agents that, for the present, seem to be a lesser threat.

A recent independent review of DoD's vaccine acquisition program recommended an integrated approach between DoD and industry and the establishment of a dedicated national vaccine production facility that allows for maximal flexibility and expandable manufacturing capability for the production of various types of vaccines. Whether the proposed facility will be government-owned and contractor-operated or contractor-owned and contractor operated is open for discussion.

Ebola virus provides a useful paradigm for how a molecular-level understanding of the pathogenesis of a virus can be used to develop a new vaccine for an infectious agent that would otherwise be difficult to tackle. This type of molecular genetics approach can reveal possible targets for antiviral drugs as well. For example, recent studies have shown that one of the domains of the ebola virus forms a coil-to-coil structure that is similar to structures found in other viruses, including HIV and influenza. This similarity suggests that the approach being used to develop products for antiviral use against HIV may also be useful for targeting the coil-to-coil region of ebola virus. In fact, targeting this coil-to-coil structure may prove to be a useful general antiviral strategy against many different viruses.

Other vaccine issues that were raised during this session include:

- Improving the usefulness of DNA vaccines, which work well in rodents but not primates.
- Consideration of combination vaccines, for example can we use what we have learned from ebola to make a combination vaccine for use against all hemorrhagic fevers?
- Application of genomics to vaccine research could have, for example if we could use the new high throughput technology to identify genomic biomarkers for vaccine efficacy, then we could use these biomarkers in the future to move forward more quickly toward licensure.
- The need for a strong infrastructure to receive the intense flow of resources that would be expected with a rapid deployment of vaccines in response to outbreaks.
- The need for ways to accelerate vaccine FDA licensure without compromising product safety, for example use of the proposed animal efficacy rule for products that are either not feasible or ethical for human efficacy trials.

VACCINES FOR THREATENING AGENTS: ENSURING THE AVAILABILITY OF COUNTERMEASURES FOR BIOTERRORISM

Philip K. Russell,[*] M.D.
Special Advisor on Vaccine Development and Production
Office of the Secretary
Department of Health and Human Services

Recent events have brought the subject of vaccines as a defense against bioterrorism into very sharp focus. We have been forced to take action in an area that, for the civilian sector, had previously been largely an academic debate and planning exercise with inadequate definitive action. We have changed from a nation of skeptics concerning the threat of bioterrorism to a nation of believers. Several government agencies are working at unprecedented speed to acquire the needed vaccines and put the correct policies into place for utilization.

However, the use of vaccines for defense against bioterrorism presents multiple challenges that are quite different from the traditional public health use of vaccines for protection against endemic or epidemic diseases. The issues are also quite different from those faced by the armed forces. The appropriate use of vaccines as a defense against bioterrorism presents major challenges in public policy development as well as public education. The ongoing public debates in the media highlight the complexity of the issues and reveal the widespread lack

[*] This statement reflects the professional view of the author and should not be construed as an official position of the Department of Health and Human Services.

of understanding of the limitations of the current vaccines, especially vaccinia vaccine. For example, there is a call for widespread vaccination against smallpox but, in contrast, there is much misinformation and inappropriate fear about the effects of anthrax vaccine.

Some of the challenges involved with developing vaccine policies for defense against bioterrorism lie in the uncertainty of the threats. In contrast, policies for the use of vaccines against naturally-occurring disease threats are based on a wealth of historical and current epidemiologic information about disease burden and potential. Additionally, there is extensive data available on the safety of widely used vaccines that can be used to confidently assess risk benefit and cost effectiveness. In the case of agents of bioterrorism, however, risk assessment is much more difficult. The great difficulty in obtaining timely and reliable intelligence on the threat of biologic terrorism is a major part of the problem. Critical policy decisions—such as which vaccines will be needed, how large the stockpiles should be, and how the vaccine should be used—are greatly influenced by perceptions of threat. The very nature of terrorism produces a high level of uncertainty about what to expect and prepare for, and there is a wide and varying spectrum of perceived threats.

Obtaining the vaccines that are needed to protect our military and civilian populations depends entirely on effective government action. The potential market has been too small, at least up to the present time, to encourage the vaccine industry to make the large investments needed in research, development and manufacturing facilities. This has changed dramatically in the past two months. Nevertheless, the current situation is a result of past misjudgments, which resulted in insufficient government investment in vaccine research and development, and manufacturing capacity. There is an urgent need for rapid progress in R&D, manufacturing, and licensing processes, all of which are painfully slow processes when done by the usual methodologies.

Vaccines have varying usefulness in defense against bioterrorism. At the top of the list is the need for smallpox vaccine to prevent an outbreak from becoming a catastrophic global pandemic. Both smallpox and anthrax vaccines would be very useful in the medical management of exposed individuals, if the vaccines were readily available and placed in geographic proximity to multiple centers for distribution. Less important to the civilian populations are vaccines against bacterial agents that can be managed with antibiotics and viral agents which, at least for the present, seem to be lesser threats. These include plague, tularemia, hemorrhagic fever viruses, alphavirus encephalidities, Rift Valley fever, and others. However, several of these vaccines should be available for both civilian and military use. A government-owned production facility may be the best means for meeting the needs of these lower priority vaccines which will probably, at least initially, be made in much smaller quantities than smallpox and anthrax vaccines.

Smallpox Vaccine

The acquisition of a smallpox vaccine stockpile for civilian use started in 1999, with an Acambis contract for 40 million doses which now has been increased to 54 million doses. The seed virus was developed by cloning a New York City Board of Health strain derived from Wyeth Dry Vax. Animal model studies indicate that this strain appears to be somewhat less neurovirulent than the parent virus. The clinical development plan is aggressive; the phase I clinical trial should occur, as planned, in the latter part of January 2002. A very rapid procurement action has been in progress over the past weeks. The goal is to stockpile enough smallpox vaccine to protect the entire nation within the year. The response from the vaccine industry has been very heartening and has provided excellent options for utilizing existing manufacturing capacity to meet current requirements. Every effort will be made by CDC, FDA and NIH to assure that these contractors succeed to meet goals, time lines, and regulatory requirements. This will require truly unprecedented coordination and responsiveness by both the manufacturers and the various agencies.

Although the first step in building the smallpox vaccine stockpile is to ensure that vaccine manufacturing is underway, there are several other immediate issues that need to be addressed:

- Vaccination policy issues continue to be controversial. The CDC recently sent out a draft smallpox response plan to the states for comment. The plan calls for primary reliance on ring vaccination—the traditional method—to control an outbreak. The CDC has vaccinated 140 staff members who are most likely to be involved in investigating an outbreak, but no further vaccination with potential responders or health care providers is planned at this time. Laboratory personnel working with pox viruses will, of course, continue to be vaccinated.
- There is a need for more vaccine immune globulin (VIG) or VIG substitute to deal with the consequences of vaccination in immunosuppressed or other high risk subsets of the population. An interagency working group is currently exploring options for expanding the VIG supply.
- There is a need to develop a safer vaccine for use in immunosuppressed individuals, pregnant women, and other individuals for which the current vaccine is contraindicated. This will not only be a challenging research and development problem but also a challenging regulatory problem due to the difficulties in proving efficacy.

Anthrax Vaccine

The current licensed U.S. anthrax vaccine is a filtrate of culture media that contains a high level of PA (protective antigen) absorbed to alum; it probably contains small amounts of the other factors as well. An ongoing study at CDC is testing immunization schedules that involve fewer than the currently recommended six doses for this vaccine. Conventional wisdom has it that the live attenuated vaccines used in Russia and China are too reactogenic to be licensed in the United States. Israeli scientists have published reports on animal studies of experimental vaccines engineered to over-express recombinant protective antigen, but no clinical data are available.

The problems that the manufacturer has had with meeting regulatory criteria have limited the U.S. supply. A small amount of anthrax vaccine has been made available to DHHS by DoD, but that amount is far below what will be needed. There is an urgent need for a sufficient supply of anthrax vaccine for vaccinating high risk populations and for use as post-exposure vaccination in conjunction with antibiotics.

There are several immediate issues that need to be addressed. The production method for current licensed vaccine must be scaled up. Several commercial firms have made informal proposals to do this. However, this is a high risk option because of the inherent variability of the manufacturing process and the high risk of failure when scaling up so rapidly to such a high volume. There needs to be more serious consideration of the applications of the various platform technologies—such as plasmid DNA, viral vectors, and a variety of other delivery technologies—that are being developed within the biotech industry.

Finally, we need to accelerate development of a second generation vaccine. The time to availability could be shortened by overlapping large scale production with clinical trials. It has been suggested that we might have a stockpile of IND recombinant protective antigen (PA) vaccine within 18 months. This may be an achievable goal if all involved interests work in an effective, coordinated manner. A recombinant PA vaccine produced in *E. coli* will likely be the first to enter a phase I trial.

In order to address this issue of a second generation vaccine, the National Institute of Allergy and Infectious Diseases has put together a team with contractor help. Efforts are underway to gather all available information on ongoing or planned development efforts for a second generation anthrax vaccine, and compile the information in a systematic fashion and convene several advisors to review the resulting data, findings, and policy options. This may involve a major research and development contract program similar to what exists for smallpox vaccine and which will hopefully build on the work that has been done by DoD and DoD-DHHS collaboration. It will hopefully involve some new players as well, including the large vaccine manufacturers. Although it is difficult to predict which particular options will receive aggressive support, there is nonetheless a system now in place that will hopefully pave the way for pursuing an effective strategy in a reasonable period of time. The speed at which a second generation anthrax vaccine is developed will depend on both the underlying science and the responsiveness of the vaccine industry to national needs.

THE DEPARTMENT OF DEFENSE AND THE DEVELOPMENT AND PROCUREMENT OF VACCINES AGAINST DANGEROUS PATHOGENS: A ROLE IN THE MILITARY AND CIVILIAN SECTOR?

Anna Johnson-Winegar,* Ph.D.
Deputy Assistant to the Secretary of Defense, Chemical and Biological Matters, U.S. Department of Defense

Introduction

In October 2001, the threat of bioterrorism became a reality. In support of this Forum's efforts to identify the obstacles to preparing an optimal response to bioterrorism—particularly as it relates to the complexities of interaction between private industry, research and public health agencies, regulatory agencies, policymakers, academic researchers, and the public—this paper will highlight emerging opportunities for more effective collaboration as well as scientific and programmatic needs for responding to bioterrorism. The focus of this paper is on the potential opportunities and issues related to Department of Defense (DoD) support for the research, development, and production of biological defense vaccines for the military and civilian populations to protect against bioterrorist threats. This paper will address the following topics:

- Current medical biological defense research and development efforts;
- Current biological defense vaccine capabilities;
- Proposed national biological defense vaccine production facility; and,
- Issues related to the use of biological defense vaccines.

In accordance with Congressional direction, DoD established a Joint Service Chemical and Biological Defense Program in 1994. The vision of the program is to ensure U.S. military personnel are the best equipped and best prepared force in the world for operating in future battlespaces that may feature chemical or biological contamination. The capabilities being developed for the military may have applicability to protection of civilians, especially as the military mission may increasingly support homeland security. Vaccines to protect against biological agents provide one critical capability to protect against the threat.

Medical Biological Defense Research and Development Efforts

The primary research program for the development of biological defense vaccines to protect U.S. forces is the Medical Biological Defense Research Pro-

---

* The information provided in this paper reflects the professional view of the author and not an official position of the U.S. Department of Defense.

gram (MBDRP). In developing countermeasures to biological agents, the MBDRP uses a technical approach that focuses on four areas:

- Identify mechanisms involved in disease process;
- Develop and evaluate products (vaccines or drugs) to prevent or counter effects of toxins, bacteria, viruses, and genetically engineered threats;
- Develop methods to measure effectiveness of countermeasures in animal models that predict human response; and,
- Develop diagnostic systems and reagents.

Biological defense vaccines are being developed to counter viruses, toxins, bacteria, and genetically engineered biological threat agents. Research activities start with basic research activities and proceed through the following steps, as research demonstrates successful candidates: (1) construction of the infectious clone, (2) identification of attenuating mutations, (3) construction of vaccine candidates, (4) testing in rodent models, (5) testing in non-human primates, (6) final selection, and (7) formulation. The formulated production may then become a candidate for an Investigational New Drug (IND) application for transition to advanced development and clinical trials, then ultimately licensed production.

An example of a product being developed within the MBDRP is the Next Generation Anthrax Vaccine. In cooperation with the National Institutes of Health, the next generation vaccine will provide greater or equal protection, require fewer doses to produce immunity, and have fewer adverse effects than the current anthrax vaccine. The reduced number of doses would provide greater flexibility to military forces by reducing the time constraint for developing immunity, hence accelerating the time for fielding a protected force. The next generation vaccine is based on recombinant protective antigen (rPA), which binds to the lethal factor (LF) and edema factor (EF) of *B. anthracis*. The recombinant production technology would eliminate need for spore-forming anthrax, and hence the need for a dedicated production facility. Overall, the next generation anthrax vaccine would decrease production cost, allow a greater range of potential vaccine production facilities, and potentially allow for streamlining of the regulatory approval process.

Another example of a product being developed within the MBDRP is Multiagent Vaccines (MAV) for Biological Warfare (BW) Threat Agents. The MAV project is a proof-of-principle effort to construct a vaccine or vaccine delivery approach that could concurrently immunize an individual against a range of BW threats. Bioengineered and recombinant vaccine technologies will be exploited to achieve vaccines that are directed against multiple agents, yet use the same basic construct for all of the agents. The MAV would be analogous to commercial vaccines (e.g., measles-mumps-rubella) but would exploit new approaches—naked DNA vaccines and replicon vaccines. The MAV would result in a reduced number of doses and thus provide greater flexibility to military forces by reducing the time constraint for developing immunity, hence accelerating the time for fielding a protected force. The MAV also could decrease production cost, allow for greater range of potential vaccine production facilities, and potentially allow for streamlining of the regulatory approval process.

Current Biological Defense Vaccine Capabilities

Joint Vaccine Acquisition Program (JVAP)

In order to enable the transition of candidate biological defense vaccines developed under the MBDRP or from other sources, a Prime Systems Contract was awarded in November 1997 to DynPort Vaccine Production Corporation, LLC. The JVAP was established for the purpose of developing, testing, and Food and Drug Administration (FDA) licensure of vaccine candidates, and production and storage of vaccine stockpiles. A major objective of the program is to establish a viable industrial base for vaccine production. The next generation anthrax vaccine (rPA) is one of several vaccines being investigated for development by the JVAP. Other vaccines in advanced development include smallpox, pentavalent Botulinum Toxoid, and tularemia. The Prime Systems Contract also provides options for other biological defense vaccines. Currently, all vaccines in the JVAP are in the development phase.

Anthrax Vaccine Adsorbed (AVA) and the Anthrax Vaccine Immunization Program (AVIP)

The only vaccine currently licensed for use in the United States to protect against anthrax is AVA. AVA is cell-free filtrate, produced by an avirulent strain of *Bacillus anthracis*. It is manufactured by BioPort Corporation in Lansing, Michigan and procured under a separate contract. It was licensed by the FDA in 1970. Six doses of the vaccine are required for full immunity, including doses at 0, 2, and 4 weeks, 6, 12, and 18 months, followed by an annual booster.

On December 15, 1997, the Secretary of Defense approved the decision to vaccinate all of the U.S. armed forces against anthrax, contingent on the successful completion of four conditions, which were met: supplemental testing of the vaccine; tracking of immunizations; approved operational and communications plans; and review of health and medical aspects of the program by an independent expert. Implementation is determined in accordance with DoD Directive 6205.3, "DoD Immunization Program for Biological Warfare Defense," November 26, 1993, with complete implementation of the plan contingent upon adequate supply of the licensed vaccine.

On May 28, 1998, the Secretary of Defense directed vaccination of the total force. Implementation of this directive was administered by the AVIP. As of May 29, 2001, more than two million doses were administered to more than 500,000 military personnel, with at least 70,000 completing the full six-shot regimen. Since then, there has been only a few who have received vaccinations. As outlined in a June 8, 2001 memorandum, the Secretary of the Army ordered a slowdown in immunization to accommodate delays in release of vaccine pending FDA approval. Implementation of the vaccination continues to designated special mission units, to vaccine manufacturing and DoD personnel conducting anthrax research, and others conducting Congressionally mandated anthrax vaccine research. Detailed information on the status of the AVIP is available at www.anthrax.osd.mil.

What Does Producing a Vaccine Mean?

With no vaccines currently in production under the JVAP and AVA as the only currently available FDA licensed vaccine for protection against BW threats, DoD is evaluating other mechanisms to increase and sustain vaccine production. In order to identify the status of vaccines, it is important to understand the major phases of research, development, and production through which they must proceed. Within different phases of vaccine development and production, there will be varying levels of production risk and overall risk. There are three major phases in the development and production of new vaccines—science and technology base, development and licensure, and licensed production. Following is a summary comparing different activities within each phase.

Production Approach

Within the science and technology phase, production is focused on small quantities and relies on bench top methods, which may include many different approaches, including new state-of-the-art experimental approaches. When a candidate product transitions to the next phase, a best approach is selected (or in some cases two or three promising approaches) and tested for scale up for full scale production. Following licensure, production proceeds at full scale and relies on a single, fixed method. Changes in the method typically require further testing and require approval by the FDA.

Vaccine Recipients

Perhaps the most obvious difference among the phases are the numbers and types of vaccine recipients and the purposes for which they receive the vaccine. Within the science and technology phase, recipients are primarily laboratory animals and include hundreds of animals. The primary purpose for using these recipients is to demonstrate the potential effectiveness of a vaccine candidate, that is proof-of-principle testing. During the development and licensure phase, vaccine recipients are humans, who participate in clinical trials. All recipients are volunteers, who participate in clinical trials that comply with FDA regulations. The focus of these investigations is to determine the safety and efficacy of a vaccine as well as to optimize dosing and scheduling. The final phase is production and includes providing a licensed vaccine to all individuals who may be at risk, in accordance with the FDA license and based on quantities available, for the purpose of providing protection against potential threats. The effected populations could be on the order of millions of individuals.

Production Risk

Production risk during the science and technology phase is moderate since only small quantities can be produced yet only small quantities are needed. Risk is minimized since FDA approval of the product is not required. During the development and licensure phase, production risk is usually high because of the risks involved in scaling up pilot lot product to full scale production. Overall risk is also high because of reliance on and surrogate models or biomarkers to determine efficacy, since law prohibits exposure of humans to chemical or biological agents.

Overall Risk

Overall risk for production of biological defense vaccines will vary depending on the type of vaccine being produced and the policy implemented for immunization. For example, use of a live vaccine (e.g., vaccinia live vaccine) poses risk that inoculated individual may be giving off live vaccinia viruses until scarification has occurred (2–5 days), hence potentially exposing unprotected individuals. Another risk is that low rates of adverse effects may become more apparent in a large scale immunization program than had occurred during testing. For example, if 1,000 people are tested in clinical trials and only one had a serious adverse reaction, there may be hundreds of reactions if the total military force is vaccinated.

Biological Defense Vaccine Development and Production Issues

One of the major factors limiting the availability of biological defense vaccines is the limited interest from the pharmaceutical industry in supporting the production of these vaccines. In contrast to vaccines to support public health needs (e.g., childhood diseases, influenza), most vaccine needs are fulfilled by the private sector. However, the private sector has some challenges in fulfilling public health vaccine needs. The vaccine production industrial base is nearly at full capacity to meet public health priorities. This will pose a challenge for the production of biological defense vaccines if production of biological defense vaccines results in the deferral of production of public health vaccines. Biological defense vaccines are considered specialty biologics and interest is primarily centered on a few small to mid-sized companies. Industry interest is limited in part because of requirements to conduct large, complicated clinical studies to demonstrate safety, immunogenicity, and efficacy (where possible).

Another major factor effecting the timely availability of biological defense vaccines are issues related to compliance with Chapter 21 of the Code of Federal Regulations (21 CFR), Food and Drug Administration (FDA). The specific issue relates to the ability to determine the clinical efficacy of biological defense vaccines. 21 CFR requires that for efficacy to be established, vaccines must be tested in informed, volunteer human subject who are exposed to the condition against which the vaccine is intended to protect. However, legal and ethical constraints prohibit exposing human subjects to biological agents. This constraint plus limited availability of human data for most vaccines mean that under current regulations, biological defense vaccine efficacy cannot be established. In order to address this constraint, FDA published a proposed rule on October 5, 1999 entitled, "New Drug and Biological Products; Evidence Needed to Demonstrate Efficacy of New Drugs for Use Against Lethal or Permanently Disabling Toxic Substances When Efficacy Studies in Humans Ethically Cannot Be Conducted; Proposed Rule." (FDA rules are available at http://www.fda.gov/cber/rules.htm.) The proposed rule is expected to be finalized during 2002. Under this rule, efficacy may be determined based on data from clinical testing on animals (using at least two different species with preference that non-human primates be one of the species.) Animal data would serve as a surrogate for human data, but there would need to be significant data demonstrating that the effects in animals is related to effects in humans. Without the ability to license vaccines based on surrogate test data, biological defense vaccines would remain as investigational new drugs, which would continue to limit availability.

Proposed National Biological Defense Vaccine Production Facility

Following years of research, development, and efforts to produce biological defense vaccines in sufficient quantities to meet DoD needs, a different approach is currently being planned. In July 2001, DoD submitted a report to Congress detailing biological defense vaccine efforts within DoD. Known as the "Top Report"—because it provides the results of an independent expert panel chaired by Franklin Top, M.D.—this report summarized key shortcomings of current biological defense vaccine acquisition efforts. The report made the following findings and recommendations:

- The scope and complexity of the DoD biological warfare defense requirements are too great for either the DoD or the pharmaceutical industry to accomplish alone,

- The panel recommended a combined integrated approach whereupon DoD would work closely with the vaccine industry and national scientific base, and
- The panel recommended the construction of a government-owned, contractor operated (GOCO) vaccine production facility, which would include production capacity for up to eight vaccines over the next 7–12 years and would cost an estimated $2.4–$3.2 billion over that time.

The report recognized that in order for the GOCO to be successful, it would require long-term government commitment, increased resources, innovative DoD business and program management practices, and effective participation by established pharmaceutical industry leaders in vaccine discovery, licensure, and manufacturing.

The design concept for a GOCO biological defense vaccine production facility would accommodate three bulk vaccine production suites, each with different processes: spore-forming bacteria (for which FDA requires separate facilities), microbial fermentation, and tissue culture (viral vaccines). A modular design would allow flexible and expandable manufacturing capacity for production of DoD-critical vaccines that are intended for force health protection.

The scale of the facility will be determined in part by the quantity of vaccines to be produced. The assumptions for the production capacity requirement are categorized into three tiers. *Tier 1* is the baseline requirement and reflects current production requirements, which is the same as current requirements for the JVAP and AVIP. This tier includes sufficient anthrax vaccine for the entire force (approximately 2.4 million doses). It additionally would require 300,000 Troop Equivalent Doses (TEDs) for other biological defense vaccines. (Troop equivalent dose is defined as the number of vaccine administrations to reach full immunity. Boosters are not included.) *Tier 2* would require three million TEDs (2.4 million for U.S. forces + 0.6 million for Commanders Reserve) of each vaccine to be produced to allow for total force protection plus sufficient quantities to support annual requirements due to personnel turnover. This requirement was the basis for the initial GOCO cost estimate. *Tier 3* would require approximately 300 million TEDs of each vaccine to support civilian protection for the entire U.S. population.

In order to define the requirements for vaccine production and to ensure that it addresses national, and not just DoD needs, an interagency advisory group has been established. Interagency participation has been led by DoD and the Department of Health and Human Services, with participation from several organizations (including the Office of Homeland Security) to ensure a broad perspective. Federal participation is essential since biological defense vaccine needs are not being met by private industry. No individual department has the sufficient, full-spectrum capability and capacity to support vaccine needs. A national vaccine authority may be essential to ensure interagency needs are addressed not only in the planning phase but also in implementation. The details of the national vaccine authority are being developed, though it is *not* likely to be established as a new agency.

Issues Related to the Use Of Biological Defense Vaccines

Why Vaccinate? Vaccine Use Risk Management Decisions

BW agents pose high risk to military forces and operations, and at least ten countries are pursuing offensive BW programs. Vaccines are the lowest risk, most effective form of protection against BW threats. Vaccines are more effective and have fewer adverse effects than antibiotics or other treatments following exposure. While masks may provide highly effective protection, they may impede performance and must be worn to provide protection. Vaccines enable force protection by providing continuous, long-lasting protection. In addition, there are currently no real-time BW detection systems available. While there are systems that provide the ability to detection respirable aerosols in near real-time, the best available systems today take 15–45 minutes to identify a specific BW agent.

Vaccines are unusual among medical products in that they are given to healthy people to keep them healthy. Table 3-1 shows several of the vaccines commonly given to protect against infectious diseases and contrasts them with the limited number of biological defense vaccines currently available. Biological agents that may be used as weapons may be naturally occurring but have a very low incidence of natural occurrence (at least in the United States.)

The risk assessment for using biological defense vaccines is different from naturally occurring infectious diseases (Grabenstein and Wilson 1999). Because to vaccinate is based on *potential* risk of disease outbreak rather than actual incidences. Consequently, a proper risk assessment for biological defense vaccines

TABLE 3-1 Selected infectious diseases vaccines and biological defense vaccines

Selected Infectious Diseases Vaccines		Biological Defense Vaccines
• Typhoid	• Hepatitis A virus	• Anthrax Vaccine Adsorbed (licensed)
• Yellow fever	• Meningococcal disease	• Botulinum Toxoids*
• Malaria	• Influenza vaccine	• Tularemia Vaccine*
• Diphtheria	• Measles	• Smallpox vaccine (Vaccinia Virus, Cell Culture-derived)*
• Tetanus	• Mumps	• Equine Encephalitis Virus Vaccines*
• Poliovirus	• Rubella	

* Investigational New Drug (IND) status

should not be a trade-off assessment between the actual adverse effects of a vaccine *vs.* the actual adverse effects of the disease, but the actual adverse effects of a vaccine *vs.* the potential adverse effects of the disease.

The policies on the use of biological defense vaccines will affect biological defense vaccine manufacturing. The two basic options for immunization are *stockpiling* vaccines in anticipation of a specific contingency or *routine use* immunization to ensure continued general readiness. If vaccines are stockpiled, manufacturing must address issues related to maintaining the stockpile as a result of the limited shelf life of some vaccines. Additionally, if vaccines are produced in bulk, once the required quantities are produced, manufacturers must ensure that the facilities remain capable of retaining an FDA facility license when production is not ongoing.

The assessment of potential and actual effects may effect product development. For example, as polio has been significantly reduced as a result of extensive vaccination, the Centers for Disease Control have recommended use of the inactivated polio vaccine (IPV) rather than the oral polio vaccine (OPV). While OPV has greater efficacy, it is also linked with rare occurrences of vaccine-associated paralysis. As cases of polio have been virtually eliminated in the United States, the risk of rare occurrences of adverse effects of the vaccine has exceeded the risk of the occurrences and effects of the disease.

If biological defense vaccines are produced and planned for use—especially among civilians populations—vaccine development criteria may place greater emphasis on vaccine safety than on vaccine effectiveness. Risk assessments may be complicated by the fact that the limited industrial base capacity for biological defense vaccine production will likely result in only one vaccine being available for military and civilian use.

There are other key differences between the military and civilian populations that make risk assessment difficult. One factor is that biological defense vaccines made for the military population are intended for use only in healthy adults. By contrast, the general population will also include significant subgroups for which vaccine safety, efficacy, or dosing information may not be fully understood, including pediatrics, geriatrics, pregnant women, and immune-compromised individuals. Currently there is no policy in place to immunize the civilian population absent a naturally occurring threat. If a licensed biological defense vaccine were available for use by the general population, an immunization policy for civilian use would be needed to address several issues before immunization could begin. Some of the issues that would need to be addressed are, for example, who would be vaccinated—the entire population, or a subgroup? Which subgroup(s)? Those living in specific regions? First responders? If symptoms of biological agent do not appear, would that be interpreted as the absence of a threat or the effectiveness of the defense? Paradoxically, would the demand for the vaccine diminish as the apparent threat also diminished? Civilians may also have greater concerns about the long term safety effects as a result of vaccine use. Additionally, there may be concerns regarding the unknown safety of the use of biological defense vaccines when interacting with other medical products. While there is no adequate basis to assess safety, there is no basis for extraordinary concern (Institute of Medicine, 1996).

Conclusions

The Department of Defense may bring valuable assets to bear to counter the use of biological agents by terrorists. Currently, the DoD mission is focused on responding to threats to the military. Because of DoD's experience in defending against biological threats, DoD will continue to play a role in addressing the threat to the civilian population as well. DoD will continue to work with other agencies, including the new Director of Homeland Security, to determine what role it will play in homeland security, which will be defined in The Federal Response Plan, presidential directives, and other sources.

The availability of vaccine to protect against anthrax and other biological agents is based on several factors. One key factor is sustained resources to transition products from the science and technology base to advanced development. Resources include not only adequate funding, but also trained personnel, which is a critical factor since the biotechnology and pharmaceutical industry as a whole is facing shortages of skilled personnel. A second factor limiting the availability of biological defense vaccines is that they are similar to orphan drugs. There is no commercial incentive for manufacturers to produce vaccines. Federal investment may be required to retain the services and capabilities of the biotechnology and pharmaceutical industry.

While the availability of vaccines is critical, the decisions of whether to vaccinate will remain equally important. Vaccination decisions will continue to have greater physiological consequences than non-medical measures to protection against the threat (e.g., whether to wear masks). The decision will need to weigh the risk of actual low rates of adverse effects against the potential for protecting against catastrophic effects. In making these decisions based on risk, communicating the risk decision will be at least as important as risk assessment. Failure to have a coordinated public policy decision on vaccination support for civilians may result in individuals self-prescribing treatments or failing to comply with recommended guidelines.

APPLICATIONS OF MODERN TECHNOLOGY TO EMERGING INFECTIONS AND DISEASE DEVELOPMENT: A CASE STUDY OF EBOLA VIRUS

Gary J. Nabel,* M.D., Ph.D.
Director, Vaccine Research Center
National Institute for Allergy and Infectious Diseases

In recent years, increasing attention has been focused on the Ebola virus as a potential public health problem, either from natural or deliberate outbreaks. Like the genetically related Marburg virus, Ebola is a filovirus that causes highly lethal hemorrhagic fever in humans and primates. Infection rapidly progresses from flu-like symptoms to hemorrhage, fever, hypotensive shock, and eventually, in about 50–90% of cases, death (Peters et al., 1996; Peters and Khan, 1999). The molecular mechanisms underlying the pathogenicity of the Ebola virus are not well understood, in part because it has emerged only relatively recently (for reviews see Balter, 2000; Colebunders and Borchert, 2000). There was a series of outbreaks in central Africa in the mid-1970s and again in the 1990s (i.e., the Ivory Coast in 1994, Gabon in 1994–1996, Zaire in 1995, Gulu, Uganda in 2000 and presently in Gabon and the Republic of Congo). Ebola virus infection has appeared once in the United States, in Reston, Virginia. The Reston strain is not pathogenic in humans, and the outbreak was fortunately restricted to non-human primates.

One of the reasons that Ebola is highly lethal is that this virus replicates at an overwhelming rate (Sanchez et al., 1996a). Thousands of Ebola virus particles per host cell can completely envelop the cell and take over its entire protein synthetic machinery. We have only recently begun to understand the molecular mechanisms underlying this phenomenon. Although we have a descriptive understanding of the cytopathic effects of viral replication, we lack a clear understanding of how these various changes in cell structure and viability occur. Elucidating these details will be critical for developing vaccines and other antiviral therapies.

Aside from the obvious immediate health threat that would be posed if it were introduced into the population, Ebola virus represents a useful paradigm for dissecting the molecular genetics of a virus. Most of what is known about Ebola pathogenesis is derived from genetic studies of the virus. Although Ebola is very similar to the genetically related Marburg virus, it differs in at least one important respect. The gene that encodes the viral glycoprotein in Ebola generates two gene products, whereas in Marburg, this gene encodes a single protein (Sanchez et al., 1996). One of the gene products is secreted as a soluble 50 to 70 kDa glycoprotein, whereas the other is a full-length 120 to 150 kDa glycoprotein that inserts

---

* This statement reflects the professional view of the author and should not be construed as an official position of the National Institute for Allergy and Infectious Diseases, National Institutes of Health.

into the viral membrane (Volchkov et al., 1995; Sanchez et al., 1996). The secreted form was originally believed to serve as an immunological decoy for the full-length glycoprotein, allowing the full-length glycoprotein to attach to the target cell. However, more recent evidence now suggests that this hypothesis is unlikely. Instead, the secreted form appears to inhibit early steps in neutrophil activation and thereby inhibit the host inflammatory response to the virus (Yang et al., 1998). The secreted glycoproteins have been shown to bind quite well to neutrophils, but bind poorly to endothelial cells (Yang et al., 1998). In contrast, the full-length glycoprotein interacts with endothelial cells but binds poorly to neutrophils (Yang et al., 1998). This glycoprotein enables the Ebola virus to recognize and introduce its viral contents into the endothelial cell lining of the blood vessels, as well as monocytes/macrophages, thereby resulting in the cellular damage that is associated with the devastating symptoms of Ebola infection.

Antiviral Targets

Detailed analyses of the mechanisms of viral entry, replication, and cell damage have identified the Ebola glycoprotein 2 (GP2) as a potential antiviral target. In particular, there is a region in the GP2 ectodomain of Ebola virus that forms a coiled coil, or hairpin-like structure similar to what exists in the human immunodeficiency virus (HIV), influenza, respiratory syncytial virus, and a variety of other viruses (Weissenhorn et al., 1998a, 1998b; Malashkevich et al., 1999). This coiled-coil region contributes to membrane fusion by undergoing conformational changes after the glycoprotein binds to the membrane receptor (Weissenhorn et al., 1998b; Watanabe et al., 2000). The fact that this structure is conserved in a number of different viruses suggests that it may represent a potential target for antiviral therapy. In fact, a peptide product directed at the analogous structure in HIV has potent antiviral effects and is currently being developed for the clinical treatment of AIDS. This or similar peptides could be useful against many other viruses as well, including Ebola.

Not only does the transmembrane glycoprotein direct the Ebola virus into specific cells, but the glycoprotein itself is also highly toxic to cells. For example, when full-length Ebola glycoprotein is overexpressed in cultured renal epithelial cells, it inserts into the membrane and causes morphological changes and detachment from culture dishes (Yang et al., 2000). This finding suggests that there is a genetic determinant in the glycoprotein that mediates its toxicity and, therefore, might represent another potential target for antiviral therapies. Mapping studies identified a serine-threonine-rich, mucin-like core domain region of the glycoprotein that is required for cytotoxicity in human endothelial cells (Yang et al., 2000). When the mucin-like region of the glycoprotein was deleted, its cytotoxicity was abolished, but protein expression and function remained unchanged (Yang et al., 2000). Every possible open reading frame in the Ebola virus genome has been tested for toxicity, except for the polymerase region. To date, only the glycoprotein has been shown to induce toxic cytopathic changes. However, a better understanding of the detailed molecular mechanism of virus assembly may eventually provide insight into other potential antiviral targets as well.

Vaccine Development

Not only does the glycoprotein play an important role in toxicity, increasing evidence suggests that it also plays an important role in the pathogenesis of Ebola infection. Infection of cultured cells with adenoviral vectors encoding the glycoprotein causes considerable cellular damage that correlates with toxicity. However, overexpression of a glycoprotein that is unable to insert into the cell membrane is not cytotoxic. In fact, injecting adenoviral vectors, or DNA forms of these vectors, into mice, rabbits, and primates actually protects the animals from disease by inducing an effective vaccine response. No human vaccine against Ebola is currently available. However, studies in animals suggest that DNA vaccines, together with replication-defective adenoviral vectors, may be particularly promising. In the DNA vaccination platform, a plasmid expression vector is injected into muscle, thereby enabling muscle to synthesize large quantities of proteins that stimulate the immune system to generate an effective immune response. DNA vaccination technology could greatly simplify the vaccination production process that would otherwise rely on very large-scale plants for making these complex and highly purified proteins. However, although current DNA vaccines work well in rodents, they are not as effective in non-human primates and are even less robust in humans. Thus, one of the important challenges for developing an effective DNA vaccination platform technology is to improve immune responses in non-human primates and humans.

The first successful studies of a DNA vaccine for Ebola virus were carried out in guinea pigs (Xu et al., 1999). Animals that were immunized with sufficient levels of Ebola virus glycoprotein to induce a high-titer antibody response survived infection. Guinea pigs with intermediate levels of titers exhibited an intermediate chance of survival. In contrast, none of the control animals, immunized with vector alone, survived Ebola infection (Xu et al., 1999). "Prime-boost" strategies combine DNA immunization and boosting with adenoviral vectors that encode viral proteins to specifically target dendritic cells. Such DNA vector-viral vector combinations can be very potent. Animals are first immunized with a DNA vector, and typically develop titers ranging from 1:1,500 to about 1:3,500. Following the adenovirus boost, antibody titers increase dramatically, ranging from 1:50,000 to 1:100,000. This far exceeds the minimum threshold that is considered to be necessary for an effective immune response in primates. A modified prime-boost strategy was recently used to immunize cynomolgus macaques against several strains of the Ebola glycoprotein (Sullivan et al., 2000). Several months later, animals were boosted with recombinant adenovirus expressing the Ebola (Zaire) glycoprotein. Control animals received empty vectors consisting of plasmid DNA and ADV-ΔE1 recombinant adenovirus in a parallel injection regimen. When animals were subsequently challenged with a lethal dose of the Zaire subtype of Ebola virus, all control animals (6 out of 6) exhibited rapid increases in their viral antigen levels and succumbed to infection within seven days. In striking contrast, all animals immunized with the combination DNA-adenovirus vaccine survived Ebola virus challenge (4 out of 4). The level of antibody production and the cellular proliferative response were closely correlated with immunoprotection.

It is of interest to note that vaccines are not only clinically useful, but they can also serve an important function as deterrents against bioweapons. It is much more difficult to engineer vaccine resistance than drug resistance in an organism. Having well-defined, publicly known, and effective vaccines is a critical preventive, or deterrent, strategy. Another benefit of a successful vaccine is that it opens the way for developing novel immunotherapies. In the case of Ebola virus, for example, hyperimmune serum from animals that are protected from the disease is currently being examined, to determine if it can be used during the course of infection as a possible post-exposure therapy.

Role of Genomics in Vaccine Development and Biodefense

Genomic approaches hold enormous potential for vaccine development, and these possibilities are only just beginning to be explored. For example:

- Analysis of global gene expression patterns can facilitate the early identification of both environmental and disease-associated pathogens.
- Gene expression patterns can be used to identify specific genetic susceptibility and resistance markers.
- Biomarkers for vaccine efficacy could be incorporated into the experimental design of efficacy trials, which could then accelerate approval and licensure processes.
- High throughput technology can be used to improve vaccine design, by allowing researchers to readily monitor how specific structural changes in the vaccine affect the cellular response to immunization.

It is possible that enough information will eventually be available and implemented within the technology that simply knowing the sequence of a particular open reading frame will be sufficient to understand how to generate an effective vaccine. Such technology would be useful not only as a defense measure against bioterrorism, but also for the prevention or treatment of naturally occurring outbreaks, such as influenza. The influenza virus constantly mutates, but if genetic information could be acquired quickly enough, it may become possible to develop more effective countermeasures.

In conclusion, the process of vaccine development must evolve to become more responsive to the changing needs and emerging outbreaks of society today. In other words, more agile vaccines are needed. Agility includes the ability to rapidly deploy vaccines in the event of an outbreak; to accelerate immunization regimens so that such an outbreak could be effectively managed; and, finally, new technology must be applied to develop better vaccines and to accelerate the development process.

MEETING THE REGULATORY AND PRODUCT DEVELOPMENT CHALLENGES FOR VACCINES AND OTHER BIOLOGICS TO ADDRESS TERRORISM

Jesse L. Goodman,* M.D., M.P.H.
Deputy Director, Center for Biologics Evaluation and Research, Food and Drug Administration

The FDA plays an important role in multiple stages of the product development process, from initial clinical studies through licensure, manufacturing and post-marketing studies which may be used to further evaluate safety and effectiveness. For these reasons, FDA is committed to working together with the scientific and clinical communities and with industry and the public to fulfill its regulatory and public health role in facilitating the development of biodefense biologics and therapeutics. Recent and ongoing FDA biodefense-related activities include, for example, meeting with sponsors and sister agencies and departments to encourage interest in developing safe and effective new products needed for public health biodefense, performing research that ultimately facilitates the development of these products; and providing intensive and early interactions with product sponsors to speed their availability.

As with any medical product, bioterrorism products need to be regulated to ensure consistent and objective protection of the public safety. While there is currently a sense of emergency and a set of urgent needs to address, the desire for rapid and innovative responses must not be allowed to compromise the objective assessment of safety and effectiveness. Thus we need a regulatory agency that can step back and provide a more objective perspective. If and when things go wrong in the wake of decision(s) made in a time of crisis, few people will remember the crisis and that the decision was in fact made with the best intentions. The public expects safe and effective products, and safety expectations are especially high for vaccines administered to healthy individuals. Maintaining public confidence in vaccines and medical products, in general, is critical to maintaining overall confidence in our nation's public health programs and leadership in matters extending far beyond bioterrorism. For these reasons,

* This statement reflects the professional view of the author and should not be construed as an official position of the Food and Drug Administration.

even in difficult times, we must continue to make and communicate clearly the best possible scientific and public health decisions about product development, licensure, availability and use.

Furthermore, bioterrorism is a moving target, not a single disease of predictable epidemiology, and all potential product uses may not be anticipated. This complicates many decisions about product use. For example, a vaccine, such as the licensed anthrax vaccine, which may have been originally studied and used in a limited population effectively and without major safety concerns may raise more significant public concerns about uncommon adverse events, whether coincidental or due to the vaccine, if and when it is administered for similar reasons to hundreds of thousands of people or when unanticipatedly used for post-exposure prophylaxis.

There are several factors that account for why we do not have an adequate supply of vaccines for bioterrorism defense:

Financial Disincentives

- Uncertain markets, especially for potentially more limited use products such as a tularemia or plague vaccine.
- Uncertain longevity of the needs, markets and of resources; short attention spans in government budgeting.
- The fact that vaccines are complex biological products that carry a high risk of uncertainty, unpredictability of success, and financial loss.
- The rigorous safety requirements and low public tolerance of risk—in part because they are often administered to healthy people as a preventative measure—and associated costs of developing biologics.
- The fact that preventive measures are generally undervalued, both perceptually and financially. Vaccines are often expected to be sold for very low prices, and the expected profit for the producer is therefore lower than for other products (e.g., drugs for treatment) competing for the same resources. However, while difficult to model when risks are unclear, it would be interesting to conduct more comprehensive and long-term cost-benefit analyses concerning the personal health impacts and the social and economic costs versus potential benefits of vaccine compared to treatment strategies for specific agents of interest.
- The added cost of the large clinical trials needed to address potential wide use including in diverse populations.
- The presence of advocacy groups with various points of view.
- A fair amount of concern about possible adverse effects of vaccines, ranging from specific disease issues to more general anti-vaccine sentiment on the part of a proportion of the public.
- A mistrust of government and industry.
- Product liability issues.

Scientific Challenges

- Lack of historical or recent precedents for vaccines against many pathogens, which makes it difficult to establish good surrogates.
- The potential for genetic or other manipulation of antigenic determinants. (Although this is presently more difficult in many cases to engineer than antibacterial resistance.)
- The potential complications of live vaccine administration to increasing immunocompromised populations.
- The intense flow of resources demanded by urgent perceived needs (sometimes referred to as the "disease du jour" phenomenon), in contrast to the more normal lengthy product development cycle.

The FDA Response

There are several regulatory approaches and mechanisms that the FDA has employed in an attempt to safely speed up product availability and licensure:

- Early and frequent consultation between the sponsor producing the product, the potential end users (e.g., health officials and providers in the military and civilian sectors), and the FDA is very resource-intensive but important. This kind of up-front investment can greatly improve the product development process by identifying creative study designs, recognizing factors that are normally not anticipated in developing a product, and reducing misunderstandings and the likelihood of unwelcome surprises. Early dialogue also increases accountability.
- Emergency use under IND (investigational new drug status) allows rapid access to products that have not yet completed requirements for licensure. INDs require acceptable evidence of safety; a reasonable though not necessarily formally proven scientific basis for efficacy; a favorable risk:benefit ratio; and an intent to license. While allowing availability of potentially lifesaving products, a disadvantage to emergency use under this rule is that the product is not licensed, which not only reflects the true scientific limitations of the data but also raises important issues about public perception.
- Fast track processes can speed up the review process for products that will provide meaningful therapeutic benefits compared to existing therapies for serious or life-threatening illnesses. Fast track allows the FDA to review information as it becomes available and as the sponsor submits it.
- Accelerated approval through the use of surrogate end points to demonstrate benefit. The use of CD4 cells for assessment of antiviral treatment of HIV was one of the first surrogates to be approved under this rule. For bioterrorist agents, protective antibody levels for a vaccine or immunoglobulin could serve as potential surrogate end points. Clinical end points can also be utilized. There still must be good post-licensure studies to demonstrate the effects on disease outcomes and to collect additional safety information, and the FDA can place restrictions on use and promotion and even withdraw the product if agreements are violated or the product proves unsafe or ineffective. Thus far, this process has worked fairly well

although, once a product is licensed, or if a disease is rare, it may be difficult to obtain patients for studies, and sponsors sometimes are unable or unmotivated to fulfill their commitments. But because most accelerated approval products also receive priority review, this process can allow for rapid approval of a product based on more limited and simpler-to-obtain clinical data than may be the case with large, randomized control trials and/or longer-term endpoints.

- Priority review is applied when a product is considered a significant advance or will be used for serious or life-threatening illness.
- Approval under the forthcoming "Animal Rule" has very important biodefense implications. In fact, the rule is specifically oriented to drugs or biologics that reduce or prevent serious or life threatening conditions caused by exposure to lethal or disabling toxic, chemical, biologic, or nuclear threats. The products should be expected to provide a meaningful therapeutic benefit over existing treatments. Human efficacy trials should either be not feasible or unethical, and the use of the animal efficacy data should be scientifically appropriate. In this proposed rule, the end point should be related to the desired benefit in humans, usually a significant outcome such as mortality or major morbidity. Clinical studies in representative populations are still needed, however, both for establishing pharmacokinetics (including, in the case of many vaccines, immunogenicity) and for assessing safety. Such studies are critical because civilian populations often include vulnerable or pharmacokinetically variable subsets. Finally, similar to the fast track and accelerated approvals, the animal rule has post-marketing and labeling commitments and restrictions. It does not apply if the product could be approved based on any other standard in FDA's regulation. It is a rule of last resort, but it certainly would be applicable to many of the situations that have been described in this workshop.

In addition to its regulatory responsibilities, the FDA's Center for Biologics conducts a significant amount of biodefense-related research, supporting approximately sixty ongoing projects that are directly relevant to identified high threat agents. The general goal is to meet otherwise unmet research needs, often with regulatory implications. Examples include how to better determine potency; defining immune and other correlates of protection; how to make safer and purer products (including characterization of the safety of cell substrates and detection of adventitious agents); better assessment of adverse events and efficacy under conditions of use, and studies which allow the agency to make regulations more scientific and less "defensive." These types of research can benefit not only the public, but also multiple companies across industry, but are often not performed by a given sponsor as they may not provide a direct and/or immediate benefit. Furthermore, through its research and related scientific interactions, the center maintains the type of cutting edge expertise that is increasingly needed for dealing intelligently and proactively with evolving products and their underlying biotechnology. This expertise and confidence fosters the science-based objectivity necessary for anticipating and/or reacting appropriately to the issues raised during the development of a product which, ultimately, accelerates the regulatory and licensure process.

By maintaining its scientific, objective regulatory stance, the FDA can increase confidence in the likely efficacy of products primarily approved based on surrogate/animal data and reduce the likelihood of serious adverse events. The FDA brings several other unique attributes to the product development process as well, including:

- Knowledge of scientific and industrial capabilities, which is very helpful when it is necessary to identify people with specific expertise. This includes knowledge of emerging technologies which are cross-cutting among diverse products that nobody else may have the opportunity to see; knowledge of manufacturing capabilities; and knowledge of potential new uses of both licensed and investigational products, for example anti-sepsis and immune modulators.
- Day-to-day participation in what it takes to develop a product, including clinical trials, quality assurance, adverse event monitoring, timelines, etc.
- A unique ability to match product needs to industrial and academic capabilities. Much of this is informal, but it can be very helpful in getting the job done well.

However, there are several things that the FDA cannot do. FDA cannot

- provide monetary or tax incentives;
- assure that anyone will make a product;
- sponsor or directly assume the burden of product development, since this would be a conflict of interest;
- provide indemnification or compensation for injuries;
- Furthermore, while the prelicensure process can provide reasonable assurances about the degree of safety and effectiveness, FDA cannot
- *guarantee* absolute safety;
- *guarantee* human efficacy under field conditions based on non-human data such as animal studies or surrogate endpoints (or, for that matter, based on efficacy observed in the controlled setting of a clinical trial).

In addition to expedited regulatory pathways, as well as orphan drug status, there are several potential incentives—both push and pull—which are outside the mission of FDA but that could be evaluated with respect to their potential to stimulate product development. Push incentives, which could be considered where markets are small or uncertain, could include:

- direct financial awards or contracts;
- tax credits;
- enhanced exclusivity;
- partnerships in product development; and

- research and development assistance to reduce the financial sting and risk of product development.

Possible pull incentives, which are probably more valuable, include:

- known markets;
- longer term financial contracts;
- defining prices that more accurately reflect known and potential public health benefit, which will require more economic discussion and modeling; and
- where possible, developing dual or multiple use products/concepts which can used not just for meeting bioterrorism needs but also for enhancing general public health and medical care.

In summary, FDA and CBER are highly committed to working with multiple partners in and outside of government to help in meeting the challenges posed by bioterrorism. Especially in times of threat and crisis, there is a need for a responsive, yet independent and science-based regulatory process. Relevant research and expertise remains critical in meeting the challenge. Existing laws and regulations can help facilitate product development in a timely manner. There are significant financial disincentives which have and may continue to impede the industrial development of some needed products where markets may be small or uncertain. Careful and open communication with the public about what is and is not known about proposed bioterrorism responses using new and existing products is critical not only in responding to specific threats and protecting the public but also in maintaining confidence in and support for the public health system as a whole.

MOVING THE VACCINE AGENDA FORWARD: OBSTACLES AND OPPORTUNITIES

Stanley Plotkin, M.D.
Medical and Scientific Consultant
Aventis Pasteur, Inc.

The vaccine industry is highly concentrated with only four major manufacturers providing more than three quarters of the market. Even within these four main manufacturers production capacity may be insufficient for unseen circumstances or urgent need, as has been demonstrated by shortages of DT Acellular pertussis vaccine and pneumococcal conjugate vaccine. Following September 11th, all four major manufacturers expressed interest in biodefense/military vaccines. But how long will patriotism sustain this interest? The industry is both high risk (e.g., the rotavirus vaccine which had to be withdrawn from use) and risk adverse (e.g., certain vaccines for pregnant women have yet to be developed). There are several factors that impede vaccine product development:

- The vaccine industry is market-driven, and the major manufacturers are simply not interested if the market is insufficient.
- The industry is highly regulated, and perceived regulatory hurdles can impede product development.
- Intellectual property conflicts can prevent companies from developing products.
- Negative marketing assessment impedes vaccine production. Marketing departments often make unchecked predictions.
- An uncertain or dubious proof of concept creates reluctance to develop a particular new product.
- Biohazard to personnel, especially with regards to biodefense vaccines, can create reluctance.

The production of an adequate biodefense vaccine supply will depend on many factors:

- Development of biodefense vaccines must be in collaboration with DoD, NIH, CDC, and other agencies with national interests. With regard to the development of a new smallpox vaccine, for example, efficacy is an important issue that requires the use of monkey models for which the vaccine industry must turn to DoD or NIH.
- Development of biodefense vaccines requires better adverse reaction surveillance. With the military anthrax vaccine, for example, initially there were no data to support the claim that the vaccine was indeed safe. New vaccines must have built-in surveillance in order to discredit unsubstantiated claims about adverse reactions based on anecdotal data.
- New vaccine production requires that liability indemnification be guaranteed. One possible solution would be to place these vaccines on a list of compensable vaccines. We need to create a no-fault system which indemnifies companies against non-negligent harm but also provides some relief for injured individuals.
- Access to new technology would likely stimulate more interest in vaccine development. For example, a number of companies have pursued DNA vaccines because they are an interesting new technology which could likely be applied to a number of different targets. After all, the main role of the vaccine industry is not to conduct basic research but apply basic research to the development of new products.
- Finally, vaccine production priorities need to be set and somebody authorized to say "this is the top priority." For example, Gary Nabel has presented some very elegant work on an Ebola vaccine. But what does intelligence say is the real risk of Ebola? Should a virus that cannot be spread as an aerosol be considered a priority?

In summary, the major vaccine manufacturers will not be able to provide all of the needed biodefense vaccines. However, they should be asked to play a significant role. That role will be facilitated by clear directions, clear priority setting, and a tight collaboration between industry and government in order to move animal and clinical testing forward.

REFERENCES

A. Johnson-Winegar:

Grabenstein, J.D., and Wilson, J.P. (1999), "Are Vaccines Safe? Risk Communication Applied to Vaccination," *Hospital Pharmacy*, Vol.34, No. 6, pp. 713–729.

Institute of Medicine. (1996), *Interaction of Drugs, Biologics, and Chemicals in U.S. Military Forces*. Washington, D.C.: National Academy Press.

Morbidity and Mortality Weekly Report. (December 15, 2000), 49(RR15);1–20 "Use of Anthrax Vaccine in the United States—Recommendations of the Advisory Committee on Immunization Practices."

G. Nabel:

Balter, M. 2000. *Science* 290: 923–925.

Colebunders R, Borchert M. 2000. *Journal of Infections*. 40: 16–20.

Malashkevich VN, Schneider BJ, McNally ML, Milhollen MA, Pang JX, Kim PS. 1999. *Proceeding of the National Academy of Sciences, U S A* 96:2662–2667.

Peters CJ, Sanchez A, Rollin PE, Ksiazek TG, Murphy FA. 1996. In: Fields BN, Knipe DM, Howley PM, eds. *Fields Virology* 1161–1176 (Lippincott-Raven, Philadelphia).

Peters CJ, Khan AS. 1999. In: Klenk, HD, ed. *Current Topics in Microbiology and Immunology* 235:85–95 (Springer-Verlag, Berlin).

Sanchez A, Khan AS, Zaki SR, Nabel GJ, Ksiazek TG, Peters CJ. 1996a. In: Fields BN, Knipe DM, Howley PM, eds. *Fields Virology* 1279–1304 (Lippincott-Raven, Philadelphia).

Sanchez A, Trappier SG, Mahy BWJ, Peters CJ, Nichol ST. 1996b. *Proceedings of the National Academy of Sciences, USA* 93: 3602–3607.

Sullivan NJ, Sanchez A, Rollin PE, Yang ZY, Nabel GJ. 2000. *Nature* 408: 605–609.

Volchkov VE, Becker S, Volchkova VA, Ternovoj VA, Kotov AN, Netesov SV, Klenk HD. 1995. *Virology* 214: 421–430.

Watanabe S, Takada A, Watanabe T, Ito H, Kida H, Kawaoka Y. 2000. *Journal of Virology*. 74: 10194–10201.

Weissenhorn W, Calder LJ, Wharton SA, Skehel JJ, Wiley DC. 1998a. *Proceedings of the National Academy of Sciences, USA* 95: 6032–6036.

Weissenhorn W, Carfi A, Lee KH, Skehel JJ, Wiley DC. 1998b. *Molecular Cell*. 5: 605–616.

Xu L, Sanchez A, Yang Z, Zaki SR, Nabel EG, Nichol ST, Nabel GJ. 1998. *Nature Medicine* 4: 37–42.

Yang Z, Delgado R, Xu L, Todd RF, Nabel EG, Sanchez A, Nabel GJ. 1998. *Science* 279: 1034–1037.

Yang ZY, Duckers HJ, Sullivan NJ, Sanchez A, Nabel EG, Nabel GJ. 2000. *Nature Medicine*. 6: 886–889.

4
The Research Agenda: Implications for Therapeutic Countermeasures to Biological Threats

OVERVIEW

As with vaccines, not only are therapeutics an integral component of our biodefense arsenal, but making it publicly known that we are producing a constant stream of new, innovative antimicrobials would serve as a very strategic form of deterrence. Several issues related to antibiotic, antiviral, antitoxin, and antibody research and development were identified and discussed during this session of the workshop. In light of the plethora of bioterrorist agents that could be used against us, of utmost importance is deciding whether we should focus our efforts on the development of broad-spectrum or agent-specific antimicrobials. For example, one possible antiviral strategy is the development of family-specific antivirals. Increasing evidence suggests that common antiviral targets exist.

Our antibiotic arsenal is limited to only a handful of old antibiotics. Unfortunately, the general confidence in existing antibiotics and the complacency that was associated with infectious diseases in the 1960's resulted in a lag in producing new classes of antibiotics. There are about twenty-five antibiotics currently in the early phases of clinical development. However, none of these are new classes of antibiotics, and none are broad-spectrum. In fact, there has been only one new class of antibiotic developed in the past two decades, and resistance to it emerged before it came to market. This is alarming given the increasing accessibility of the tools and knowledge needed to develop antibiotic-resistant strains of bioterrorist agents.

There is concern that the situation will become ever worse with the recent FDA changes in clinical trial design requirements. It is expected that the increased cost associated with larger clinical trials will discourage companies from pursuing new antibiotic development, especially when there are other therapeutic interests vying for the same resources. Although the FDA attempts to balance the demands of a public health emergency with their needs as a regulatory agency and offers several accelerated routes to licensure, including the proposed animal efficacy rule, there is still a sense that these regulatory processes need to be streamlined even more in order to accelerate drug discovery and development efforts and provide more incentive for the pharmaceutical industry.

Our antiviral amamentarium is even more limited than our antibiotic arsenal. Cidofovir, for example, can only be administered intravenously and is highly nephrotoxic, making it unsuitable for mass casualty use. The clinical utility of ribavirin as an antiviral drug strategy for bioterrorist agents remains unclear.

Antibiotics and antivirals are not the only potential therapeutic defense against bioterrorist agents. Basic research on the anthrax toxin system has led to some exciting prospects for antitoxin targeting. The most promising are the dominant negative inhibitors (DNIs), mutant forms of the protective antigen that block translocation of the virulence factors across the plasma membrane. Currently, DNIs are a very late stage product. If they can be proven efficacious in infected animal models, they could be produced and deployed very rapidly. There are several other approaches in much earlier stages of development.

The use of recombinant monoclonal antibodies is another option which has been implicated for use against several biothreat agents, including anthrax, smallpox, and botulinum neurotoxins. For example, recent research has shown that a small mixture of recombinant monoclonal antibodies provides complete protection in mice against botulinum neurotoxin type A. Antibodies have several advantages as a bioweapons defense tool: they have been shown in multiple studies to be safe; ten have already been approved by the FDA and seventy more are in clinical trials, so their route to licensure is known; the technology and knowledge needed for production are readily available; their overall course through the discovery and approval process is much quicker than those of other types of therapeutics; and the technology platform used to produce and manufacture antibodies could be applicable to multiple agents.

Finally, scaling up research and development of all of these various potential therapeutics will require an evaluation of the availability of and need for additional laboratory capacity. In particular, there are a very limited number of BSL-3 and 4 labs where nonhuman primate studies can be conducted. Hope was expressed that in the future the FDA will accept rodent data in lieu of nonhuman primate data, if it can be demonstrated that the efficacy is the same in rodents as in nonhuman primates. This would allow for more testing in a greater number of facilities, although it would still require at least BSL-3 capability. Aerosol testing requires BSL-4 capabilities, as well as trained, vaccinated personnel.

COUNTERMEASURES TO BIOLOGICAL THREATS: THE CHALLENGES OF DRUG DEVELOPMENT

Gail H. Cassell,* Ph.D.
Vice President, Infectious Diseases
Drug Discovery Research and Clinical Investigation
Eli Lilly and Company

The Problem

The diversity of existing biological weapons and the ever increasing possibilities preclude simple therapeutic countermeasures to bioterrorism. Furthermore, response possibilities are rather limited even for known threats. Although there are 13 viruses on the current list of potential biothreats, there is only one indicated antiviral—cidofovir—and it both requires intravenous administration and is highly nephrotoxic. More broadly, there are no truly broad spectrum antivirals, and only a limited number of antivirals for routine pathogens like influenza, herpes, hepatitis-B, and HIV.

The situation is somewhat better but still worrisome with respect to antibiotics. There has been only one new class of antibiotics developed in the past two decades, and resistance emerged to this class before it entered the commercial market. This is a clear challenge to developing an armamentarium against biological pathogens.

At first glance, the situation with respect to antibiotics currently in clinical development looks encouraging. About 25 antibiotics are in the first 3 stages of development, with several billion dollars devoted to their development. However, there are no new classes being pursued, nor are new broad spectrum antibiotics. Furthermore, most are quinolones, and 50 percent or more of the strains of E. coli in Beijng are resistant to quinolones as are many foodborne pathogens. In addition, quinolones are contraindicated for children, and neither quinolones nor tetracycline are acceptable for pregnant women.

The Challenges

So it would be mistaken to be sanguine about current antibiotic therapies to counter bioterrorism. Nor can one be optimistic about near-term prospects. Eli Lilly recently conducted a competitive analysis revealing that most of the large pharmaceutical companies, with the possible exception of Pfizer, are reducing their investment in antibiotic development. There are probably several reasons.

A decade ago, we looked at new technologies like high throughput screening, combinatorial chemistry, and microarray assays, and anticipated a golden

* The information provided in this paper reflects the professional view of the author and not an official position of Eli Lilly and Company.

age of antibiotics. But today we have no new classes of antibiotics as a result of those efforts.

It has become painfully apparent that discovering new antibiotics is not as easy as once believed. We have, for example, a plethora of targets. Numerous targets have been found with documented *in vivo* expression of antigens. But they are not necessarily what are called "drugable targets." A target can be validated and essential for bacterial viability, but if there is not a chemical entity that will penetrate the bacterial cell wall and inhibit growth, you don't have a real target. In addition, the chemical entity must be safe and not highly toxic. There is a 90 percent failure rate from the discovery of a target to the launch of a new antibiotic. This lack of success has likely damped further spending in this area.

Moreover, there has been increased investment in antivirals. And, with the sequencing of the human genome, competition for resources within pharmaceutical companies has turned to other therapeutic areas where there are tremendous opportunities and great unmet medical needs with bigger market opportunities. In fact, infectious diseases, specifically those requiring antibiotic therapy, do not fare too well in financial analyses.

Conclusion

In short, our antibiotic armamentarium is limited, there is growing concern about an increasing number of potential new weapons, and there has been a marked increase in resistance to existing antibiotics.

It seems clear that no public health response to bioterrorism is likely to prove effective without addressing the overall problem of antimicrobial resistance and the challenges of drug discovery and development.

Finally, the best deterrent against the use of a biological weapon of mass destruction may be a constant stream of new, innovative antibiotics and antivirals. Knowledge of such commitment and successful developments would surely dissuade the efforts of our enemies in such an arena.

THE FDA AND THE END OF ANTIBIOTICS

David M. Shlaes, M.D., Ph.D.
Vice President for Infectious Diseases
Wyeth-Ayerst Research
Robert C. Moellering Jr., M.D.
Physician-in-Chief, Beth Israel Deaconess Medical Center
Herrman Blumgart Professor of Medicine
Harvard Medical School

Antibacterial research has, for almost two decades now, been the "Cinderella" area in the pharmaceutical industry. The market for these products, used to

TABLE 4-1

Cure Rate	*Delta*
90%	10%
80–89%	15%
< 80%	20%

treat acute, not chronic disease, is modest. There are many products available including many generics. For the most part, the market is growing only slowly and, except for the problem of resistance, is largely satisfied.

Recently, when presenting proposals for Phase III trial designs for antibacterial compounds to the FDA and European regulatory bodies, a number of companies, both small and large, were told that the designs for equivalence studies had to target a 10% delta for the lower limit of the confidence interval. This requirement, seemingly innocent and technical, threw the pharmaceutical industry into a panic and probably contributed to the recent withdrawals by Lilly and Bristol-Myers-Squibb from the antibacterial discovery business. Why?

Most clinical trials leading to approval of antibacterial drugs are based on equivalence studies. The delta statistically defines the boundary for equivalence. In the past, based on the 1992 points to consider document from the FDA (FDA, 1992), a step function for the delta has been used (Table 4-1).

These boundaries have meant the study results must show there is a 95% (97.5 % one sided) probability that true cure rate for the new drug is not more than 10 to 20 % lower than the cure rate for the approved drug. In most reasonable sized studies the new drug has had to be as good as or better than the old one to be successful.

The European regulatory authorities and the FDA are now suggesting that a 10% delta be used routinely in drug development (FDA, 2001b), and the FDA has now "disclaimed" the old step function on their web site (FDA, 2001a). It is not completely clear upon what data this suggestion is based, other than purely statistical considerations. In fact, a quick calculation will show that two independent trials successful at a 15% delta would result in approving a drug inferior at the 10% delta only 2% of the time. The concern that the FDA has expressed is over something called "biocreep." In this concept, a slightly inferior experimental drug becomes the comparator for the next generation of compounds and so on until the experimental drugs of the future asymptotically approach the efficacy of placebo. However, one must wonder whether, for serious infections, this is any more than a theoretical concern, especially when most recent approvals (Synercid, Zyvox) have been based on comparison with standard therapy using older agents. One of the regulatory agency's best weapons against biocreep is their control over the choice of comparators in clinical trials.

The deltas used in the step-function of the 1992 points to consider document were chosen, not based on scientific reasoning, but based on the size of trial that would be required given the cure rate (Pharmaceutical Research and Manufacturers of America [PhRMA], personal communication). The trial size required is very sensitive to efficacy rate, evaluability, β error (power) and the delta. For typical trials with an injectable antibiotic, the patient numbers under various assumptions are shown in Table 4-2. The increased numbers have implications for the ability to run a trial in a reasonable length of time, time of availability of the new drug to patients and physicians, time to market and overall costs of development. For example, Bristol-Myers-Squibb was told that they would be required to run a trial in acute bacterial meningitis at a 10% delta requiring enrolling over 700 patients (Roger Echols, BMS, personal communication). That would be the largest meningitis trial ever done, require about five years to accomplish this enrollment and would require enlisting over 90% of the patients from outside the United States. They declined based on the impracticality of the design and the fact that in the later years of the study, it was not clear that their comparator would still be considered the standard of care in the medical community. Similar concerns exist regarding our ability to run trials at a 10% delta for infections caused by resistant bacteria like vancomycin-resistant enterococci.

TABLE 4-2 Number of patients for each indication with a one-sided 97.5% CI (assumes 75% evaluability)

Indication	Cure Rate	90% Power 10% delta	90% Power 15% delta
A Number of Studies: 2	85%	1532	688
B Number of Studies: 2	80%	2248	1000
C Number of Studies: 2	70%	2948	1316
D Number of Studies: 1 Related to indication C	65%	1598	710
TOTAL		8326	3714
		80% Power 10% delta	80% Power 15% delta
TOTAL		6226	2770

Table 4-2 also shows the increase in numbers required if a 10% delta is required for all indications. Costs of a Phase III trial are directly related to the number of patients enrolled. Therefore, in the scenario above, the costs increase more than 100% going from a 15% to a 10% delta design and much more if one was starting at the old step function plan. This can be ameliorated to some extent by decreasing the β power to 80% from 90%. However, doing that results in a 32% risk of falsely concluding that the experimental compound is inferior to the comparator—a risk not acceptable for most companies.

One might argue that large pharmaceutical companies can easily absorb these costs and, that if they want to sell a product, they should do so. However, just as in government agencies like NIH, proposed research in the pharmaceutical industry is subject to prioritization. In the case of the industry, business considerations play a large role in the process. Therefore, in most companies, programs with modest potential markets and large costs are automatically deprioritized unless there is some other, overriding strategic issue to be considered. Thus, one unintended result of promulgating these guidelines will be a decrease in the number of companies carrying out antibacterial research as was seen in the late 1980s and is occurring again now.

PhRMA has suggested a number of alternate approaches to the FDA and the industry is more than willing to work with FDA, IDSA and other interested parties to address their concerns regarding clinical trial design in antibacterial development. However, the attempt by regulatory authorities to implement an across-the-board requirement for 10% delta trial designs has already wreaked irreparable damage to our ability to provide a reliable pipeline of new antibiotics for serious infections. We hope that the advisory committee of the FDA will understand these concerns and act appropriately. We would also ask that the European regulatory authorities reconsider their stance for the same reasons.

THE ROLE OF ANTIVIRALS IN RESPONDING TO BIOLOGICAL THREATS

C.J. Peters, M.D.

Professor, Departments of Microbiology and Immunology and Pathology

University of Texas Medical Branch at Galveston

Many available viruses could be used to cause harm to others under many different scenarios. It is important to try to focus on some specific priorities to attempt to limit the problem to a tractable scope, yield maximum benefit in the short term and develop more comprehensive goals that we can hope to attain in the longer term. It is important to consider vaccines and drugs together as part of an overall antiviral strategy.

The Threat

To ameliorate the adverse consequences of a bioterrorist (BT) attack, the scenario is everything and, equally, the scenario for every possible attack is unknowable. However, aerosol attacks have the greatest potential to cause mass casualties and also lend themselves to stealthy application (see chapter on aerosols). Only a limited number of viruses are known that grow to high titer and are stable and infectious in aerosols and thus lend themselves to this form of attack. Tables one and two list a number of human-pathogenic viruses that have often been considered as aerosol threats (Alibek, 2001; Ferguson, 1999; Alibek, 1998). The viral hemorrhagic fevers (VHF) (Table 4-3) are among the most dangerous (Peters, 2000). The VHF agents are lipid-enveloped RNA viruses with a genome size of around 1–2 million Daltons belonging to four different virus families (Peters and Zaki, 1999). They are zoonotic viruses and all are aerosol-infectious, either shown through formal studies in the laboratory and/or by the observation of frequent "unexplained" laboratory infections. As might be expected from their taxonomic diversity, they differ in individual strategies for maintenance in nature and in their pathogenesis of human disease. Several of these viruses were developed by the Soviets for use as strategic weapons for mass destruction, including Machupo, Lassa, Rift Valley fever, and Marburg viruses. At this time, technical difficulties may limit the prospects for weaponization of Crimean Congo HF and the hantaviruses, but like most problems, these are subject to solution.

TABLE 4-3 Viral hemorrhagic fevers commonly mentioned in association with biological warfare or biological terrorism

PRIMARY HEMORRHAGIC FEVERS (HF)		
ARENAVIRIDAE		
	Lassa Fever	
	South American HF (Argentine, Bolivian, etc)	
BUNYAVIRIDAE		
	Phlebovirus	*Rift Valley fever*
	Nairovirus	*Crimean Congo HF*
	Hantavirus	*HF with renal syndrome*
		Hantavirus pulmonary syndrome
FILOVIRUS		
	Marburg HF	
	Ebola HF	
FLAVIVIRUS		
	Yellow fever	
	Tick-borne HF (Kyasanur forest disease, Omsk, etc)	

TABLE 4-4 Other viruses suggested to have potential in biological warfare or biological terrorism

- Smallpox
- Monkeypox
- Nipah
- "Viral encephalitides"
  - -Venezuelan equine encephalitis
  - -Other alphaviruses
  - -Tick-borne encephalitis virus
- "Eradicated": polio and measles
- Influenza A: 1918 strain, Hong Kong H5N1, others

The other candidate aerosol infectious viruses (Table 4-4) also are largely zoonotic with the exception of the very important BT agent, smallpox. The zoonotic agents can spread to close family contacts and to health care staff, but continuing chains of transmission are not a threat. Smallpox is quite different because its natural history is one of continuous inter-human transmission. The lack of a reservoir outside human-kind, the moderately higher transmissibility, and the existence of a highly effective vaccine that can be efficiently delivered combined to allow the eradication of the virus as a cause of natural disease. Monkeypox is another poxvirus which shares high aerosol stability and infectivity with smallpox but which has a much lower interhuman transmissibility and case fatality (Jezek and Fenner, 1988).

Nipah virus is representative of a newly proposed genus of a very well-established family, Paramyxoviridae. Before the other human pathogen in this genus, Hendra virus, emerged in Australia in 1995 (Murray et al., 1998) the existence of the genus was unsuspected, but now its members are seen to be widely distributed among flying foxes (Macrochiroptera) and at least these two members have the potential to cross over into domestic animal species and also infect humans (Chua et al., 2000). Their future behavior is unpredictable, but analysis of the state of emerging infections in the world and the recent recognition of two serious episodes in the recent past suggest they are highly dangerous (Peters, 2001). The spread of Nipah virus among swine in Malaysia was progressive and could have resulted in an enormous economic, human, and political disruption if it had extended into Thailand and China. Stopping the march of the virus entailed destruction of more than one million pigs after it caused 265 human cases with a 40% case fatality and significant residual morbidity among survivors (Parashar, et al., 2000). The aerosol properties of this virus are unknown, but it appears to spread among pigs by small-particle aerosols or droplets. The great majority of human cases had close contact with living swine, but the occurrence of a small number of cases in persons living near pig farms but without actual contact with pigs also raises the question of aerosol infection of humans. Where will the next unexpected virus come from and what taxon will it belong to?

Several alphaviruses causing viral encephalitis have been regarded as potential biological agents. The best established is VEE which was weaponized by the US and the Soviets. VEE also has the potential to establish itself as an endemic mosquito-borne disease in North America based on past epidemiological evidence (Weaver, 1997).

Other less well-established threats will undoubtedly become more important in the future. These could even include common viruses such as polio and measles should they be eradicated (previous Forum). The problem is exemplified by smallpox virus. When vaccine was widely used and substantial immunity was maintained in the population, the threat of smallpox epidemics arising from isolated cases was less than it is today. Polio epidemics would be highly disruptive, but measles would probably be the worst threat. For example, measles epidemics during the U.S. Civil War were among the greatest impediments to expanding armies because of their heavy impact in both morbidity and mortality among recruits and staging centers for the two armies (Steiner, 1968).

Influenza virus is infectious by aerosols and is capable of propagating efficiently among humans by aerosols even though other routes are also important (Kilbourne, 1975). In general, the impact of aerosol spread on control is huge. In the US we can deal effectively with fecal-oral, fomite, and large droplet transmission through our general level of sanitation or by using simple mask, eye protection, and hand washing measures. However, aerosols must be controlled with efficient filters on breathing air and the filters must be well-fitting and in use during the time of risk; any society would have difficulty dealing with an aerosol spread epidemic. The ability to produce recombinant influenza strains from natural strains or using synthesized genes is an accomplished feat (Neumann et al., 1999). The prediction of which viruses will be highly transmissible and lethal will come in the near future. Whether such viruses would be produced and could actually spread among human populations is another matter.

The Solutions

Biological warfare (BW) and terrorism present different challenges. BW scenarios could involve the use of biological weapons on the battlefield and would target a selected group which could therefore be immunized using their training schedules, different operational missions, and on-going military service to select and prioritize.

A strategic biological warfare effort could also be directed against civilian populations, as was envisaged by the Soviets, and this scenario could possibly be countered through vaccine protection. However, a civilian population facing an ill-defined bioterrorist threat would be much harder to protect by immunization because of the problems of vaccine coverage and the inevitable adverse events associated with all vaccines. The example of smallpox vaccine would be illustrative. Use of this vaccine in the defined and limited group of military persons in basic training was accomplished in the 1970's with little morbidity among the vaccine recipients and little risk to non-military. If this program had been expanded to the entire civilian population with the consequent adverse effects (Koplan and Hicks, 1974), there would have been a huge backlash. The "swine flu " vaccination program in 1976 provides an excellent example of the likely outcome (Neustadt and Fineberg, 1978). Widespread vaccination against an ill-defined threat would be associated with the adverse effects that will ensue from any vaccine and would bring the effort would be brought to a stop with serious negative medical and political consequences. Even a perfect vaccine would be tarred by the unfortunate events that occur by chance in a large, healthy population receiving no treatment at all.

Thus, vaccines against viruses could be very useful but would likely only be used on a large scale in civilian populations in the face of a clear and present danger. Nevertheless, they are an important part of an over-all antiviral strategy because they provide protection for particular groups, including those studying the virus in the laboratory, antiviral drug developers, and those working with the virus in regions where it occurs naturally. Furthermore, availability of attenuated strains can be essential to expanding research activities, including antiviral drug development, to laboratories with lower levels of containment.

Antiviral drugs could provide protection, subject to all the same problems of stockpiling and delivery as antibacterial agents now considered for use against such threats such as anthrax, plague, and tularemia. It is now clear that, with an adequate molecular and structural biological base, drug development capability, and financing, effective antivirals can be discovered and indeed even designed. Antivirals inhibiting enzymes active in nucleic acid synthesis or protein cleavage have been highly effective although with some price in toxicity. Other targets exist which might be less closely allied to host cell constituents and which might provide a greater therapeutic index, just as drugs such as penicillin inhibit gram positive bacterial cell wall synthesis, which has no counterpart target in mammalian cells. Monoclonal antibody strategies are also possible, but many of the protective targets are the neutralizing antibody epitopes, which are commonly virus specific. This can be overcome with multi-virus cocktails, but this in turn demands development and production efforts of multiple antibodies. Fortunately, the neutralizing epitopes are highly conserved and usually linked to virulence on the individual viruses of interest. The success of any antiviral strategy—drugs (including antibodies), vaccines, or combinations—will depend critically on the context in which it is used. Stockpiles of remedies will be essential, as well as expectant use of some vaccines. Equally important will be plans to deliver emergency vaccines or drugs in a timely fashion where they are needed. And most importantly, it will be incumbent on the physician to recognize the diseases and request definitive laboratory evaluation of the provisional diagnosis. Finally, the definitive laboratory diagnosis must be quickly available.

Both antiviral vaccines and drugs suffer from major development problems. They would require an expensive developmental effort that has never been able to attract industrial support based on disease activity in endemic areas, even when the U.S. Department of Defense has expressed an interest and provided an additional market. A large number of viruses are involved (Table 4-3, Table 4-4), further multiplying the problem. Finally, the utility of these medical countermeasures is severely limited if they cannot be tested in adequate numbers of humans for efficacy and safety. The safety testing would be extremely important in prophylactic use or if large numbers of people are to receive the substance.

The only feasible solution is to provide public support for the discovery and testing of BT related antiviral vaccines and drugs. As a practical matter, the approach would have to be directed toward finding solutions which will apply to the most important individual viruses or to broad groups of agents, perhaps at the generic or family level. We should look very carefully, without being shackled by our past attitudes, at new technologies for translating the advances of biotechnology into human products; the way we test, develop, and approve drugs and vaccines for different uses; and the actual increment in safety derived from additional requirements on manufacturers.

A Case Study

A vaccine against Argentine hemorrhagic fever (Junin virus) provides insight into some of the obstacles. This vaccine was desired by DoD because of suspicions, later confirmed, that the Soviet Union was developing Junin and the related Machupo virus as biological weapons. Immunity against Junin virus was shown to protect against Machupo infection in experimental animals, thus providing the expectation that a Junin vaccine could protect against both arenaviruses.

A laboratory attenuated strain of Junin virus had been shown to be safe and immunogenic in at-risk laboratory workers in Argentina. This strain was too reactogenic and insufficiently characterized for general use, but it provided a very important bench-mark for development of a further-attenuated vaccine for humans. The availability of limited testing of prototype vaccines in humans before undertaking definitive work can be an extremely important step in facilitating vaccine development with the potential to save years or decades in development time and costs.

Vaccine development was pursued as a joint US-Argentine project with participation of Argentine colleagues at each step and with efforts to brief Argentine public health officials at regular intervals over the decade-long period. This was extremely important when the time for field-testing of the vaccine arrived because charges of "scientific imperialism" could have blocked the actual evaluation of the vaccine against the disease even though it only occurred naturally in

Argentina. After the vaccine met FDA standards and had undergone preliminary evaluation for safety and immunogenicity in US volunteers it was shown to be protective in double-blind, placebo-controlled trials in the endemic area for Argentine hemorrhagic fever (Maiztegui et al., 1998; Barrera-Oro and McKee, 1991). Subsequent use in Argentina brought the number of inoculated humans above 200,000, further establishing the safety in large numbers of recipients.

Among the lessons is that useful drugs and vaccines can be developed in the US outside the private sector if there is a substantial long-term financial and institutional investment, if they can be tested overseas, and if there is forethought and participation of the target test nation's scientific and political establishment. Other issues such as change in political and economic circumstances in the participating country and temporary or permanent shifts in disease incidence during vaccine development unpredictable and difficult to control hurdles to success.

Existing Measures for Specific Agents

The number of potentially deployable antiviral measures for recognized viral threats is limited (Table 4-3). All the viruses in the arenavirus family are inhibited in cell culture by the antiviral drug ribavirin and there are substantial data to support the use of intravenous ribavirin in the treatment of Lassa fever (McCormick et al., 1986). Preclinical and anecdotal human data for the use of the drug in other arenavirus infections strongly suggests it is efficacious (Peters, 2002). Emergence of resistance under therapy has not been seen (Kenyon et al., 1986). The drug has undergone extensive preclinical evaluation and is licensed in the U.S. for aerosol and oral use in respiratory syncytial virus pneumonia and hepatitis C respectively. Its use in pregnant women and children is restricted because of preclinical data suggesting teratogenicity and growth retardation. Because of the lack of a market and the expense of New Drug Applications, additional studies or licensure of intravenous ribavirin for use in arenavirus infections have not been pursued. The drug has also been used in Crimean Congo HF (family Bunyaviridae, genus *Nairovirus*) in South Africa and a modest clinical experience supports the positive preclinical data for its efficacy.

Rift Valley fever (family Bunyaviridae, genus *Phlebovirus*) presents an important problem (Peters, 2000). In addition to its potential use as a BW agent, it is also a threat for importation. It is a mosquito-borne disease that usually depends on the presence of sheep or cattle to support its transmission, but humans may well be able serve as substitute amplifiers in the proper circumstances. Aerosol-mediated human infection is the basis for the BW potential of the virus and also the infections commonly occurring among laboratory workers. Thus, both human and veterinary vaccines would be desirable for Rift Valley fever control if the virus were introduced into the U.S. by bioterrorists or natural spread. Its propensity for spread outside its natural range in sub-Saharan Africa has been demonstrated by introductions into Egypt, Saudi Arabia, and Yemen in past years. The veterinary involvement brings the additional dimension of economic disruption of the livestock industry; export of meat products would be interrupted for months or years and the movement of animals within the US would be greatly restricted. There is no satisfactory veterinary vaccine, but two human vaccines have been developed. The inactivated, Salk-type vaccine has a long lag before induction of immunity and is available in limited supply, not positive attributes to deal with a rapidly moving disease such as Rift Valley fever has proven to be (Pittman et al., 1999). In addition, the infrastructure for producing additional inactivated vaccine does not exist in the U.S. any longer. The other human Rift Valley fever vaccine is a live-attenuated product that has been tested in 60 persons with a good safety profile and the elicitation of antibodies expected to be protective, based on preclinical studies and the experience with the inactivated vaccine in laboratory workers (Pittman PR; Morrill J, Peters CJ, unpublished observations). There are no on-going efforts in further development and testing of this product.

Yellow fever (family Flaviviridae) presents a hazard either through direct aerosol delivery, spread by artificially infected vectors, or by importation into cities where human-mosquito cycles could support further transmission. The US presumably has a low receptivity to mosquito amplification of the virus, but many tropical metropolises are thought to be highly vulnerable. The vaccine is one of the best known and best studied viral vaccines and has been in use for more than 60 years; but in recent months has been associated, for the first time with fatal adverse reactions in several older vaccinees (Martin et al., 2001). Furthermore, the vaccine is not manufactured in the U.S. and world stocks are not sufficient in some areas to meet today's needs, let alone an unforeseen surge (Nathan, et al, 2001).

Tick-borne encephalitis is another flavivirus disease that has been considered as a potential BW threat. Inactivated vaccines against the virus are widely available in Europe and are thought to be safe and protective against the virus as delivered by tick bite. None are available in the U.S., even in investigational status, nor are they protective against other tick-borne flaviviruses that cause VHF.

Smallpox is one of the most serious viral threats (see chapter on BW threats or on smallpox). Although there is no information as to who possesses the agent beyond the two authorized laboratories, it has to be noted that the virus could be disseminated directly by aerosol, based on its stability and infectiousness by that route; and it can also spread from person-to-person. The extent of this interhuman spread is controversial but modeling using selected sets of parameters (O'Toole, et al, 1999; Meltzer, et al, 2001) suggests that the outcome of even a limited dissemination could be disastrous. The amount of vaccine to contain an outbreak of smallpox in our mobile world could be substantial, given the need to ring-vaccinate contacts, medical staff, and many others. This is best met by production of large quantities of vaccine above current stocks. The safety and efficacy of the inexpensive calf-produced vaccine was excellent and hopefully the desire to manufacture the newer vaccine stocks in cell culture will not impede the availability of the older, proven modality of control.

In the case of smallpox, the vaccine has some of the attributes of an antiviral drug because vaccination within 4 days or perhaps even longer after exposure prevents smallpox. Alternate approaches are discussed in the smallpox chapter, but it is worth pointing out that no further-attenuated protective vaccine can be evaluated for efficacy in either normal or immune deficient hosts, no vaccine can be proven to be safer without inoculating literally millions of normal persons, no vaccine can be shown to be safer in immunosuppressed hosts without performing unacceptable experiments, and no smallpox drug can be shown to be effective in humans. Surrogate markers are just surrogates and the mainstay of protection of the world's population must remain the proven vaccine. An effort to provide a perfect solution to protection and treatment of all possible exposed persons may well siphon off resources needed for protection from other threat agents.

The classical smallpox vaccine faces additional hurdles for its effective use today because of the much larger number of immunosuppressed persons. Accidental vaccination as well as transmission from vaccinees pose hazards, and the management of an immunosuppressed patient exposed to smallpox would be a challenge. An effective drug against vaccinia virus would be life-saving for immunosuppressed persons. Drug efficacy could be shown in normal humans receiving smallpox vaccine and its relative efficacy in immunocompetent and immunodeficient animals studied in the laboratory.

Research Directions

It would be desirable to have two things in our portfolio for all the viruses known to pose BW threats. First, we should understand how to make vaccine candidates that would be protective in realistic animal models against all the threat agents and that would utilize a technology feasible for human vaccine production. Second, we should have broadly reactive antiviral drugs or other therapeutic approaches that would be effective in families or at least genera of threat agents. The priorities for advanced development would depend on the perceived level of threat and would be leavened by the natural occurrence of the target diseases, which would determine both the availability of a test bed and the additional benefit to society from developing the viral countermeasure.

In the field of vaccines, we need a revolutionary approach to vaccine development that attempts to move research on human immunogens forward to another quantum level. The history of dengue vaccine research is an example of where we do not want to be: time line stretching over decades; under-funded and marginally focused efforts; candidate vaccines are classical live-attenuated and poorly characterized in modern terms; and some promising results but no end in sight. Approaches to live vaccines could be greatly simplified with a flexible, standardized, safe vector; research dedicated to resolving this problem in humans could be critical to our ability to respond to the threat of viral terrorism. Alphavirus replicons provide one of several hopeful possibilities (Polo et al., 2000; Smerdon and Liljestrom, 1999). An alternative would be a protein sub-unit based vaccine, and the best immunogenicity is generally obtained with a self-assembling particle such as with hepatitis B surface antigen or core particles. Purification of proteins could be improved through variations of approaches such as the histidine-tagged proteins employed for laboratory immunogens. The immunology of most of the viral agents listed is not well-understood and should have more work, particularly in primates. Among the viruses listed, Ebola stands out as the single agent that requires much more in-depth understanding of its immunology. To date no simple approach has protected primates against death after exposure to the Zaire subtype of Ebola virus (Sullivan et al., 2000).

In the field of antivirals, there is a good deal of basic research needed on the viruses themselves, although structural and molecular biology are providing increasingly well-defined opportunities among the better-studied taxons. Both alphavirus and flavivirus structure is understood at a high level of resolution (Pletnev, et al.; Perera, et al., 2001) and the construction of the particles provides opportunities to intervene in surface assembly or capsid-nucleocapsid interactions that would be completely independent of viral polymerases. Inhibition of receptor binding through structural knowledge is another approach. The fusion peptides of filoviruses and arenaviruses are now better understood (Gallaher, 1996; Gallaher et al., 2001; Weissenhorn et al., 1998), and it may be possible to prevent infection with inhibitors designed using the same principles as for an apparently successful HIV drug (LaBranche et al., 2001).

Among the VHF there is often extensive endothelial cell involvement and modulation of their metabolism; taking advantage of insights gained through extensive work on septic shock and the "sepsis syndrome" may be useful. One promising approach has been the design of inhibitors of NF kappa B intranuclear binding (Yang et al., 1999) to prevent the over exuberant inflammatory reaction thought to be a major pathogenetic mechanism of Lassa fever and other arenavirus diseases (Mahanty et al., 2001; Marta, et al., 1999).

Conclusions

The approach to civilian biodefense against viruses consists of three elements. In the short-term we should obtain adequate stocks of smallpox vaccine, as well as the ancillary materials to deploy it. We also need to develop both policies for immunosuppressed persons and treatment strategies if they become infected with vaccinia. In the medium vista we should bring to fruition the projects that are already well along the way but are not actually ready for deployment (see Table 4-5): modest stockpiles of intravenous ribavirin, assuring availability of yellow fever vaccine, and further testing of a live-attenuated Rift Valley fever vaccine. These goals are both good bioterrorism policy and good public health preparedness. Other medium term goals can be addressed depending on priorities and resources. Long-term goals should be focused on developing better ways to translate the findings from molecular virology into drugs and vaccines for human use; to understand the immunology, structure, and molecular biology of viruses that are threats; and to apply basic science to development of broad spectrum solutions.

TABLE 4-5 Existing potentially deployable antiviral measures

VIRUS	MODALITY	COMMENTS	NEEDED IF TO BE USED
Arenaviruses	Junin vaccine, live-attenuated	Not licensed in U.S. but preclinical studies met FDA requirements for IND. Extensive experience on safety and efficacy in Argentina. DoD has master seed virus and has manufactured in the past. Current stocks limited.	Re-manufacture from existing master seed. Additional testing of new lots, licensure.
	Ribavirin, intravenous	Aerosol and oral forms licensed in U.S. DoD holds IND. Overseas studies support efficacy in Lassa fever. Preclinical and anecdotal human data support use in other arenavirus HF. Intravenous form not readily available.	Assessment of amount of powdered drug actually available. Manufacture of additional stocks of intravenous solution. Licensure of iv formulation.
Rift Valley fever	Vaccine, inactivated	DoD holds IND. Effective in preventing laboratory infections and safe in a few thousand humans. Requires 2–3 weeks for protection as well as annual boosters. Limited availability.	No facility to manufacture new lots; containment and vaccinated staff required to work with live virus.

continued

TABLE 4-5 *continued*

	Vaccine, live-attenuated	Under IND. Administered to >60 humans with safety and immune response suggesting efficacy. Not under active development. Limited availability. DoD has master seed.	Needs additional phase II testing. Manufacture from existing master seed. Will need decision as to whether to license based on preclinical data, human immune response (compared to experience with inactivated vaccine in lab workers) or field trial.
Crimean Congo HF	Ribavirin, intravenous	Drug not licensed in U.S. for iv use. Preclinical and anecdotal human data support its use but no definitive data. Employed routinely in South Africa. DoD has limited stocks and holds IND.	Needs more clinical data, presumably a trial. As for arenaviruses.
Yellow fever	Vaccine, live-attenuated	Licensed vaccine. Currently not manufactured in U.S. Stocks, surge capacity are in doubt.	Manufacture, stockpiles.
Tick-borne encephalitis	Vaccine, inactivated	Not licensed or produced in U.S. but widely available in Europe	Development of manufacturing capability.
Smallpox	Vaccine, live attenuated	U.S. and world stocks substantial but in face of multifocal epidemic would be inadequate. Manufacture of additional vaccine underway	Stockpile underway.
Monkeypox	Vaccine, live attenuated	Same vaccine as smallpox. Threat mainly direct delivery of virus; limited interhuman transmission.	Smallpox vaccine stockpile underway.

NEW RESEARCH IN ANTITOXINS

R. John Collier, Ph.D.
Maude and Lillian Presley Professor of Microbiology and Molecular Genetics, Harvard Medical School

Inhalational anthrax is a deadly disease. Based on recent events, the case fatality rate with supportive care and appropriate antibiotics is approximately fifty percent. However, in the case of massive attack and according to previous knowledge, the case fatality rate could easily approach one hundred percent. Although an anthrax vaccine exists, there are currently no plans to implement mass immunization. At least for the foreseeable future, the civilian population is completely susceptible to infection.

Currently, antibiotic therapy is our only therapeutic countermeasure. Because death from anthrax is largely, if not solely, due to the action of the anthrax toxin, antitoxins may prove to be a valuable ancillary treatment. An added advantage of antitoxins is that they not only inhibit toxin action and the progression of symptoms but, because toxins are aimed at immune system cells, antitoxins also boost the immune system by protecting those cells.

There are several viable antitoxin options based on what we have learned in these last few years about the structure and action of anthrax toxin. Three antixoxin approaches in particular appear very promising: dominant negative inhibitors (DNI), which are mutant forms of the protective antigen that block translocation; polyvalent inhibitors (PVI), which are chemically synthesized inhibitors that block toxin assembly; and soluble forms of the toxin receptor, ATR, which block toxin attachment to cells. There is also a fourth approach that is based on the fact that, because lethal factor (LF) is a zinc protease, some of the number of metallic protease inhibitors that are already in use in other drugs may be effective inhibitors of LF as well. At least one major pharmaceutical company is currently screening their large library of metallic protease inhibitors for activity in this regard.

Anthrax toxin consists of three large proteins: edema factor (EF), protective antigen (PA), and lethal factor (LF). None of these proteins alone is toxic, but a combination of EF and PA induces an edematous response, and a combination of PA and LF causes lethality. LF and EF are enzymes that act on target molecules in the cytosol of mammalian cells. As is commonly known, proteins generally do not penetrate membranes. But *B. anthracis* has devised a way to do this. In particular, PA serves as the vehicle for delivering EF and LF into the cytosol.

The model diagrammed in Friedlander, Chapter 2 p. 50, Figure 2-1 provides a basis for understanding each of the three inhibitors that have been developed. First, the growing bacteria release EF, LF and PA as monomeric proteins into its environment. Then, the proteins diffuse to the cell surface and undergo a rather intricate assembly process which results in the formation of toxic complexes.

These complexes pass into the cell and then into the endosomal compartment, where LF and EF are released into the cytosol.

The toxic complex that enters the cell is formed by the assembly of an activated PA heptamer which begins when monomeric PA binds to the toxin receptor, ATR. When the bound PA encounters cell-bound feron, a protease, a subdomain of the PA molecule is released as fragment PA-20, leaving only the PA-63 fragment bound to the receptor. There are about 10,000 receptors per cell, which means that there are about 10,000 PA molecules being activated on the cell surface, where they diffuse, collide, recognize each other, and spontaneously form a heptameric ring known as the pre-pore, or pre-insertion, form. Unlike the native PA molecule, heptameric PA is capable of binding EF and LF with very high affinity.

The crystallographic structures of the native PA, heptameric PA, EF, and LF are all known. Native PA has a four-domain structure. Part of domain 1 is cut off by furin, and the remainder of domain 1 presents a site to which EF and LF bind; domain 2 is the pore-forming domain; domain 3 is involved in the ligamerization; and domain 4 is involved in receptor binding.

Dominant Negative Inhibitors (DNIs)

DNIs are mutant forms of PA that block EF and LF from entering the cytosol through the membrane. Normally, there are loops on the subunits of the pore-forming domain 2 that undergo massive confirmational changes in acidic conditions and move in such a way that they form an aqueous pore across the membrane. The lumen of the pore may be the passageway for EF and LF, although this has yet to be proven. It is clear, however, that the translocation of EF and LF bound to the heptameric PA occurs in concert with pore formation. There are sites in PA, in the loops in contact with the lumen of the pre-pore, that when mutated result in an inactive form of PA. It is not completely understood how these mutated residues actually block translocation, but clearly they are potent inhibitors of toxin action. In cell culture, a one-to-one ratio of mutant to wild-type PA almost completely inhibits toxin action. This has also been demonstrated in rat models. Even with a wild-type PA/LF mixture that is ten times lethal dose, if as little as a quarter as much of the dominant negative form of PA is coinjected into the rat, the animal survives indefinitely with no symptoms whatsoever.

Of all of the approaches described here, DNIs are likely to prove the most effective. It is remarkable that by introducing a small change in only one amino acid out of some 700, it is possible to convert a toxin subunit into a very effective inhibitor of toxin action. The dead and inactive complexes that are formed by DNI action are probably channeled off to the lysosomes where they are destroyed. So this mode of action creates a sink, or destruction pathway, for the normal toxin subunits that the bacteria produces. Because many toxins are oligomeric and involve pore formation, this approach could also be generalized and applied to other toxins as well.

Currently, DNIs are a very late stage product. If they can be proven efficacious in infected animal models, they could be produced and deployed very rapidly. The product should be safe since PA is nontoxic and acts at very low levels relative to the toxin. Also, these mutations do not appear to affect the immunogenicity of PA. With regard to post-exposure prophylaxis, because it would not be possible to identify individuals had inhaled a lethal dose of spores, it would not be unreasonable to inject all exposed individuals with DNIs. For individuals who are in danger of experiencing a delayed form of anthrax, the antitoxin in conjunction with appropriate adjuvants would hopefully induce an active immunity to PA and thereby stop infection and the progression of disease.

Polyvalent Inhibitors (PVIs)

PVIs block toxin action by prohibiting EF and LF from even binding to heptameric PA in the first place. PVIs are polyacrylamide polymers that act at multiple sites on the heptamer. This is a very early stage product, although proof of principle experiments have validated the approach. PVIs have been shown to inhibit toxicity in cell culture as well as rescue rats that have been injected with lethal doses of toxin. This approach could be generalized and used in the development of inhibitors for other oligomeric virulence factors. However, much more work needs to be done before PVIs could ever seriously be considered for therapy.

Soluble Forms of the Toxin Receptor, ATR

This type of inhibitor is still on the drawing board. It is a logical ramification of the discovery of the identity of ATR, a type 1 membrane protein with an extracellular domain that binds directly to PA. Not much is known about ATR's function except that it is closely related to the TEM8 protein that is upregulated on colorectal cancer endothelial cells. A soluble version of one of the domains of ATR, generated by eliminating the transmembrane part of the protein and its cytoplasmic tail, has been shown to protect cells in culture by competing with the normal receptor for PA. The potency of the soluble form of ATR has yet to be tested in vivo. But once positive results are achieved, development could proceed rapidly.

RECOMBINANT HUMAN ANTIBODY: IMMEDIATE IMMUNITY FROM BOTULINUM NEUROTOXIN AND OTHER CLASS A BIOTHREAT AGENTS.

James D. Marks, M.D., Ph.D.
Departments of Anesthesia and Pharmaceutical Chemistry
University of California, San Francisco

Botulinum Neurotoxins as Biothreat Agents

The spore forming bacteria *Clostridium botulinum* secrete botulinum neurotoxin (BoNT), the most poisonous substance known (Gill, 1982). The protein toxin consists of a heavy and light chain, which contain three functional domains (Simpson, 1980; Montecucco and Schiavo, 1995; Lacy et al., 1998). The C-terminal portion of the heavy chain (H_C) comprises the binding domain, which binds to cellular receptors on presynaptic neurons, resulting in toxin endocytosis (Dolly et al., 1984, Montecucco, 1986). The N-terminal portion of the heavy chain (H_N) comprises the translocation domain, which allows the toxin to escape the endosome. The light chain is a zinc endopeptidase that cleaves different members of the SNARE complex, depending on serotype, resulting in blockade of neuromuscular transmission (Schiavo et al., 1992; Lacy and Stevens, 1999).

There are 7 BoNT serotypes (A-G) (9), four of which (A, B, E, and F) cause the human disease botulism (Arnon et al., 2001). Botulism is characterized by prolonged paralysis, which if not immediately fatal requires prolonged hospitalization in an Intensive Care Unit (ICU) and mechanical ventilation. The potent paralytic ability of the toxin has resulted in its use in low doses as a medicine to treat a range of overactive muscle conditions including cervical dystonias, cerebral palsy, post-traumatic brain injury, and post-stroke spasticity (Mahant et al., 2000). BoNTs are also classified by the Centers for Disease Control (CDC) as one of the 6 highest-risk threat agents for bioterrorism (the Class A agents), due to their extreme potency and lethality, ease of production and transport, and need for prolonged intensive care (Arnon et al., 2001). It is likely that any one of the seven BoNT serotypes can be used as a biothreat agent. Both Iraq and the former Soviet Union produced BoNT for use as weapons (United Nations Security Council, 1995; Bozheyeva et al., 1999) and at least 3 additional countries (Iran, North Korea, and Syria) have developed or are believed to be developing BoNT as instruments of mass destruction. Iraq produced 19,000 L of concentrated BoNT of which 10,000 L were weaponized in missile warheads or bombs (United Nations Security Council, 1995; Zilinskas, 1997). The 19,000 L are not fully accounted for and represent an amount of toxin capable of killing the world's population three times over. The Japanese cult Aum Shinrikyo attempted to use BoNT for bioterrorism by dispersing toxin aerosols at multiple sites in Tokyo (Arnon et al., 2001).

Exposure of even a small number of civilians would paralyze the health care delivery system of any metropolitan center. Treatment of botulism requires prolonged ICU hospitalization and mechanical ventilation for up to six weeks. With the downsizing and closing of hospitals, most ICUs run at 80–100% occupancy. In San Francisco, for example, there are approximately 210 ICU beds, with an average occupancy rate of greater than 90%. As few as thirty cases of botulism would fill all empty ICU beds and occupy them for up to 6 weeks. This would eliminate availability of ICU beds for post-operative patients requiring ICU care, such as organ transplantation, neurosurgery, cardiac surgery, and traumatic injuries. Patients requiring such operations would represent 'collateral damage', with necessary surgery postponed, or transferred to outlying hospitals. Major civilian exposure to BoNT would have catastrophic effects. It has been estimated that aerosol exposure of 100,000 individuals to toxin, as could occur with an aerosol release over a metropolitan area, would result in 50,000 cases with 30,000 fatalities (St. John et al., 2001). Such exposure would result in 4.2 million hospital days and an estimated cost of $8.6 billion. In this study, the most important factors reducing mortality and cost were early availability of antitoxin and mechanical ventilation (St. John et al., 2001). Such treatment could reduce deaths by 25,000 and costs by $8.0 billion.

Prevention and Treatment of Botulism

No specific small molecule drugs exist for prevention or treatment of botulism, but an investigational pentavalent vaccine is available from the CDC (Siegel, 1988) and a recombinant vaccine is under development (Byrne and Smith, 2000). Regardless, mass civilian or military vaccination is unlikely due to the rarity of disease or exposure and the fact that vaccination would deny subsequent medicinal use of BoNT. Post-exposure vaccination is useless, due to the rapid onset of disease.

Antibodies for Prevention and Treatment of Botulism

Neutralizing antibody (Ab) can be used for both prevention and treatment of botulism. Historically, such Ab has been made by hyperimmunzing either horses (equine antitoxin) or human volunteers (human botulinum immune globulin). After immunization, the serum is collected and antibody prepared. The resulting antibody is polyclonal, consisting of hundreds to thousands of different antibodies that bind to many different parts of the toxin. Equine antitoxin has been shown to protect against the development of botulism in multiple animal models when administered prior to or after exposure to toxin (Franz et al., 1993). Antibody therapy is also the standard of care for botulism in humans, with equine anti-toxin and hyperimmune human globulin used to treat adult (Black and

Gunn, 1980; Hibbs et al., 1996) and infant botulism (Arnon, 1993) respectively. The best evidence for the value of antibody in treating botulism comes from a prospective randomized comparison of human botulinum immune globulin to non-immune globulin (Arnon, 1993). In this study, infants treated with human botulinum immune globulin had their ICU stays reduced by 2 weeks and their hospital stay reduced by 3 weeks compared to treatment with non-immune globulin. Treatment with immune globulin was beneficial even when administered 1 week after hospitalization.

Polyclonal antibodies typically neutralize toxin with high potency. Their manufacture, however, requires immunization and plasmapheresis, making large-scale production not feasible. As a result only a minimal number of doses of equine antitoxin and human immune globulin exist. These supplies would be inadequate to treat a major neurotoxin exposure. In addition, polyclonal antibodies suffer from batch to batch variability and potential infectious risk associated with an animal or human product. Horse antitoxin is also immungenic when administered to humans and can cause serum sickness and anaphylactic shock.

Monoclonal Antibody Technologies

Since 1975 it has been possible to make antibodies derived from a single antibody producing B-lymphocyte, so called monoclonal antibodies (mAbs) (Kohler and Milstein, 1975). Mabs consist of a single Ab and are made by fusing a B-cell to an immortal cell line, which can be expanded without limit. Mabs can be manufactured in unlimited supply, do not require a source of immune donors, are consistent batch to batch, and have no infectious risk. Initial mAbs were derived from mice or other rodents and were immunogenic when administered to humans, limiting their clinical development. Unfortunately, hybridoma technology has not proven generally adaptable to making human antibodies by fusing B-cells from immunized humans to an immortal cell line. With the advent of modern molecular biology techniques, however, it has become possible to make monoclonal antibodies that are far less immunogenic. Chimeric antibodies are made by grafting human antibody constant domains onto the murine variable domains, yielding antibodies which are approximately 75% human in sequence (Morrison et al., 1984). Humanized antibodies are made by grafting only the antigen binding antibody loops from the variable domains onto human variable domain frameworks, yielding an antibody which is 90% human in sequence (Jones et al., 1986). Such antibodies are far less immunogenic in humans than murine mAbs. The human constant domains also impart a long serum half life of up to one month. As a result, a single dose of antibody can provide 3 to 6 months of protection against pathogens.

Recently, it has proven possible to make antibodies that are entirely human in sequence either by immunizing mice transgenic for the human immunoglobulin locus and making hybridomas (Mendez et al., 1997) or by using phage display (Marks and Marks, 1996). For phage display, repertoires of antibody genes are cloned into a bacterial phage vector where the antibodies are displayed on the surface of the bacteriophage fused to one of the phage coat proteins (Marks et al., 1991). While it is not technically possible to display full length IgG antibodies on phage, it is possible to display smaller single chain Fv (scFv) or Fab antibody fragments. Such antibody fragments contain the antigen recognition of the IgG (the variable domains), but lack the constant regions.

To perform phage display, repertoires of antibody heavy and light chain variable domain genes are assembled and cloned into a phage vector to create libraries of scFv or Fabs displayed on the phage surface. The source of the variable region genes can be any species, including immunized humans. Once phage libraries are constructed, binding phage antibodies can be isolated from non-binding phage by a variety of types of affinity chromatography. Binding phage antibodies can be detected by ELISA, and the antibodies characterized with respect to affinity, epitope recognized, sequence, and biologic activity. Given the high transformation efficiency of bacteria, it is possible to make libraries of millions to billions of different antibodies, allowing immortalization of the entire immune response to an antigen. As a result, hundreds to thousands of antibodies are generated, allowing isolation of high affinity antibodies to rare epitopes. In contrast, generation of hybridomas captures only a fraction of the immune response, due to the inefficiency of the fusion process. Other advantages of antibody phage display include the ability to make antibodies from immunized humans and to engineer antibody affinities to values not achievable by hybridoma technology (Schier et al., 1996).

Neutralization of Botulinum Neurotoxin by Monoclonal and Oligoclonal Antibody

Recombinant mAb could provide an unlimited supply of antitoxin free of infectious disease risk and not requiring a source of human donors for plasmaphoresis. Such mAb must be of high potency in order to provide an adequate number of doses at reasonable cost. In some instances, the potency of polyclonal antibody can be recapitulated in a single mAb (Lang et al., 1993). In the case of BoNT, potent neutralizing mAb have yet to be produced: single mAb neutralizing at most 10 to 100 times the 50% lethal dose (LD_{50}) of toxin in mice (Pless et al, 2001; Hallis et al., 1993).

To generate mAb capable of neutralizing BoNT serotype A (BoNT/A), we generated scFv phage antibody libraries from immunized humans, mice, and mice transgenic for the human immunoglobulin locus (Amersdorfer et al., 1997; Amersdorfer et al., in press). After screening more than 100 unique mAb from these libraries, three groups of scFv were identified which bound non-overlapping epitopes on the BoNT/A binding domain (H_C) and which neutralized toxin *in vitro* (prolonged the time to neuroparalysis in a murine hemidiaphragm model) (Amersdorfer et al., 1997; Amersdorfer et al., in press). *In vitro* toxin neutralization increased significantly when two scFv binding non-overlapping epitopes were combined. The small size of the scFv, however, precluded study of *in vivo* toxin neutralization, due to the rapid clearance of the 25 kDa scFv from serum (Colcher et al., 1990).

To evaluate *in vivo* BoNT neutralization, IgG were constructed from the variable domain genes of three BoNT/A scFv that neutralized toxin *in vitro*. No single IgG significantly neutralized toxin *in vivo*, but combining mAb led to potent toxin neutralization. The most potent mAb pair protected mice challenged with 1500 LD_{50}s of toxin, while combining all three mAb protected mice challenged with 20,000 LD_{50}s of toxin (per 50 μg of antibody administered) (Nowakowski et al., 2002). The potency of the three antibody combination (oligoclonal Ab) was formally titered using the standard mouse neutralization bioassay and was determined to be 45 International Units/mg of Ab. One International Unit (IU) neutralizes 10,000 LD_{50}s of BoNT/A toxin, yielding a potency of 450,000 LD_{50}s /mg of Ab. This is 90 times more potent than the hyperimmune human globulin used to treat infant botulism (Arnon, 1993) and approaches the potency of hyperimmune mono-serotype horse type A anti-toxin (Sheridan et al., 2001).

The increase in potency appears to result primarily from a large increase in the affinity of the oligoclonal Ab for toxin compared to the individual mAb (Nowakowski et al., 2002), and also to greater blockade of the toxin surface which interacts with cellular receptors (Mullaney et al., 2001). Such mechanisms may be generally applicable to many antigens in solution, suggesting that oligoclonal Ab may offer a general route to more potent antigen neutralization than mAb. The precise contribution of these two mechanisms to the increase in potency is unknown. It is also unknown as to whether engineering the affinity of one of the mAbs to a value approaching that of the oligoclonal Ab would yield a similar increase in potency as combining mAbs.

Conclusions, Obstacles, and Future Research Needs

Recombinant oligoclonal Ab offers a safe and unlimited supply of drug for prevention and treatment of BoNT/A intoxication. With an elimination half-life of up to 4 weeks, Ab would provide months of protection against toxin. Since the current oligoclonal Ab consists of either chimeric or human IgG, production could be immediately scaled to produce a stockpile of safe anti-toxin. Alternatively, we have already replaced the chimeric S25 IgG with a fully human IgG and increased potency of the oligoclonal Ab more than 2 fold. Work is ongoing to replace chimeric C25 with a fully human homologue. Chimeric, humanized, and human mAb represent an increasingly important class of therapeutic agents whose means of production are known. The high potency of the oligoclonal Ab makes it possible to manufacture millions of doses of antitoxin from a single manufacturing facility which could be stockpiled for future use. Ten mAb have been approved by the FDA for human therapy and more then 70 other Mab therapeutics are in clinical trials (Reichert, 2001). As a result, the process of scaling production and manufacturing, as well as the necessary toxicology and clinical safety testing requirements are known. This should result in a rapid development timeline, especially compared to vaccines or small molecule drugs. The major challenges and obstacles to development are FDA regulatory issues related to combining mAbs and a predicted worldwide shortage of IgG manufacturing capacity (Reichert, 2001).

Oligoclonal Ab would be applicable to the other BoNT toxin serotypes and these antibodies should be generated as rapidly as possible. Oligoclonal antibody may also be able to potently neutralize other class A agents as well. Anthrax toxicity is toxin mediated, and polyclonal Ab has been shown to be protective for this agent (Little et al., 1997; Beedham et al., 2001). Vaccinia immunoglobulin can be used to prevent or treat smallpox or complications arising from vaccination of immunocompromised hosts (Feery, 1976). Ab may also be useful for plague and the hemorrhagic fevers (Hill et al., 1997; Wilson et al., 2000). Given the threats posed by these agents, rapid generation and evaluation of oligoclonal Ab for their neutralization is warranted.

MEETING THE REGULATORY AND PRODUCT DEVELOPMENT CHALLENGES TO ADDRESS TERRORISM

Andrea Meyerhoff,* M.D.
Director, Anti-terrorism Programs
Office of the Commissioner, Food and Drug Administration

FDA's mandate in anti-terrorism warrants a balance between its requirements as a regulatory agency and the demands of a public health emergency. We attempt to achieve this balance by facilitating the availability of safe, effective drugs, vaccines, and medical devices in a manner that is consistent with our legal responsibilities as a regulatory agency.

Organization of FDA Anti-Terrorism Programs

The FDA is divided into five centers which are organized based on the type of products that are regulated. Three of these centers regulate products that deal with medical care: CDER (Center for Drugs); CBER (Center for Biologics), which regulates vaccines; and CDRH (Center for Devices and Radiation Health), which regulates a range of medical devices from diagnostic assays to mechanical ventilators.

---

* This statement reflects the professional view of the author and should not be construed as an official position of the Food and Drug Administration.

The Director of Anti-terrorism Programs is housed in the Office of the Commissioner, which is not housed in any particular center but rather coordinates across these centers. There is a designated anti-terrorism point of contact (POC) within each center and with whom the director liaisons on antiterrorism issues.

For many products under development, there is already a relationship established with the appropriate regulating center. New products that are seeking regulatory guidance and old products that may have an anti-terrorism application are often routed through the Anti-terrorism Programs first and then passed on to the appropriate POC. Anti-terrorism Programs works with the POC to coordinate all efforts that are relevant to each particular stage of product development.

Existing Regulatory Mechanisms for Enhanced Product Availability

There are several existing regulatory mechanisms that can be invoked to address issues of anti-terrorism product availability. They apply to a number of different phases of product development, from the early pre-IND (i.e., before the drug is introduced into human trials) to the review of the NDA (i.e., new drug application for marketing approval):

- Pre-IND meeting is an attempt to begin early dialogue between the sponsor and the review division and provide regulatory guidance in preparing IND (investigational new drug) applications. IND applications include a set of data that are shown to the agency before the product is used for the first time in humans. The entire body of data are reviewed by all of the various disciplines that are brought to bear at that stage, and missing pieces of data are identified. Pre-IND meetings are regarded as resources for developers. There is no set period in the pre-IND phase when this meeting must occur. Some sponsors approach the FDA quite early; others meet immediately before they submit the IND just to make sure that everything is okay.
- IND regulations refer to the set of regulations that determine how a product will be used when it is initially introduced into a human population. IND regulations may also be viewed as a mechanism for making an investigational product available. IND regulations have three basic components: an informed consent form; review of the protocols for planned use by an institutional review board; and a plan for the collection of safety and efficacy data from the human population in which the product is going to be used.
- "Streamlined" IND is not an official regulatory term, but it serves the purpose of addressing the requirements of the IND regulations while simultaneously making a product available to a large population in an emergency setting. Streamlined INDs are in place for both biologic and drug products, and the template can be used for other products as well if the need should arise.
- The animal efficacy rule was proposed and published in the Federal Register in October 1999. It is intended to apply when a disease cannot be studied in

humans, that is when the disease is either very rare or it would be unethical to introduce the disease into a human population. Clearly, diseases due to biological agents of intentional use would fit into this category. This rule provides the framework for which efficacy data could be derived from an animal model of disease and is intended to address efficacy only. The safety of the drug still needs to be studied in the human population. The rule is based on the use of a scientifically valid animal model and generally requires the use of two species. In cases where a well-established species is already recognized as a scientifically valid model for disease, it would be decided on a case-by-case basis whether efficacy data is needed from a second species as well. Currently, this rule is still only proposed and has not been used (the approval of ciprofloxacin for anthrax invoked accelerated approval, not the animal rule); finalization is anticipated within the next few months.

- Accelerated approval (sometimes referred to as subpart H regulation) refers to a set of regulations that permit the use of a surrogate marker for the purposes of demonstrating efficacy of a product if the product is considered reasonably likely to provide an improvement in mortality or serious morbidity. Still, post-marketing data would need to be collected to verify the surrogate. This is the regulatory approach that was taken for ciprofloxacin for anthrax which was initially approved for human use in the mid-1980s, so there was already a fairly well-developed set of human pharmacokinetic data and a very large safety database. The surrogates in this case were human serum levels of ciprofloxacin which have been shown to be associated with improved survival in monkeys that have been exposed to aerosolized *B. anthracis* spores. Serum level in humans have been shown to reach or exceed weight-adjusted levels in monkeys. The labeled regimen for post-exposure inhalational anthrax is a sixty-day dosing period. Safety databases of patients who received the drug for more than sixty days, patients who received the drug for sixty days, patients who received it for less fewer than sixty days, and patients who received other antibiotics all show similar adverse event rates. GI events are the most common, with a slightly incidence of higher abdominal pain and rash in the sixty-day group. However, patients who received the drug for sixty days showed no previously unidentified adverse events associated with the shorter, more usual seven to fourteen day dosing periods. There is also a substantial pediatric safety database which supports the approval of ciprofloxacin for post-exposure inhalational anthrax indication in pediatric patients.
- Priority review is a request that is made by the applicant at the time of NDA filing. It is generally used for products that are considered to have special public health significance and results in a review process that is shortened to six months rather than the usual ten or twelve. In addition to accelerated approval, ciprofloxacin for anthrax also received priority review.

Recent FDA Anti-Terrorism Initiatives: Drug Development

Recent FDA antiterrorism-specific initiatives, most of which involve anthrax, include:

- In early November, 2001, the FDA published a Federal Register notice recognizing that doxycycline and penicillin are also approved for anthrax. The Federal Register notice was published because product labels do not contain specific dosing information for post-exposure inhalational anthrax, even though scientific data support this labeling. The Federal Register notice states this, provides the dosing recommendations, and invites applications from manufacturers of these drugs to request labeling supplements. This was done as a way to expand the options of products available to manage what was clearly a growing population of people who had been exposed to aerosolized spores of *B. anthracis*. Because of potential side effects, drug intolerance, other medications, and any of a number of other reasons why people cannot take a particular class of drugs, having more available options expands our ability to manage large populations of exposed individuals.
- There are a number of ongoing efforts among several government agencies to provide regulatory guidance for the development of animal models to be used in the evaluation of drugs specifically for diseases related to bioterrorism. There has been much ongoing collaboration with DOD laboratories and the NIH to establish guidelines and goals for studying these products in animal models.

The FDA has been considering other products besides antimicrobials that could be made available for the treatment of clinically apparent inhalational anthrax.

REGULATION AND PRODUCTION OF RECOMBINANT HUMAN ANTIBODIES

Kathryn E. Stein,* **Ph.D.**
Director, Division of Monoclonal Antibodies
Center for Biologics Evaluation and Research
Food and Drug Administration

The FDA is fully prepared to deal with the issue of monclonal antibody cocktails and, in fact, has had relevant policies in place since 1994. At that time, it was anticipated that manufacturers might develop antibody cocktails directed at either different epitopes on a particular antigen or different antigens on a particular organism. From a safety and efficacy perspective, these policies consider

* This statement reflects the professional view of the author and should not be construed as an official position of the Food and Drug Administration.

cocktails as single products. However, there must be a rationale for the use of each component in the cocktail as well as a means for determining the dose of each component. These data could come from preclinical animal models, for example, or *in vitro* neutralization or other tests.

The real question is, what kinds of antibodies should be included in the cocktails? For example, because the murine Fc region is the most immunogenic part of a monoclonal, both chimeric and humanized antibodies, with human Fc regions, have been engineered and shown to exhibit much less immunogenicity in humans than whole murine antibodies. With regards to how antibodies are produced, there is some concern that phage display may create combinations of heavy and light chain genes that would raise unusual issues regarding immunogenicity.

One could envision a mixed antibacterial and antiviral cocktail comprised of antibodies to a diverse assortment of potential bioterrorist agents. However, more research is needed to identify protective factors and determine which virulence factors the antibodies should target. (See Marks for further discussion on antibody options.)

There are many antibodies currently being researched in academe that could be developed into cocktails. There needs to be greater partnerships among academe, government, and industry in order to bring the intellectual property to the antibody engineers so that these products can be developed.

There are ways to lyophilize monoclonal antibodies such that the cocktails could be stable at room temperature and on the shelf of every emergency room, although formulation needs to be further studied. The FDA is willing to consider any proposed products.

Limitations on monoclonal cocktails pertain mostly to production. The worldwide capacity for mammalian cell culture has reached its maximum. In order to increase production to build a stockpile for prophylaxis or treatment in the event of a large-scale bioterrorist attack, we must either build more manufacturing facilities or buy or renovate already existing facilities. Such large-scale production will likely require government assistance and funding.

REFERENCES

D. Shlaes:

Food and Drug Administration (FDA). Division of Anti-Infective Drug Products. Points to Consider, Clinical Development and Labeling of Anti-Infective Drug Products. 1992. Online. www.fda.gov/cder/present/anti-infective798/biostats/tsld005.htm

Food and Drug Administration (FDA). Division of Anti-Infective Drug Products. Points to Consider, Clinical Development and Labeling of Anti-Infective Drug Products, Disclaimer of 1992 Points to Consider Document. March 08, 2001a. Online. www.fda.gov/cder/guidance/ptc.htm.

Food and Drug Administration (FDA). International Conference on Harmonization (ICH) of Technical Requirements for Registration of Pharmaceuticals for Human Use, Guideline, Choice of Control Group and Related Issues in Clinical Trials. May 2001b. Online. www.fda.gov/cder/guidance/4155fnl.htm

C.J. Peters:

Alibek K. 2001. Biological weapons: past, present, and future. 177–185. In: Layne SP, Beugelskijk TJ, Patel CKN, eds. Firepower in the Lab. Automation in the Fight Against Infectious Diseases and Bioterrorism. Washington, D.C.: Joseph Henry Press.

Alibek K, Handelman S, eds. 1999. Biohazard. New York: Random House.

Barrera Oro JG, McKee, Jr., KT. 1991. Toward a vaccine against Argentine hemorrhagic fever. *Bulletin of the Pan American Health Organization*, 25:118–126.

Chua KB, Bellini WJ, Rota PA, Harcourt BH, Tamin A, Lam SK, Ksiazek TG, Rollin PE, Zaki SR, Shieh WJ, Goldsmith CS, Gubler DJ, Roehrig JT, Eaton B, Gould AR, Olson J, Field H, Daniels P, Ling AE, Peters CJ, Anderson LJ, Mahy BWJ. 2000. Nipah virus: a recently emergent deadly paramyxovirus. *Science* 2888:1432–1435.

Ferguson JR. 1999. Biological weapons and U.S. law. Pp. 81–91. In: Lederberg J, ed. Biological weapons: limiting the threat. Cambridge: The MIT Press.

Gallaher WR. 1996. Similar structural models of the transmembrane proteins of Ebola and avian sarcoma viruses. *Cell* 85:477–478.

Gallaher WR, DiSimone C, Buchmeier MJ. 2001. The viral transmembrane superfamily: possible divergence of Arenaviruses and Filovirus glycoproteins from a common RNA virus ancestor. *BMC Microbiology* 1:1.

Jezek Z., Fenner F, eds. 1988. Human monkeypox. *Monographs in Virology* 17:140. New York: Karger.

Kenyon RH, Canonico PG, Green DE, Peters CJ. 1986. Effect of ribavirin and tributylribavirin on Argentine hemorrhagic fever (Junin virus) in guinea pigs. *Antimicrobial Agents and Chemotherapy* 29:521–523.

Kilbourne ED. 1975. Epidemiology of influenza. Pp. 483–538. In: Kilbourne ED, ed. The Influenza Viruses and Influenza. New York: Academic Press, Inc.

Koplan JP, Hicks JW. 1974. Smallpox and vaccinia in the United States—1972. *Journal of Infectious Diseases* 129:224–226.

LaBranche CC, Galasso G, Moore JP, Bolognesi DP, Hirsch MS, Hammer SM. 2001. HIV fusion and its inhibition. *Antiviral Research* 50:95–115.

Mahanty S, Bausch DG, Thomas RL, Goba A, Bah A, Peters CJ, Rollin PE. 2001. Low levels of interleukin-8 and interferon-inducible protein-10 in serum are associated with fatal infections in acute Lassa fever. *Journal of Infectious Diseases*183:1713–1721.

Maiztegui JI, McKee KT, Barrera Oro JG, Harrison LH, Gibbs PH, Feuillade MR, Enria DA, Briggiler AM, Levis SC, Ambrosio AM, Halsey NA, Peters CJ, AHF Study Group. 1998. Protective efficacy of a live attenuated vaccine against Argentine hemorrhagic fever. *Journal of Infectious Diseases* 177:277–283.

Marta RF, Montero VF, Hack CE, Sturk A, Maiztegui JI, Molinas FC. 1999. Proinflammatory cytokines and elastase-alpha-lantitrypsin in Argentine hemorrhagic fever. *American Journal of Tropical Medicine and Hygiene* 60:85–89.

Martin M, Weld LH, Tsai TF, Mootrey GT, Chen RT, Niu M, Cetron MS, and the GeoSentinel Yellow Fever Working Group. 2001. Advanced age a risk factor for illness temporally associated with yellow fever vaccination. *Emerging Infectious Diseases* 7:945–951.

McCormick JB, King IJ, Webb PA, Scribner CL, Craven RB, Johnson KM, Elliott LH, Belmont-Williams R. 1986. Lassa fever. Effective therapy with ribavirin. *New England Journal of Medicine* 314:20–26.Meltzer MI, Damon I, LeDuc JW, Millar JD. 2001. Modeling potential responses to smallpox as a bioterrorist weapon. *Emerging Infectious Diseases* Nov–Dec;7(6):959–969.

Modeling Potential Responses to Smallpox as a Bioterrorist Weapon. EID vol 7. 2001. Online. http://www.cdc.gov/ncidod/eid/vol7no6/meltzer.htm.

Murray K, Eaton B, Hooper P, Wang L, Williamson M, Young P. 1998. Flying foxes, horses, and humans: a zoonosis caused by a new member of the Paramyxoviridae. Pp. 43–58 in Emerging Infections I, W.M. Scheld, D. Armstrong, and J.M. Hughes, eds. Washington, D.C.: ASM Press.

Nathan N, Barry M, Van Herp M, Zeller H. 2001. Shortage of vaccines during a yellow fever outbreak in Guinea. *Lancet* 358:2129–2130.

Neumann GT, Watanabe H, Ito S. ,Watanabe H, Goto P, Gao M, Hughes DR, Perez R, Donis E, Hoffman G, Hobom, and Y. Kawaoka. 1999. Generation of influenza A viruses entirely from cloned cDNAs. Proceedings of the National Academy of Sciences USA, 96:9345–9350.

Neustadt RE., Fineberg HV. 1978. The swine flu affair. In Decision-making on a slippery disease, US Department of HEW, ed. Washington, DC: US Government Printing Office.

O'Toole T. 1999. Smallpox: an attack scenario. *Emerging Infectious Diseases*. 5:540–546.

Parashar UD, Sunn LM, Ong F, Mounts AW, Arif MT, Ksiazek TG, Kamaluddin MA, Mustafa AN, Kaur H, Ding LM, Othman G, Radzi HM, Kitsutani PT, Stockton PC, Arokiasamy J, Gary Jr., H, Anderson LJ. 2000. Case-control study of risk factors for human infection with a new zoonotic paramyxovirus, Nipah virus, during a 1998–1999 outbreak of severe encephalitis in Malaysia. *Journal of Infectious Diseases* 181:1755–1759.

Perera R, Owen KE, Tellinghuisen TL, Gorbalenya AE, Kuhn RJ. 2001. Alphavirus nucleocapsid protein contains a putative coiled coil alpha-helix important for core assembly. *Journal of Virology* 75: 1–10.

Peters CJ, Zaki SR. 1999. Viral hemorrhagic fever: an overview. Pp. 1180–1188. In: Guerrant RL, Walker DH, Weller PF. Tropical Infectious Diseases: Principles, Pathogens, and Practices. New York: W.B. Saunders.

Peters CJ. 2000. Are hemorrhagic fever viruses practical agents for biological terrorism? Pp. 203–211 In: Scheld WM, Craig WA, Hughes JM, eds. Emerging Infections. Volume 4, Washington, D.C.: ASM Press.

Peters CJ. 2001. The viruses in our past, the viruses in our future. Pp. 1–32 in SGM Symposium 60: New Challenges to Health: the Threat of Virus Infection, G.L. Smith, and others, eds. United Kingdom: Cambridge University Press.

Peters CJ. 2002. Arenaviruses. In: Richman DD, Whitley RJ, Haydon FG, eds. Clinical Virology. 2nd edition, in press.

Pittmann PR, Liu CT, Cannon TL, Makuch RS, Mangiafico JA, Gibbs PH, Peters CJ. 1999. Immunogenicity of an inactivated Rift Valley fever vaccine in humans: a 12-year vaccine. *Vaccine*. 18:181–189.

Pletnev SV., Zhang W, Mukhopadhyay S, Fisher BR, Hernandez R, Brown DT, Baker TS, Rossmann MG, Kuhn RJ. 2001. Locations of carbohydrate sites on alphavirus glycoproteins show that E1 forms an icosahedral scaffold. *Cell*. 105:127–136.

Polo JM, Gardner JP, Ji Y, Belli BA, Driver DA, Sherrill S, Perri S, Liu MA, Dubensky Jr., TW. 2000. Alphavirus DNA and particle replicons for vaccines and gene therapy. *Developmental Biology* 104:181–185.

Smerdou C, Liljestrom P. 1999. Non-viral amplification systems for gene transfer: vectors based on alphaviruses. *Current Opinions in Molecular Therapy* 1:244–251.

Steiner PE. 1968. Disease in the Civil War. P. 243 in Natural Biological Warfare in 1861–1865, Charles C.Thomas, ed. Springfield, IL.

Sullivan NJ, Sanchez A, Rollin PE, Yang ZY, Nabel GJ. 2000. Development of a preventive vaccine for Ebola virus infection in primates. *Nature* 408:605–609.

Weaver SC. 1997. Convergent evolution of epidemic Venezuelan equine encephalitis viruses. Pp. 241–249 In: Saluzzo JF, Dodet B, eds. Factors in the Emergence of Arbovirus Diseases. Paris:Elsevier.

WeissenhornW, Calder LJ, Wharton SA, Skehel JJ, Wiley DC. 1998. The central structural feature of the membrane fusion protein subunit from the Ebola virus glycoprotein is a long triple-stranded coiled coil. Proceedings of the National Academy of Sciences USA, 95:6032–6036.

Yang X-B, Fennewald S, Luxon BA, Aronson J, Herzog N, Gorenstein DG. 1999. Aptamers containing thymidine 3'-O-phosphorodithioates: synthesis and binding to nuclear factor-κB. *Bioorganic and Medicinal Chemistry* 9:3357–3362.

J. Marks:

Amersdorfer P, Wong C, Chen S, Smith T, Desphande S, Sheridan R, Finnern R, Marks JD. 1997. Molecular characterization of murine humoral immune response to botulinum neurotoxin type A binding domain as assessed using phage antibody libraries. *Infection and Immunity* 65: 3743–3752.

Amersdorfer P, Chen S, Wong C, Smith T, Deshpande S, Sheridan R, Marks JD. 2002 (In press). Genetic and immunologic comparison of anti-botulinum type A antibodies from immune and non-immune human phage libraries. *Vaccine* 20:1640–1648.

Arnon SS. 1993. in *Botulinum and Tetanus Neurotoxins: Neurotransmission and Biomedical Aspects* (DasGupta, B. R., Ed.) pp 477–482, Plenum Press, New York.

Arnon SA, Schecter R, Inglesby TV, Henderson DA, Bartlett J G, Ascher MS, Eitzen E, Fine AD, Hauer J, Layton, M, Lillibridge S, Osterholm MT, O'Toole T, Parker G, Perl TM, Russell PK, Swerdlow DL, Tonat K. 2001. Botulinum toxin as a biological weapon. *Journal of the American Medical Association* 285, 1059–1070.

Beedham RJ, Turnbull PCB, Williamson ED. 2001. Passive transfer of protection against *Bacillus anthracis* infection in a murine model. *Vaccine* 19, 4409–4416

Black RE, Gunn RA. 1980. Hypersensitivity reactions associated with botulinal antitoxin. *American. Journal of Medicine* 69, 567–570.

Bozheyeva G, Kunakbayev Y, Yeleukenov D. 1999. *Center for Nonproliferation Studies, Monterey Institute of International Studies, Monterey, CA*. Former soviet biological weapons facilities in Kazakhstan: past, present, and future. June 1999: 1–20. Occasional paper No. 1

Byrne MP, Smith LA. 2000. Development of vaccines for prevention of botulism. *Biochimie* 82, 955–966.

Colcher D, Bird R, Roselli M, Hardman KD, Johnson S, Pope S, Dodd SW, Pantoliano MW, Milenic DE, Schlom J. 1990. In vivo tumor targeting of a recombinant single-chain antigen-binding protein. *Journal Of The National Cancer Institute* 82, 1191–1197.

Dolly JO, Black J, Williams RS, Melling J. 1984. Acceptors for botulinum neurotoxin reside on motor nerve terminals and mediate its internalization. *Nature* 307: 457–460.

Franz DR, Pitt LM, Clayton MA, Hanes MA, Rose KJ. 1993. In: DasGupta BR, ed. 1993. *Botulinum and Tetanus Neurotoxins: Neurotransmission and Biomedical Aspects* Plenum Press, New York pp 473–476.

Feery BJ. 1976. The efficacy of vaccinial immune globulin, a 15 year study. *Vox sanguinis*. 31: 68–76.

Gill MD. 1982. Bacterial toxins: a table of lethal amounts. *Microbiology Review* 46: 86–94.

Hallis B, Fooks S, Shone C, Hambleton P. In: DasGupta BR, ed. 1993. *Botulinum and Tetanus Neurotoxins: Neurotransmission and Biomedical Aspects* Plenum Press, New York 473–476.

Hibbs RG, Weber JT, Corwin A, Allos BM, Abd el Rehim MS, Sharkawy SE, Sarn JE, McKee KTJ. 1996. Experience with the use of an investigational F(ab')2 heptavalent botulism immune globulin of equine origin during an outbreak of type E origin in Egypt. *Clinical Infectious Diseases*. 23:337–340.

Hill J, Leary SE, Griffin KF, Williamson ED, Titball RW. 1997. Regions of Yersinia pestis V antigen that contribute to protection against plague identified by passive and active immunization. *Infection and Immunity* 65:4476–4482.

Jones PT, Dear PH, Foote J, Neuberger MS, Winter G. 1986. Replacing the complementarity determining regions in a human antibody with those from a mouse. *Nature* 321: 522–525.

Kohler G, Milstein C. 1975. Continuous cultures of fused cells secreting antibody of predefined specificity. *Nature* 256: 495–497.

Lacy DB, Tepp W, Cohen AC, DasGupta BR, Stevens RC. 1998. Crystal structure of botulinum neurotoxin type A and implications for toxicity. *Nature Structural Biology* 5:898–902.

Lacy DB, Stevens RC. 1999. Sequence homology and structural analysis of the Clostridial neurotxins. *Journal of Molecular Biology* 291:1091–1104.

Lang AB, Cryz SJ, Schurch U, Ganss MT, Bruderer U. 1993. Immunotherapy with human monoclonal antibodies: Fragment A specificity of polyclonal and monoclonal antibodies is crucial for full protection against tetanus toxin. *Journal of Immunology* 151: 466–472.

Little SF, Ivins BE, Fellows PF, Friedlander AM. 1997. Passive protection by polyclonal antibodies against Bacillus anthracis infection in guinea pigs. *Inf. Immun.* 65:5171–5175.

Mahant N, Clouston PD, Lorentz IT. 2000. The current use of botulinum toxin. *Journal of Clinical Neuroscience* 7: 389–394.

Marks JD, Hoogenboom HR, Bonnert TP, McCafferty J, Griffiths AD, Winter G. 1991. By-passing immunization. Human antibodies from V-gene libraries displayed on phage. *Journal of Molecular Biology*. 222:581–597.

Marks C, Marks JD. 1996. Phage libraries-a new route to clinically useful antibodies. *New England Journal of Medicine* 335: 730–733.

Mendez MJ, Green LL, Corvalan JR, Maynard-Currie CE, Yang XD, Gallo ML, Louie DM, Lee DV, Erickson KL, Luna J, Roy CM, Abderrahim H, Kirschenbaum F, Noguchi M, Smith DH, Fukushima A, Hales JF, Klapholz S, Finer MH, Davis CG, Zsebo KM, Jakobovits A. 1997. Functional transplant of megabase human immunoglobulin loci recapitulates human antibody response in mice. *Nature Genetics* 15:146–156.

Montecucco C. 1986. How do tetanus and botulinum toxins bind to neuronal membranes? *Trends in Biochemical Science* 11: 315–317.

Montecucco C, Schiavo G. 1995. Structure and function of tetanus and botulinum neurotoxins. *Quarterly Review of Biophysics* 28, 423–472.

Morrison SL, Johnson MJ, Herzenberg LA, Oi VT. 1984. *Proceeding of the National Academy of Sciences, USA*. Chimeric human antibody molecules: mouse antigen-binding domains with human constant region domains. 81:643–646.

Mullaney BP, Pallavicini MG, Marks JD. 2001. Epitope mapping of neutralizing botulinum neurotoxin A antibodies by phage display. *Infection and Immunity* 69:6511–6514.

Nowakowski A, Wang C, Powers D, Amersdorfer P, Smith T, Montgomery V, Sheridan R, Blake R, Smith L, Marks JD. (In press) . Deconvoluting the immune response: Potent botulinum neurotoxin neutralization by oligoclonal antibody.

Pless DD, Torres ER, Reinke EK, Bavari S. 2001. High-affinity, protective antibodies to the binding domain of botulinum neurotoxin type A. *Infection and Immunity* 69:570–574.

Reichert J.M. 2001. Monoclonal antibodies in the clinic. *Nature Biotech* 19: 819–822.

Schiavo G, Benfenati FBP, Rossetto O, Polverino DLP, DasGupta BR, Montecucco C. 1992. Tetanus and botulinum-B neurotoxins block neurotransmitter release by proteolytic cleavage of synaptobrevin. *Nature* 359, 832–835.

Schiavo G, Rossetto O, Catsicas S, Polverino DLP, DasGupta BR, Benfenati F, Montecucco C. 1993.. Identification of the nerve terminal targets of botulinum neurotoxin serotypes A, D, and E. *Journal of Biological Chem* 268, 23784–23787.

Schier R, McCall A., Adams GP, Marshall KW, Merritt H, Yim M, Crawford RS, Weiner LM, Marks C, Marks J. D. 1996. Isolation of picomolar affinity anti-c-erB-2 single-chain Fv by molecular evolution of the complementarity determining regions in the center of the antibody binding site. *Journal of Molecular Biology* 263, 551–567.

Sheridan RE, Deshpande SS, Amersdorfer P, Marks JD, Smith T. 2001. *toxicon* Anomalous enhancement of botulinum toxin type A neurotoxicity in the presence of antitoxin. 39, 651–657.

Siegel LS. *J. Clin. Microbiol.* Human immune response to botulinum pentavalent (ABCDE) toxoid determined by a neutralization test and by an enzyme-linked immunosorbent assay. *Journal of Clinical Microbiology* 26: 2351–2356.

Simpson LL. 1980. Kinetic studies on the interaction between botulinum toxin type A and the chloinergic neuromuscular junction. *Journal of Pharmacological Experimental Therapy* 212:16–21.

St. John R, Finlay B, Blair C. 2001. Bioterrorism in Canada: An economic assessment of prevention and postattack exposure. *Canadian Journal of Infectious Diseases* 12:275–284.

United Nations Security Council. 1995 Tenth report of the executive committee of the special commission established by the secretary-general pursuant to paragraph 9(b)(I) of security council resolution 687 (1991), and paragraph 3 of resolution 699 (1991) on the activities of the Special Commision., New York, NY.

Wilson JA, Hevey M, Bakken R, Guest S, Bray M, Schmaljohn AL, Hart MK. 2000. Epitopes involved in antibody-mediated protection from Ebola virus. *Science* 287, 1664–1666.

Zilinskas RA. 1997. *Journal of the American Medical Association.* Iraq's biological weapons: the past as future? 278, 418–424.

5

Assessing the Capacity of the Public Health Infrastructure

OVERVIEW

A strong public health system is an integral component of bioterrorism defense. Even if the contents of our biodefense arsenal were sufficient to treat any and all disease caused by a bioterrorist agent, we would still need a rapid detection and response system for the delivery of therapeutics or prophylaxis to all exposed individuals. However, there are many critical gaps in the public health infrastructure and many lessons to be learned from our response to the recent anthrax events. These gaps exist at every level—federal, state, and local—and in nearly every realm of public health, from federal laboratory diagnostic capacity to local first responder education.

The anthrax outbreak was a relatively small-scale situation. Had we experienced a massive release, the CDC and Laboratory Response Network (LRN) would have been stretched beyond capacity. Consequently, the CDC is developing more consolidated bioterrorism guidelines and recommendations to aid the federal-level response and strengthen the LRN. Local and community level bioterrorism response preparedness is equally important. Indeed, distribution to the site of the event, for example, will likely not be a problem. The greater challenge will be distributing it within the community once it arrives at its destination. There is a very strong and urgent need to strengthen local public health capacity, not just in terms of available resources but also, and perhaps more importantly, trained and organized personnel. The U.S. public health work force consists of about half a million people, most of whom have never received any formal public health training. It is essential that this work force be well-prepared and understand their role in a bioterrorism emergency. Local capacity building will probably also require recruiting more expertise, which requires both money and time. Local medical care surge capacity—including personnel, training, space, supplies, and equipment—must be strengthened. Hospitals are nowhere near being prepared to take on the tens to hundreds of thousands of mass casualties expected in the event of a large bioterrorist event. Equally important is the frontline health care responder who will likely be the one to sound the alarm. To this end, first responders must be adequately trained to recognize the symptoms of the various bioterrorist infectious agents. Real-time response role-playing exercises based on probable biological attack scenarios would be helpful in such planning.

Coordinating bioterrorism operational planning among jurisdictions, including with every hospital, will be a significant challenge since state level health departments have limited leverage to make this happen. One suggestion is to apply a model plan to be disseminated to local jurisdictions where it can then be adopted and exercised. It was recommended that jurisdictions share best practice information and new systems be integrated with systems that are already in place.

LESSONS BEING LEARNED: THE CHALLENGES AND OPPORTUNITIES

Julie L. Gerberding,* M.D., M.P.H.
Acting Deputy Director, National Center for Infectious Diseases, Centers for Disease Control and Prevention

We are learning many lessons from the recent anthrax events. By reviewing the behind-the-scenes processes as the investigation unfolded, we can identify conspicuous gaps and evaluate what needs to be done to strengthen our biodefense response capabilities.

The response to the recent events can be divided into overlapping stages. The first stage was initial detection of the threat and the immediate response to what was happening. This included case detection by astute clinicians, presumptive laboratory diagnoses, and evaluation of suspicious powders. Laboratory confirmation rapidly ensued, both in laboratories within the LRN and at CDC. One of the strengths of our response was the rapid deployment of personnel, antimicrobials, and other assets in response to requests from state and local health departments, which occurred within hours of detection or confirmation of events.

In the next stage, full-scale investigation and prevention interventions were priorities. This included post-exposure prophylaxis, building closure, environmental sampling and criminal investigation in addition to traditional epidemio-

---

* This statement reflects the professional view of the author and should not be construed as an official position of the Centers for Disease Control and Prevention.

logic and clinical evaluation. Interim guidance and recommendations were developed nearly in sync with these activities. We were very careful to label these recommendations "interim," knowing that we would probably update or consolidate them as more information and data became available.

The current stage of response is that of recovery and regrouping. Priorities now include optimizing post-exposure prophylaxis, promoting adherence, monitoring the short and long-term safety of the prevention interventions, building remediation, and recalling personnel and assets back to CDC. Re-entry is a very important component of the current stage of response. In other words, coping with the transition from a crisis state to a more proactive and reflective state. This includes considering ways to improve adherence to the sixty-day antimicrobial therapy, evaluating treatment, and understanding the overall impact and cost of this situation. Evidence-based guidelines and recommendations also are being developed. Most importantly, input is being sought from a variety of consultants to identify strengths of the response as well as gaps that require action to improve our capacity to respond to future events.

There are several lessons to be learned regarding the CDC and federal response in particular:

- In terms of competency, we need new paradigms and skills at CDC. Forensic epidemiology is a new discipline for us; one that requires new perspectives and investigative methods. Working side-by-side with the FBI and other law enforcement agencies is something we have done before but certainly not on this scale or with this degree of ongoing involvement.
- We must learn to make adaptive decisions, that is decisions that must be made in real time with very little data and that require an experimental approach. Inducing explanations or policy decisions from immediate situations, instead of from the more extensive databases that normally frame most public health decisions, learning from new information, adjusting guidelines and policies in response, and building the science as one moves forward are necessary parts of this process.
- We need enhanced environmental microbiology expertise. Most of the investigations, including those that are still ongoing, are highly focused on environmental evaluation. Evidence-based air, water, and surface sampling strategies, risk assessment, decontamination methods, and re-entry criteria are needed for *B. anthracis* and other potential agents of bioterrorism.
- We must be able to quickly access expertise in a range of specialized fields, such as small particle physics, ventilation systems, and building engineering.
- Response capacity must be expanded. For example, laboratory capacity must be sufficient to support a large-scale event. The CDC intramural laboratory response was outstanding. Extramural public health laboratories, including those in the Laboratory Response Network, were similarly challenged and also performed extraordinarily well. Laboratorians rose to the occasion, and a number of

surge capacity needs were identified and fairly quickly met. Measures that were taken during the response include:

- CDC expanded the BSL-3 space so that anthrax typing could continue simultaneously with ongoing diagnostic studies.
- CDC created a new Level A laboratory in less than 72 hours that processed 600 environmental specimens per shift. On the first day of full-scale operation, the new lab processed 1,000 samples.
- CDC implemented a new integrated data management system which coordinated all of the laboratory results from the many participating laboratories across the center and linked with relevant patient data at CDC. Thus, we could enter a single data set and find both laboratory and clinical information about specific patients. This was a ground-breaking accomplishment which greatly facilitated daily laboratory coordination.

To date, at CDC we have processed about 5,400 anthrax-related specimens, and there is still more work to be done. The number of specimens managed outside CDC exceeds 70,000. Nevertheless, this was a relatively small-scale situation. Had it been a massive release, we would have been stretched beyond capacity.

Collaboration is essential. The CDC has a great deal of experience with collaboration, but we have never had to collaborate with so many partners for so long and so intensely. Key partners included state and local health departments, clinicians, health care facilities and organizations, and numerous federal agencies including other agencies in DHHS, the FBI, U.S. Postal Service, EPA, DoD, USAMRIID, and the U.S. State Department.

Coordination is crucial; an effective response depends on knowing who is in charge. All layers of government and public health must be synchronized to make this kind of collaborative effort work, but how should this effort be coordinated? Internally, the CDC Emergency Operations Center was supported by a series of teams, such as environmental, postal, clinical, information technology, communications, the personnel team, the telephone hot line team, etc., that were led by senior personnel. Field investigations in each locale were directed by an on-site senior field team leader and co-leader who were linked to a corresponding support team at CDC. The operations center facilitated information flow from the field teams to decision-makers and deployment of resources and personnel to support the field teams and implement decisions.

Coordinating activity outside of the CDC was especially challenging. The National Security Council (NSC), the Office of Homeland Security, DHHS, governors, and health commissioners all played important roles in coordinating CDC's response.

Communication is key. Our communications capacity was not the strength of this investigation, to say the least. We did field hundreds of phone calls during the peak times of this investigation and provided a great deal of information to those who needed it most, including critical partners in the investigation. Immediately after the attack in New York and Washington, DC, there was almost no communication from CDC because we were operating under federal emergency response management plans. But as the investigation unfolded, we were allowed to carefully communicate a limited amount of information. Later, it became clear that more information from CDC was desperately needed, but by then we were in a reactive phase where we were trying to catch up with information needs. Clearly, a proactive information management plan is a critical priority for future response efforts.

Consultation is also very important, so that we can learn as we go forward. For example, we have conducted nine consultations at CDC to solicit input from experts about how we can improve our response capacity. These have included partners from affected areas, clinicians, professional organizations, communications experts, research scientists, environmental scientists, and many others with relevant expertise. Future meetings are planned to examine other aspects of the federal response, such as how to scale up for other scenarios and how to use new detection systems to rapidly identify an event.

THE RESPONSE INFRASTRUCTURE: INVESTIGATING THE ANTHRAX ATTACKS

Bradley Perkins,* M.D.
Division of Bacterial and Mycotic Diseases
Centers for Disease Control and Prevention

From October 4 to November 2, 2001, the first ten confirmed cases of inhalational anthrax caused by intentional release of *Bacillus anthracis* were identified in the United States. Epidemiologic investigation indicated that the outbreak—in the District of Columbia, Florida, New Jersey, and New York—resulted from intentional delivery of *B. anthracis* spores through mailed letters or packages. These are the first U.S. cases of intentional inhalational anthrax that we know about and the first inhalational cases in the U.S. since 1976; only eighteen other cases have been reported through the last century.

The median age of patients was 56 years (range 43 to 73 years) which, in contrast to cutaneous cases, is slightly older than expected. Seventy percent of the cases were male and, except for one, all were known or believed to have processed, handled, or received letters containing *B. anthracis* spores. The median incubation period of the first six cases (i.e., known cases) from the time of exposure to onset of symptoms was 4 days (range 4 to 6 days).

At initial presentation, symptoms included fever or chills (n=10), sweats (n=7), fatigue or malaise (n=10), minimal or nonproductive cough (n=9), dysp-

---

* The information provided in this paper reflects the professional view of the author and should not be construed as an official position of the U.S. Department of Health and Human Services or the Centers for Disease Control and Prevention.

nea (n=8), and nausea or vomiting (n=9). The drenching sweats and the extent of fatigue, malaise and GI symptoms were quite dramatic and unlike what had been reported in previous literature, although some of the Sverdlovsk autopsy data showed GI symptoms in a number of the inhalational cases. On the other hand, the laboratory tests were not very remarkable. The median white blood cell count was only moderately elevated at 9.8 x 10^3 /mm^3 (range 7.5–13.3), often with increased neutrophils and band forms. Nine patients had elevated serum transaminase levels, and six were hypoxic. All 10 patients had abnormal chest X-rays; abnormalities included infiltrates (n=7), pleural effusion (n=8), and mediastinal widening (seven patients). Importantly, the mediastinal widening was a fairly subtle feature that was missed on a couple of initial interpretations. The pleural effusions were hemorrhagic pleural effusions which were recurrent and a predominant feature of the clinical illness. Computed tomography of the chest was performed on eight patients, and mediastinal lymphadenopathy was present in seven. With multidrug antibiotic regimens and supportive care, survival of patients (60%) was markedly higher (<15%) than previously reported. Two of the deaths were probably due partly to the fact that patients did not receive antibiotics or appropriate antibiotics when they came for medical attention.

For most cases, the infection was identified from blood cultures. Positive CSF cultures were remarkable in terms of the number and the striking morphology of *B. anthracis*; and on blood agar, they reached confluent growth in about 6 hours, which again is remarkable. Blood cultures grew for all patients whose cultures were tested prior to receiving antibiotics. The three patients who did not have positive cultures had all been treated with antibiotics; their diagnosis was made instead by a combination of immunohistochemical staining for the capsule and cell wall of *B. anthracis*, DNA for *B. anthracis* or, in one case, a serology that showed a four-fold rise for anti-PA IgG. Importantly, the diagnosis was based on specimens from the pleural cavity, pleural biopsy, transbronchial biopsies, and actual cytology blocks from pleural fluid.

The eleventh inhalational anthrax case, which was not included in the above summary data, was a 94-year old Connecticut woman who was admitted on November 16. She had a 3- to- 5-day prodromal illness that was fairly vague in its manifestations and was complicated by the fact that some people were attributing her illness to depression as a result of a recent death of a friend. Although her chest X-ray was within normal limits, blood cultures obtained on the day of admission grew gram-positive rods and, over the next several days, clinical progression was rather remarkable with development of bilateral bloody pleural effusions, hypotension requiring vasopressor support, intubation and then death. Although mediastinal adenopathy had not been detected on her admission chest X-ray (probably because of dehydration), it was present at the time of autopsy. Also at the time of autopsy, her gross and microscopic findings were consistent with inhalational anthrax.

THE CENTERS FOR DISEASE CONTROL BIOTERRORISM INVESTIGATION

Kevin Yeskey,* M.D.
Acting Director of Emergency and Environmental Health Services, Centers for Disease Control and Prevention

The bioterrorism program at the CDC began in 1999. The program had two initial components: an intramural capacity development component intended to enhance CDC's bioterrorism response capacity and an extramural cooperative agreement program that served to develop state and local public health preparedness for a bioterrorist event. CDC's intramural activities included hiring of subject matter experts in priority areas of bioterrorism; expanding and enhancing the laboratory capacity to handle biological and chemical agents; development of specific communications technologies using the Internet; enhancing CDC's surveillance and epidemiology capacity; and developing and managing the National Pharmaceutical Stockpile.

The extramural cooperative agreement is a five-year program that has four focus areas: preparedness and planning; surveillance and epidemiology; laboratory capacity; and communications. Every state has received funding for at least one component of the cooperative agreement, but not all states received funding for each component. In the preparedness focus area, sites receiving funding have begun to develop public health bioterrorism preparedness plans. In the epidemiology and surveillance focus area, states have hired personnel to enhance their bioterrorism surveillance and reporting capacity. Additionally, several special projects have been initiated that utilize alternative sources for surveillance, such as medical examiners and poison control centers. State and some municipal health department laboratories have developed the capacity to provide initial screening for several of the biological agents most likely to be used as biological weapons. These labs are part of the Laboratory Response Network that enables them to have access to a secure communications system, order reagents, receive new protocols, obtain proficiency testing, and receive training from the CDC. One of the main communications activity of the cooperative agreement is the Health Alert Network (HAN). The HAN offers a means to provide rapid communication to health departments via high speed Internet access. Another, more secure, communications system is the Epidemic Information Exchange (Epi-X). This system offers a more secure mechanism to communicate to a more directed audience.

There have been challenges in the development of the CDC bioterrorism program and the recent response to the terrorist events since September 11, 2001. In the area of preparedness, activity must extend beyond the creation of a

* The information provided in this paper reflects the professional view of the author and not an official position of the U.S. Department of Health and Human Services or the Centers for Disease Control and Prevention.

written response plan. Preparedness involves assessment of a community's vulnerabilities, resources, and threats. Bringing all concerned parties to the assessment process is one of the main challenges facing public health responders. Clinical providers, treatment facilities, and first responders must develop integrated and coordinated preparedness plans. Inclusion of non-traditional public health partners, such as law enforcement, is essential to this process. Surveillance challenges include the implementation of "real-time" surveillance methodologies that are not as labor intensive as those used during the response to the anthrax incidents. Laboratory challenges include expanding the screening capacity for biologic and chemical agents to the local level. Additional challenges include having the capacity for accurate field-testing to assess hoaxes at the scene of an event, rather than performing all testing at the state health laboratory, which often overloads them. Specimen transport can also be difficult as some air couriers refused to transport environmentally contaminated samples during the recent anthrax incidents. Communications challenges include providing accurate information on a timely schedule. Field teams must also have standardized pre-developed data management tools so that others who need to evaluate these data can easily access data gathered in the field.

VA CAPABILITY TO ENHANCE THE MEDICAL RESPONSE TO A DOMESTIC BIOLOGICAL THREAT

Kristi L. Koenig,[*] M.D., FACEP
National Director, Emergency Management Strategic Healthcare Group (EMSHG)
Veterans Health Administration
Department of Veterans Affairs

VA Missions and Organization

The Department of Veterans Affairs (VA) is a cabinet-level department that has the care of veterans as its primary mission. VA manages and controls the vast medical care assets of the largest integrated healthcare system in the country. Currently, VA has 163 medical centers nationwide, in addition to approximately 450 community-based outpatient clinics, 130 nursing homes, 73 home care programs, and 206 counseling centers. VA personnel across the nation include about 15,000 physicians and more than 1,000 dentists, 58,000 nurses, 36,000 pharmacists, and 130,000 ancillary staff. Thus, VA is the *federal presence in the local community*, with facilities and personnel in virtually every neighborhood in the country.

---

[*] The information provided in this paper reflects the professional view of the author and should not be construed as an official position of the U.S. Department of Veterans Affairs.

VA is composed of three Administrations: Veterans Health Administration (VHA), Veterans Benefits Administration, and National Cemetery Administration. VHA is the largest of the three administrations and has four statutory missions. In addition to its primary mission of "medical care," VHA has affiliations with most of the nation's teaching institutions (Education Mission) and a vast research program (Research Mission). It is the so-called "Fourth Mission" or Contingency Support that is least known.

The executive agent for the Fourth Mission is the Emergency Management Strategic Healthcare Group (EMSHG). EMSHG is currently authorized 86 FTEs that include a headquarters staff of 24 and 62 out-based personnel consisting of District Managers, Area Emergency Managers (AEMs), and Management Assistants located at field offices throughout the nation. AEMs serve as liaisons to VHA's 22 Veterans Integrated Service Networks (VISNs) by providing emergency consultation and support in the development and implementation of VISN and VA medical center emergency management plans.

Comprehensive Emergency Management (CEM) Programs

Emergency Management Missions

EMSHG coordinates emergency management programs that ensure health care for eligible veterans, military personnel, and the public through the Federal Response Plan and the National Disaster Medical System (NDMS) during Department of Defense (DoD) contingencies, national security emergencies, and disasters.

VHA's Fourth Mission consists of the following six functions:

- VA Contingencies
- DoD Contingencies
- Federal Response Plan
- National Disaster Medical System (NDMS)
- Radiological Hazards
- Continuity of Operations/Continuity of Government

EMSHG plans and coordinates VA's role as the primary backup to DoD during war or national emergencies. It responds to taskings received by VA under the Federal Response Plan and Federal Radiological Emergency Response Plan to provide support to veterans and non-veterans alike. EMSHG also supports VA's continuity of operations plans through maintenance of relocation sites, and operates the VHA's Emergency Operations Center. VA, via EMSHG, assists in the implementation of the NDMS to supplement state and local medical resources in the event of a major domestic disaster or emergency. EMSHG's AEMs provide support for VA healthcare facilities designated as Federal Coordinating Hospitals for NDMS. Additional information on VA's role in emergency management can be found on the EMSHG website at www.va.gov/emshg.

VA Role in Bioterrorism

VA's primary focus is the protection of its own veteran patients and staff. In addition, Presidential Decision Directive (PDD) 62 directs VA to support the Department of Health and Human Services (DHHS) in providing adequate stockpiles of pharmaceuticals and training of personnel in civilian NDMS hospitals.

VAMC Preparedness

VA uses an all-hazard, CEM approach. The four phases of CEM—mitigation, preparedness, response, and recovery—are incorporated into each emergency management plan. This approach is consistent with that required by the Joint Commission on Accreditation of Healthcare Organizations (JCAHO) Standards for hospitals across the country as of January 2001. EMSHG contributed significant content to the 2001 JCAHO Environment of Care Standards on Emergency Management and has been invited to develop training standards for JCAHO surveyors.

Counterterrorism funding and preparations are part of the overall all-hazard, CEM approach. Under CEM, a hazard vulnerability analysis is performed in each location with assistance from the EMSHG AEMs assigned to VAMCs throughout the country. This approach provides the flexibility to meet any contingency. For example, a pandemic influenza event would have essentially the same effects on the healthcare infrastructure as biological terrorism with a contagious agent such as smallpox or pneumonic plague. While the bioterrorism event would be unique in terms of the involvement of law enforcement, VA facilities use an Incident Management System for all emergencies that coordinates all appropriate participants both within a VA facility and from outside agencies such as law enforcement. VA has a robust exercise and training program (> 400 exercises per year) that includes specific attention to bioterrorism as part of its comprehensive approach.

Presidential Decision Directive 62

VA supports the Department of Health and Human Services (DHHS) in two roles under Presidential Decision Directive (PDD) 62: 1) management of pharmaceutical stockpiles, and 2) training of personnel working in civilian NDMS hospitals.

VA procures, rotates and manages four pharmaceutical caches for the DHHS Office of Emergency Preparedness' (OEP) National Medical Response Teams (NMRT). These caches are mostly geared towards management of chemical casualties. In addition, VA manages a "Special Event" cache that is staged at National Security "high-risk" events upon request.

Through its Prime Vendor System, VA also purchases the contents of the CDC's National Pharmaceutical Stockpile. These stockpiles are larger than the NMRT caches and contain equipment and antibiotics suitable for treatment of biological terrorism casualties.

At the end of FY2001, OEP transferred $832,000 to VA EMSHG to begin training of NDMS hospital personnel. The first part of the project will be to perform a "needs assessment." VA already has a robust education and training program. One example is the partnership between VA and Soldier Biological Chemical Command (SBCCOM) that resulted in hosting the Domestic Preparedness Program Hospital Provider Course at more than 40 VA facilities across the country last fiscal year.

Federal Response Plan (FRP)

When the President declares a disaster, the Federal Response Plan (FRP) is activated. VA provides personnel, pharmaceuticals and supplies upon request from DHHS under Emergency Support Function ESF #8: Health and Medical Services. In fact, since the FRP was promulgated by the Federal Emergency Management Agency (FEMA) in 1992, VA's assistance has been requested in every major disaster that has occurred in the United States, its territories or possessions. More than 1,000 clinical personnel have been deployed along with large quantities of supplies.

In addition to providing resources for presidentially declared disasters, VA has provided emergency managers to assist with the staging of medical personnel and supplies at sites of various "high-threat events" (e.g., NATO 50, Olympics, Inauguration, Papal Visit, and Economic Summit of the Eight).

Disaster Emergency Medical Personnel System (DEMPS)

DEMPS is a nationwide registry or database of full-time VA employees who wish to volunteer to deploy if needed to assist with a disaster. The database is maintained at EMSHG headquarters and is currently being populated. Skills, professional qualifications and credentials, and documentation of appropriate training are being collected. Some volunteers have special training in the management of terrorist attacks.

VA Unique Resources

- In addition to its national infrastructure with *personnel* and *facilities* across the nation, VA has several other unique resources.

• There are robust *research* and *education* programs with affiliations with most of the country's medical schools.

• *Telemedicine* or telehealth is another resource that might be especially useful in a bioterrorism event with a contagious agent so that experts could view patients from a distance.

• The *Prime Vendor System* allows economic and efficient purchase of pharmaceuticals and supplies.

• National alerting and *communications* systems are in place and routinely tested.

• VA also has extensive expertise in the identification and treatment of victims experiencing Post Traumatic Stress Disorder and counts large numbers of *stress counselors* among its assets.

Current Initiatives

EMSHG Technical Advisory Committee

In July of 2000, EMSHG formed a Technical Advisory Committee (TAC). The TAC is a multidisciplinary group of internal VA leaders and emergency management experts from federal partner agencies such as DHHS, FEMA, DoD, CDC, and the Federal Bureau of Investigation. The first goal of the TAC was to advise on preparing VA medical centers for any type of weapons of mass destruction threat.

The TAC was divided into the following task forces: Organizational Support; Basic Training; Decontamination and Personal Protective Equipment; Accessing Stockpiles; Surveillance; Quarantine; VA's Role in the Local Community; and Research and Assessment.

Emergency Management Academy

In the fall of 2001, EMSHG initiated the Emergency Management Academy (EMA). The EMA has a focus on emergency management and *healthcare.* Target audiences will be trained using web-based technology, satellite videoconference, and hands-on sessions. A knowledge management library is posted on the EMSHG website as part of the EMA. VA is capable of granting continuing medical education credits. Accreditation is provided not only for VA personnel, but also for attendees at the annual NDMS conference and other major disaster-related conferences and programs.

Emerging Issues

One of the major issues that our nation faces is the lack of surge capacity. This is due in large part to the influence of managed care and the resultant shift to ambulatory services and decreases in in-patient beds. The national nursing shortage contributes to this lack of excess patient care capacity. Further, if we have a concurrent overseas conflict and U.S. disaster, reservists will be mobilized further draining the health care system. While we currently count beds to assess capacity, beds are no longer a good surrogate marker. We need to develop methods to assess *patient* care capacity and explore creative solutions to the lack of surge capacity.

Summary

VA is the largest national integrated healthcare system with facilities and personnel across the country. The initial response to any disaster is local—it will take some time before state and federal resources can be mobilized. Therefore, VA is uniquely positioned to assist with both local and federal counterterrorism efforts. While our primary focus remains protection of our ability to care for veterans, VA also provides the federal medical presence in the local community since we are "in and of that community."

THE PROGRESS, PRIORITIES, AND CONCERNS OF PUBLIC HEALTH LABORATORIES

Mary J. R. Gilchrist, Ph.D.
Director, University of Iowa Hygienic Laboratory
President, Association of Public Health Laboratories

The Laboratory Response Network (LRN) was instituted in 1999 in preparation for the U.S. response to bioterrorism. The LRN consists of public health laboratories that form linkages to the private hospital, clinical, and referral laboratories, which refer isolates to the public health laboratories for confirmation of identity of suspect microorganisms. The system was not fully operational when the October anthrax events began but, even so, it functioned relatively well. One can only guess at the difficulties that would have occurred without the embryonic phase of the LRN. Human illness was limited so the system was not fully tested for dealing with a large outbreak of disease. However, it did appear to work even though training had not been widely and intensively administered to the local laboratories. In several cases the isolates of *Bacillus anthracis* were presumptively identified at the local laboratory and referred to the state lab for full identification. It is apparent that without widespread knowledge of the diagnosis of illness, some of these cases might have been missed completely.

Because the event played out over several weeks in the form of envelopes being delivered to offices, the primary challenge to the public health laboratories was to rule out other suspect powders. At the peak, the public health laboratories in the LRN were testing over 1,200 powders per day. These were those powders that were deemed a credible threat by the law enforcement community so it is likely that some ten-fold or greater numbers were evaluated and rejected as non-credible threats. Even following this filter by law enforcement, the impact of the public health laboratory testing was significant. If you assume that 5 to 10 people were exposed to each of the credible threats, on average, then 6,000 to 12,000 people per day were saved from fear and/or the need for prophylaxis. In some states the management of the program was smooth and in others capacity was exceeded and turnaround time prolonged. Obviously fine-tuning is necessary. The good news is that the private laboratories rapidly became motivated to learn to test for bioterrorism agents and that public health laboratories developed greater ties with the emergency management community so that the next event can be more readily managed even if it involves greater numbers of human illnesses.

A full evaluation of the testing capacity of the LRN should be conducted. The CDC has been collecting data that should be subjected to scrutiny and projections made about future capacity. The various types of tests should be scaled for their relative effort. With the chain of custody and evidence documentation procedures that we used at University Hygienic Laboratory (UHL), as well as the complex packaging, the suspect powders were extremely labor intensive. Environmental and nasal swabs were more easy to open and required no evidence documentation but the environmental swabs had more spore formers to evaluate than did nasal swabs. An approximation of relative effort for powders:environmental swabs:nasal swabs would be 10:4:1. Thus, if a facility claims to have conducted a large number of tests, the relative effort demands of the various types of tests should be considered. Most of the 1,200 tests cited in the paragraph above were powders but a few facilities also reported swab samples.

After the events of October and November have played out, a full assessment of the LRN is indicated. Evidently, it will be imperative to provide real-time communication of testing capacity throughout the system. This would allow for redistribution of effort where needed, to identify sites not at capacity that could relieve those exceeding capacity. Originally, a proposal for a protected website to perform this function was entertained but not implemented due to concerns about security. If a website cannot be adequately secure, the rapid development of an alternative is strongly recommended.

The testing of samples during the fall bioterrorism outbreak was primarily conducted using conventional methods, culture and staining. These tests are relatively accurate but speed is often important and the culture requires 2–3 days to produce a final result. More rapid tests are being adopted. Many states have purchased advanced instrumentation for real time amplification (PCR-type) testing, for example, Smartcyclers, Lightcyclers, and ABI 5700 or 7700 TaqMan devices. The reagents for use with these devices are being produced at CDC. The proficiency tests to confirm the adequacy of the reagents and devices were underway in many states when the anthrax outbreak began. The CDC should be recognized for its prescience in developing these assays and supported in furthering this cause. The rapid identification of organisms is very important and should be advanced. Reagent production should be supported fully so that it is adequate and timely. Manufacture, storage, and distribution of reagents should be protected from other types of terrorism by consideration of off-site locations and stockpiles. States should have redundancy in instrumentation so that when an instrument fails, there is backup testing capacity available. The validity of the tests should be fully demonstrated. FDA approval should be sought so that specimens from humans can be fully evaluated in a rapid fashion.

Smallpox diagnosis should be part of the system available for PCR-type testing in the states. The specimen would be decontaminated prior to testing, obviating the need for BSL-4 facilities, so safety is not a limiting issue in states where BSL-2/3 facilities are available. Although it is true that most true cases of smallpox may be readily identifiable clinically, there will be skin rashes that cannot be ruled out. Moreover, it is possible that atypical rashes may not occur in those who were previously vaccinated, as we encountered in the atypical measles cases of the latter part of the last century. Rapid specific etiologic identification will minimize panic and preclude excess use of limited supplies of vaccine. We must have widely dispersed capability for rapid diagnosis. The bioterrorism events of the last two months have fully demonstrated the accuracy of this imperative.

The advent of bioterrorism this fall allowed the recognition of other needs. Security of the laboratory building became a priority in Iowa when the "Ames" strain was said to be the cause of the first case of anthrax. The media stated that the organism was "man made" in a laboratory in Iowa. The National Guard visited the UHL that evening and they stayed to guard our building for several weeks. Our building is a multiuse building that was built in 1917 to serve as a tuberculosis sanitarium. There are many security challenges that we face as we prepare to bridge the interval preceding occupancy of the new building that we seek. Security arrangements range in the other state public health laboratories from sophisticated in California to minimal in other states. The CDC has reinforced its security following the events of September 11 when it was closed due to the threat of terrorism. It remains a potential target for future events and should duplicate functions elsewhere whenever they constitute a critical path. Public health laboratories should be subject to critical infrastructure protection not unlike transportation and energy structures.

Among the needs that we recognize are specimen transport and communications capability enhancements. To fulfill these needs, courier systems and communication devices, respectively, are now recognized as critical. Finally, safety remains an issue in many laboratories. Although Biosafety Level 3 laboratories are present, or being planned and built in most state public health laboratories, they may not be situated in close approximation to all areas where they are needed. Optimized safety facilities will decrease turnaround time and maximize surge capacity.

Among the problems that were recognized during the last two months were those of communication between the Level A (hospital or clinical laboratories) and the Level B/C (public health laboratories). The National Laboratory System

(NLS) is the concept currently under consideration for implementation to link public health and private laboratories. It is being evaluated in four states that are employing prototype methods to form these linkages. Prior to September 11, 2001, many private laboratories were not convinced of the need for bioterrorism training. Thus, such training was shunned when it was offered. Now, the laboratory community is anxious for training and for linkages. The National Laboratory System (NLS) should be fully implemented throughout the U.S. When implemented, it will provide reciprocal communications links, feedback of results to the private laboratories and of needs to the public health laboratories.

The LRN should be linked with other laboratory networks that have complementary functions. The veterinary diagnostic laboratories in the states have recently formed a network with the National Veterinary Services Laboratory (NVSL) at the hub. This veterinary laboratory network is comparable to its human diagnostic counterpart, the LRN with the CDC at the hub. Linkages between the two networks would facilitate the efforts of both networks. Zoonotic microbial agents that are familiar to veterinarians and their laboratory communities cause many primary bioterrorism disease threats, e.g., anthrax, tularemia, plague, and brucellosis. Indeed, if used by terrorists the organisms may strike both humans and animals. For these reasons, greater linkages between human and veterinary laboratories are strongly indicated. Other potential linkages would include those with military, food, and environmental and international labs. Many of these linkages are already being discussed. Funding and organizational planning will breathe life into the linkages and make them function.

Chemical terrorism laboratory capacity should be distributed widely throughout the states. At present there are only some five states that have such funding and are developing capacity. None of these are in the plains states so Iowa would not be well served should there be a chemical terrorism event. It would be particularly problematic should air traffic be simultaneously shut down and no specimen delivery could occur. The bioterrorism events of the past two months have definitely taught us that laboratory capacity should be widely distributed and under local control. Similarly, testing for chemical terrorism must not be limited to a few sites.

With some $8 million annually, the CDC administers funding and coordination of the LRN. Although these limited funds were instrumental in supporting the planning and organizational phase, they are inadequate for the operational phase of the LRN. The amount being distributed to the states is inadequate to fully fund laboratories that must continue routine operations and also sustain efforts directed at identification of bioterrorism agents. The UHL is currently receiving $100,000 per year for its bioterrorism grant. This amount is insufficient to support a sustained effort and would strain the capabilities of the current staff beyond the breaking point. Federal funding should be enhanced so that the system can operate smoothly.

The concept of "dual use" should be widely celebrated. The PCR instrument at UHL that was purchased with bioterrorism funding is also used for detection of West Nile Virus, whooping cough and a host of other agents. Our staff members work on detection of other agents during times of quiescence in bioterrorism. They must do so to retain full proficiency in identification of all agents. It is not possible to be proficient at identification of bioterrorism agents unless one can identify all other agents. The government should embrace the concept of "dual use" as necessary as well as economical. The emergence of infectious diseases occurs either naturally or intentionally and the means of emergence cannot be ascertained before the agent is detected and identified. From time to time, governmental agencies drift toward compartmentalizing funding and restricting dual use as though it were an evil. In truth, it is an inevitable, necessary and highly laudable use of funding.

A major concern to the bioterrorism response community is the prospect of false positive and false negative tests that may be produced by devices currently owned or being sold to first responders and to citizens. The "Smart Ticket" is a hand held device originally used around military bases to continuously monitor air for presence of incoming agents. When used in this fashion and backed up by confirmatory tests, it has worked adequately. However, in recent years it was purchased by the first responder communities for delivery of instantaneous results. Unfortunately, the device gave false positive signals in numerous instances. These false positive signals may have subsequently depleted capacity to manage real events. Organisms such as *Bacillus subtilis* and *Bacillus cereus*, present in soil and dust, are common sources of false positive signals. *Bacillus thuringiensis*, the agent used on vegetables as an insecticide, is also a source of false positive signals. Testing some green peppers with the "Smart Ticket" led to the fear that foods were contaminated with *B. anthracis* spores in at least one incident. Not only are false positive results a problem with this device. Unfortunately, the device requires as many as 10,000 spores to yield a positive result. Thus, it also may cause many false negative results. The CDC published an admonition against the use of the "Smart Tickets" for decision-making. Stronger constraints against the use of the device should be considered.

An ominous situation may be at hand with regard to a new group of test devices. There are several new devices being marketed for home testing. One is potentially hazardous to health. It involves a growth medium said to be selective for *Bacillus anthracis* that also claims to provide some indication of the identity of the species. Although one might argue that the isolation of the anthrax bacillus would be rare and thus the device not hazardous, it remains to be proven that the device would not grow other hazardous organisms. The device has a sticker warning the curious that an anthrax culture is in progress but this is not a convincing safety element. Of greater concern than safety, perhaps, is the potential for production of false positive results that will confound the public health community and unduly deplete resources. Testing devices should be regulated even when they are not for use on human clinical specimens. If these devices are not regulated, test results may mislead investigators and divert attention from critical needs. Worse, the results may lead to unprotected exposures to hazardous agents, followed by disease and death.

Although not fully functional, the LRN was a relative success story during the last two months. This fact should not cause us to conclude that the LRN has been fully tested for functionality in upcoming bioterrorism attacks. The LRN was not subjected to the greater stresses that it would have to endure if/when there is an event involving large numbers of ill individuals. Post event evaluation of the LRN should be conducted to ensure that weaknesses are identified and repaired and strengths are amplified and systematized. The Association of Public Health Laboratories is currently composing a questionnaire to distribute to the public health laboratory community to identify unmet needs that were identified during the fall anthrax outbreak. Good planning should be based on credible data. Funding should follow the need for a better system and should support areas of greatest need.

THE ROLE OF THE CENTERS FOR PUBLIC HEALTH PREPAREDNESS

Stephen S. Morse, Ph.D.
Columbia University
Mailman School of Public Health

The public health system is at the forefront of our defenses against bioterrorism, as it is against infectious diseases in general. As demonstrated by the recent anthrax events, the first indication of an attack may well be the appearance in emergency rooms or doctors' offices of people sick with an unexpected illness. Conceptually, many of the steps that the public health system needs to take in order to strengthen our national biodefense are very similar to what needs to be done to prepare for an unexpected, naturally occurring outbreak of infectious disease (call it "emerging infections plus"). Not only is public health an important component of biodefense, it may perhaps be the only component in the earliest phases of a response to a bioterrorist attack. Sustaining this capacity between crises poses a difficult but essential challenge to recognizing early warnings and saving as many lives as possible. The U.S. public health work force consists of about half a million people, many of whom have never actually had any formal training in public health. Ensuring that the entire work force is well-prepared and understands its role in an emergency is an important need.

Showing great foresight in this regard, last year the CDC set up a new program (through a cooperative agreement with the Association of Schools of Public Health) to build a network of centers for public health preparedness that links academe and public health practice. The goal was to develop competency-based curricula in public health that would be appropriate to local needs and could also serve as models for national replication and national testing. It is also a goal to provide materials through technology-supported learning, which means that the information should be available on the web or by other means such that people can access it at any time for training purposes.

Currently, there are seven academic centers in the system. Columbia University was one of the first academic centers in the system to receive funding; Columbia's center at the Mailman School of Public Health is partnered with the New York City Department of Health to focus on emergency preparedness, including bioterrorism and infectious disease preparedness. Members of the center include people with public health backgrounds as well as specialists in curriculum and distance learning. Development of the preparedness plan required initially looking at already existing preparedness plans for New York City to see how they could be revised and strengthened. Dr. Kristine Gebbie (a Center member) and her colleagues at the Columbia University School of Nursing had developed a set of emergency preparedness competencies for public health, to serve as a basis for developing training. The core competencies for public health workers include such fundamental items as "describe the public health role in emergency response" and "describe the agency chain of command in an emergency." Focus groups and discussions with the health department had indicated that all employees could benefit from a general orientation to emergency preparedness, keyed to these competencies. The next step was to work with the city's health department to develop a competency-based training program, with a specific portion of the workforce, as an example.

The first training program was designed (by a curriculum team led by Dr. Marita K. Murrman) for public health nurses in school health. There are about 800 New York City school nurses who represent a large pool of professional, clinically trained people that could assist in an emergency. Their first primary emergency role is opening and staffing Red Cross shelters. In fact, this is exactly what happened after September 11. After a pilot run in June, the training program (given in conjunction with the American Red Cross) had been completed by late August, and a number of the nurses who participated in it were called in to staff shelters or volunteer in other capacities. Because they had received this type of training, they said they were able to hit the ground running.

Since September 11, the Center has continued to work closely with New York City on identifying continuing preparedness and education needs and, in response to their request, helping to evaluate the city's response to the disaster. The Center also provided the city, as well as CDC, with a database of faculty members of the School of Public Health who could help the city in an emergency or help elsewhere if surge capacity is needed. We have also been working with the New York Presbyterian hospital system, which has thirty-one units and covers about one-quarter of the patients in the greater New York City area, to develop improved surveillance and plans for enhancing the interface between public health and the hospital system. Recognizing that public and provider education is a key need, the Center developed training programs on bioterrorism response for clinicians (coordinated by Dr. Robyn Gershon, in partnership with the New York Academy of Medicine, the state medical society and other organizations) and provided lay language information through community forum presentations and the School's website. There are continuing needs in these areas, as well as in the closely related area of hospital preparedness, which the Center is continuing to address.

Shortly before September 11, we met with the Office of Emergency Management (OEM) of New York City. We learned that the incident command side of emergency response—fire, police, FEMA, FBI, etc.—speak a very different language than does the public health community. The discussions with OEM suggested that it would be useful to train their first responders not only in terms of core competencies but also to orient them to the role of public health in general (as well as doing the reciprocal task with the public health community, to help bridge these cultural and vocabulary differences). Currently, the Center is developing content for this purpose.

For further information, the Columbia University's Center for Public Health Preparedness website (http://cpmcnet.columbia.edu/dept/sph/CPHP/index.html) provides information on the Center's programs and links to other websites. Further information on CDC's overall network of Centers for Public Health Preparedness can be found on the CDC website:

http://www.phppo.cdc.gov/owpp/centersforPHP.asp.

THE ROLE OF COORDINATED INFORMATION DISSEMINATION: THE CASCADE SYSTEM IN THE UNITED KINGDOM

John Simpson, M.B.B.S., M.F.P.H.M.
Head of Emergency Planning Co-ordination Unit
Department of Health, United Kingdom

In 1994 some issues concerning the provision of timely information for healthcare professionals developed in the U.K., particularly after the notice of a vaccine withdrawal appeared in the media before the Department of Health's official information to doctors was delivered. It was decided by the then Chief Medical Officer that a system should be set up to enable urgent messages cascade down to all doctors and other relevant health professionals in a timely manner. Before 1994 messages were sent in the form of a Chief Medical Officer letter by first class (next day) post.

This system originally used, and which to a considerable extent still is used, is a closed computer network system called EPINET which was developed by the Public Health Laboratory Service (PHLS). This system was developed for transfer of communicable disease notification information and outbreak intelligence. The Chief Medical Officer's office was connected to the system which allowed a message to be sent to all Directors of Public Health and Consultants in Communicable Disease in local Health Authorities. The Health Authority could then, by using either a courier system or fax (when available) send copies of the letter to hospitals and General Practitioner (family physician) surgeries. A simplified structural diagram of the Health Service in England is appended in Figure 5-1.

It was decided that there would be two levels of urgency for messages, 6 hours (immediate) and 24 hours (urgent). The time frame was from receipt of the message from the Health Authority to the delivery to the end user (hospital or surgery). The system was found to be useful by end users and many general practitioners who did not have a fax machine were encouraged to buy one by the scheme.

The system is used particularly for notifications of pharmaceutical withdrawals, changes in policy, notification of important new scientific/medical information, urgent notification of new/changed procedures or the need for increased vigilance for certain diseases.

An audit of the system showed that the messages were regarded by end users as useful and timely, and that decisions taken about what messages should be sent by the Chief Medical Officer's office were good. It was also commented that the immediate system should only be used when really necessary, and that some messages should be sent out by the districts at local discretion as they may not directly affect their locality. There were also comments about poor dissemination in hospitals to front-line staff and from general practitioners to community staff and suggestions on how to improve this were taken up.

Over the years as technology has improved virtually all Health Authorities also receive an e-mailed copy over the NHS net e-mail system, a secure e-mail system (or in a few cases Internet e-mail) as well. These messages can then be e-mailed by the local Health Authority via the NHS net to Hospitals and GP surgeries who are on the system or by fax if they are not (all UK GPs, except for a very few, now have a fax). Also, if a message is only thought relevant for a certain group, e.g., public health doctors, it can be marked for their attention only. Interestingly, some of these messages have been sent pre-public announcement and there has been little leaking, which has made Central Government comfortable with the system. Since September 11, the system has been used to tell the service to urgently review major incident plans and disseminate protocols for "white powder" incidents and to point out where useful information on deliberate biological and chemical release can be found on the world-wide web. For important messages, the CMO often still sends an urgent letter by post to back-up the information sent by the system.

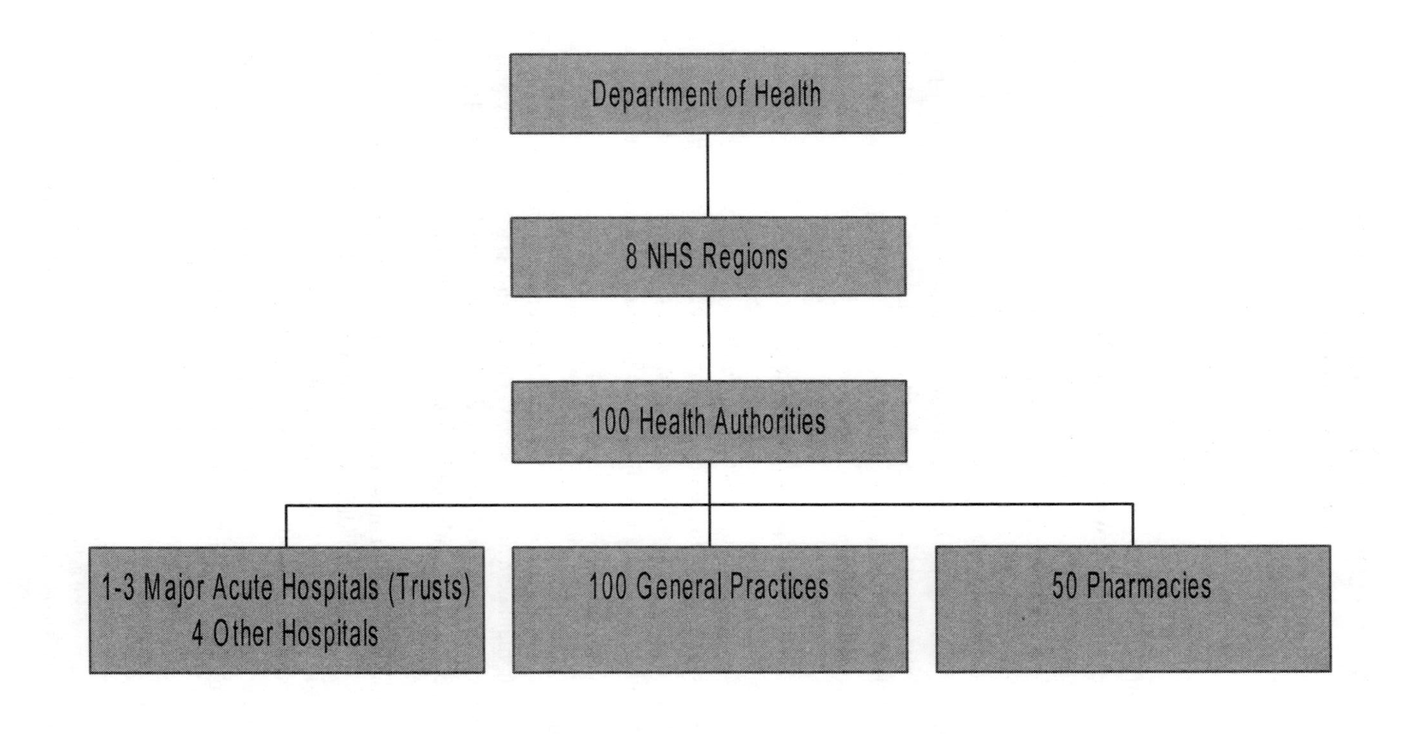

FIGURE 5-1 Simplified schematic diagram of NHS structure in England

STATE AND LOCAL NEEDS FOR RESPONDING TO BIOTERRORISM

Michael S. Ascher,* **M.D., FACP**
Office of Public Health Preparedness
Office of the Secretary
U.S. Department of Health and Human Services

Until recently, I was the lead medical officer for the State of California focussing on bioterrorism, and I will speak today from that perspective. One thing we have learned from California's many disasters is that all disasters are basically the same. Fire, flood, earthquakes—they all require the same set of skills. I thought this would not apply to public health emergencies, but I have become convinced that the logistics of responding to a smallpox outbreak or disbursing antibiotics or dispatching firefighters is almost the same. It involves different people, but the same questions: Where do you get the people? How do they get to the scene? What happens to them when they get there? How do you refresh them? How do you feed them?

Managing the Response

This suggests my first point, which is fairly simple. Someone should develop standardized instant management system "modules" for the public health response to bioterrorism. It should be modeled on things that worked well in the past, every jurisdiction should operate by the same rules, and the response elements should be vertically integrated.

The importance of the latter can be illustrated by the Oakland fire. One reason it became a full-blown disaster was that Oakland had not joined the standardized incident management system. When the fire started, there were fire trucks and men within easy distance who could have come in to help. But their radios were not compatible with Oakland's, and they couldn't attach their hoses to Oakland's fire hydrants.

A similar logistical problem is that the plans for the drug stockpiles do not extend down to the local jurisdiction. California has 61 health officers, some with larger jurisdictions than many states. But no one has determined their needs or capabilities—found out what kind of hydrants they have.

Local capacity can be vitally important. In the recent anthrax episode, there were so many instances of suspected exposures that the FBI wasn't able to determine which were credible. But we were able to get local health officers en-

* The information provided in this paper reflects the professional view of the author and should not be construed as an official position of the U.S. Department of Health and Human Services.

gaged, who could speak with authority and credibility. People stopped calling in and the situation de-escalated.

Public Health Communication

A critical deficit in our plan exists in the area of public and media communications. The original national plan included an Internet backbone, hardware, secure websites, curriculum, distance learning, public information, and media programs. Most of the money, however, went to just putting computers on health officers' desks. In fifteen California jurisdictions, the health officers didn't have computers. The main reason we didn't have a functional public information program in these recent episodes is that we ran out of money in that focus area. This is despite the fact that a successful response depends almost completely on what you tell the public. This also suggests a problem with priorities.

Building Capacity

My state faces a projected 20 percent budget shortfall this year. Since public health is not traditionally considered to be a part of the public safety system, it is not exempted from these cuts as other functions are. The threat of bioterrorism should and must change that. You do not see cuts in fire departments, and you should not see them in public health. There is some hope that this message is getting across.

One issue related to the overwhelming demand placed on diagnostic laboratories is that the majority of the tests they performed involved hoaxes. This is partly due to the fact that the original model for biologic incidents is identical to a HAZMAT response to chemical exposures. As a result we wasted a lot of energy doing work on non-credible threats. However, this did illustrate the flexibility and capacity of the existing laboratory network. It showed the potential to take on such problems without much modification.

The problem with our response, however, was that most of the testing was done at expensive and very sophisticated Level B laboratories rather than conducting initial screening at Level A laboratories. The very large capacity of Level A laboratories could not be used effectively. We had not fully thought about the need for that much Level A capacity. We met the demand by using more sophisticated laboratories. A Level A lab, for example, was set up at CDC. There are a lot of labs that could fulfill this role. There is also the issue of registration to handle special agents. Out of probably 2,500 clinical labs in California that can culture anthrax, there are no more than approximately half-a-dozen registered to handle it for analysis. We need to work out a way so that screening tests can be performed without sending the samples to Level B labs.

Level C laboratories are needed at the state or regional level to perform the sophisticated confirmation now limited to CDC or USAMRIID. Money is part of the issue, but also workforce availability. It is very hard to find and hire people with the requisite skills. One thing that has been very helpful has been the CDC Emerging Infectious Disease Fellowship Program which has assigned individuals to work with us in our labs. Expanding Level C laboratories could also help in the transfer of applied research from biotech partners to the public health network.

I happen to be an Army Reserve officer and Commander of one of two small infectious disease teams that have been organized into something called Consequence Management Medical Response Teams. This is another of the many federal assets that need to be recognized and woven into our bioterrorism defense net.

Learning Lessons

Finally, we have not given enough attention to "war games" and exercises. Those that have been conducted, for example, Dark Winter and TOPOFF, revealed vulnerabilities and were very sobering to participants and observers. Recognizing and responding to those vulnerabilities, we should conduct real-world exercise that drill down to the local response. Different variations should be tried in two or three settings. Not only would we learn a lot, we would get local responders involved in the learning process. If the real thing happens, we will be much better prepared to respond.

U.S. PUBLIC HEALTH SERVICE OFFICE OF EMERGENCY PREPAREDNESS

Donald C. Wetter,[*] P.A.-C., M.P.H.
United States Public Health Service

There was much valuable information shared in many areas during the meeting at the IOM. While briefly mentioned on occasion during the conference, the topic of surge capacity of hospitals and other medical institutions during a biological terrorism incident needs more emphasis. This includes hospital bed capacity, alternate care facility use, medical provider staffing, medical logistics and operations to name but a few areas. Laboratory capacity, public health information, vaccines, etc. were important aspects of the discussion regarding preparedness for and response to bioterrorism. This planning is incomplete without fully integrating the community that is caring for the victims of these attacks.

The assumption for this discussion is that there would be a large number of patients presenting to the healthcare system. It is difficult to define the term "large" in the context of bioterrorism because even an incident with relatively few patients ill from a bioterrorism agent could also create thousands of "worried-well" individuals presenting to a hospital. With this in mind, the healthcare system

---

[*] The information provided in this paper reflects the professional view of the author and should not be construed as an official position of the United States Public Health Service.

must be able to respond to all individuals seeking care. The current financial state of US hospitals has decreased the likelihood that a significant number of beds will be available for a sudden increase in patients. Staff shortages and closure of hospitals or sections of some facilities create a daily marginal surge capability.

The issue of staffing is not only numbers; it remains education on bioterrorism issues. Again, due to the strain on the current medical system, it is difficult to fund training for hospital staff. Competition for staff time because of the rapid pace of increasing medical knowledge also lowers the probability that a hospital or practice manager will choose terrorism for staff education over other more commonly seen diseases.

Though most emergency services use some Incident Command System (ICS), it appears that many hospitals and public health agencies are not familiar with the system or at least do not wish to adopt it. In the complex operations of terrorism response, it is essential that the healthcare institutions coordinate with the rest of the community emergency management and use some form of ICS. As discussed throughout this meeting communications and coordination are vitally important. This was reinforced to me after my six-week assignment at the New York City Emergency Operations Center during the World Trade Center incident, anthrax investigation, and American Airlines 587 crash.

Finally, the issue of funding emergency response to a biological event needs to be addressed. The Federal Emergency Management Agency is tasked with the lead in consequence management for a terrorist incident. While FEMA took the lead for the World Trade Center and much of the response funding, the agency did not do so for the bioterrorism response of the anthrax event. The policy regarding this part of disaster management is unclear. FEMA funded some states in their response to the West Nile virus, but to this date, not to the anthrax attack. Hospitals and health departments will possibly need access to additional funds to respond properly to bioterrorism events.

ASSESSING STATE AND TERRITORIAL HEALTH DEPARTMENTS

James (Jerry) Gibson, M.D., M.P.H.
Director of Disease Control, South Carolina Department of Health and Environmental Control and immediate past president of the Council of State and Territorial Epidemiologists

I will state my perceptions of the preparedness of state and local public health departments to detect, investigate, and respond to potential bioterrorist attacks and threats of such attacks. I am making one fundamental assumption that I believe most of us share here: that a major long-term goal in building an effective response to bioterrorist attacks is to re-build the American public health infectious disease control capacity which suffers currently from at least thirty years of deferred maintenance. This disease control capacity is congruent with the acute infectious disease control infrastructure, which consists of trained organized people, communication systems and laboratories. The state of disarray of these systems is summarized well in the Institute of Medicine's 1988 report *The Future of Public Health*. I have seven points to make, and then six recommendations.

First, public health organization and capability are highly varied across the 51 state and 3,000 local health departments of the United States. Some are strong, but many are very weak, some of those large population centers. However, the need to respond well to a bioterrorist threat is present in all jurisdictions. Therefore, there exist critical disparities in needed capability. Our preparedness building cannot ignore the weak departments.

Second, the public health system is very fragmented in many states. Local health departments are separate from the state health department, which has little leverage to improve them. Often they communicate minimally, and working in partnership is difficult.

Third, our task for bioterrorism preparedness is to build complex human systems that must work right the first time they are challenged. That is difficult: people are less consistent than vaccines. It implies to me that the bioterrorism response systems must be integral parts of the regular infectious disease surveillance and control systems if they are to be exercised regularly, and perform when needed.

Fourth, in the end, public health response capacity is trained people. Local health departments in particular are very short of these. Thus capacity building requires recurring funds to hire people, not a one-time capitol investment. There is no way around this need.

Fifth, to plan, organize, hire staff and train them takes time. Even in the private sector it takes time, and the ability of state and local health departments to adapt to urgent circumstances and speed up operations is also highly variable. Thus finding ways to help state bureaucracies develop a *sense of urgency*, while still maintaining their programs for HIV/AIDS, family planning, diabetes mellitus, vital statistics, etc., is essential. The key implication of this fact is that the speed at which health departments can absorb new funds is limited; capacity building does not happen overnight.

Sixth, "planning is an unnatural process… .". Getting a bioterrorism operational plan, integrated with key partners, in every county and city (not to mention every hospital) is a major challenge. Health departments have limited leverage to assure it happens. Thus this critical process of assuring local bioterrorism preparedness planning will take substantial resources.

Seventh, the most important point, most state health departments are very dependent on federal grant funds to operate their programs. In the state of South Carolina, about 38% of the integrated health department's (state and the 46 counties) budget comes from federal cooperative agreements; another 38% is earned from Medicaid and other sources, and thus is not available for bioterrorism preparedness. Essentially all the discretionary funds are federal. Therefore, I would like to propose six principles by which new federal bioterrorism grants be allocated to state and local health departments. These principles come from a new document from the Association of State and Territorial Health Officials (ASTHO) Anti-terrorism Task Force.

1. Such grants should provide for state flexibility of use. Funds should be routed in such a way that their use for public health is assured, but beyond that the state should have the discretion to spend them where they are most needed. Also, states should have the authority in emergencies to redirect federal grant funds and federally funded staff to areas of critical need, without penalties. This process should avoid cumbersome multiple layers of permission seeking from the granting agency.
2. Funding should be based primarily on state need rather than be competitive. Given the wide range of state capabilities, competitive funding will only make the strong stronger and leave the weak as vulnerable as before.
3. Funding budgets must be multi-year, to allow for the time needed for states to absorb funds. Funding will have to continue to some extent long-term, since new staff are an essential part of preparedness.
4. State and local health departments should be required to plan and submit funding proposals together, so that planning and implementation can be coordinated. Likewise, the grant process should require coordination and communication between public health and other agencies receiving bioterrorism funds.
5. A mechanism is needed for state and local health departments to share best practices and ideas that work rapidly. Possible a series of ongoing email-based surveys of innovative ideas could help do this.
6. New information, communication or surveillance systems should be built on or be integrated with existing systems such as NEDSS, HAN and Epi-X.

I was also asked to give my first three priorities for action to build state and local public health response capability to a bioterrorist attack. These are:

- Make federal cooperative agreement funds available to all states to be used primarily to build city/county public health capability for disease surveillance and investigation. In many states, what is likely to work best is to hire surveillance and epidemiology mentor/trainers on a regional basis to train, support and work with local-level infectious disease control staff to build active surveillance. These regional bioterrorism epidemiology staff can also promote local liaison between key participants and preparedness planning.
- Provide good educational materials and methods of dissemination by state health departments for primary care clinician education on detection, reporting, clinical care, and infection control of the first-line and also second-line bioter-

rorist agents. The dissemination should be via grants to states and large city public health departments, because the optimal method of disseminating and reinforcing the messages will vary locally and must be determined locally. A major purpose of such education programs should be to build relationships between state/local health departments and clinicians.

- Provide sufficient funds to complete the national, state-based system of electronic infectious disease surveillance (NEDSS) which has been begun. This is a very challenging task and will take several years to complete, test, de-bug, and optimize. It will also require significant recurrent costs for maintenance.

COUNTERING BIOTERRORISM THREATS: LOCAL PUBLIC HEALTH PERSPECTIVES

Thomas L. Milne
Executive Director
National Association of County and City Health Officials

In most communities, it is typically a local or state public health department that responds when there is a diagnosis of even a single case of a serious infectious disease. A significant outbreak often results also in the mobilization of resources from the Centers for Disease Control and Prevention. Fortunately, local, state and federal public health efforts have successfully contained most outbreaks of infectious disease in recent years to a relatively small number of cases.

The growing concern that an intentionally caused event involving a biological agent could occur in this country have prompted many activities directed toward increasing public health preparedness. Local public health agencies, along with the National Association of County and City Health Officials in Washington, DC, have been engaged in bioterrorism preparedness work since 1999, in partnership with CDC, representatives of state health departments, and representatives of local boards of health. Local health officials urge that the following principles and factors guide the work ahead:

Principles

1. There is significant likelihood that terrorism events involving biological agents will occur. Such events will take place in communities and will affect people living in communities.
2. While state and national level preparedness is important, it is very important that communities be prepared as well, with preparedness plans and necessary capacities to respond in place.
3. Preparedness plans, to be effective, must be developed through broad collaboration including all significant stakeholders in communities, including hospitals, emergency responders, fire, law enforcement, public health, physicians, and elected officials.

4. There is a significant under-investment in public health, particularly at the local level, where capacities have been declining in recent years.

5. Virtually no local public health agency, regardless of size and level of budget, has the full range of capacities needed to assure an adequate public health response to bioterrorism events. Significant resources are needed from the federal government to assure that the local and state infrastructure is adequate to the tasks required.

6. Investment in public health preparedness for bioterrorism will have multiple benefits because the capacities and competencies required are directly applicable to the general daily responsibilities of public health departments at the local and state levels.

Local Public Health Infrastructure

There is no such thing as a consistent local public health system. There are approximately 3,000 local public health departments in the U.S. Most are small and serve small populations. The median size health department employs 13 staff while the mean size is 67 employees. About 70 percent of local health departments serve populations of 50,000 or smaller. Most are agencies of local government, with county health departments the most common form. However, in 16 states (primarily in the east and southeast U.S.), local health departments are mostly or entirely local offices of the state department of health. About 160 counties in the country have no form of local public health services. Services, authorities, and staffing levels vary widely among health departments, and no two are the same.

The increased and much needed emphasis on public health preparedness should prompt discussions of the need to build a more consistent system of local public health, assuring that the necessary capacities and competencies are available to serve all residents of the country. Strategies to address this need should begin with local and state initiatives and, only if and where needed, include federal mandates.

Needed Capacities

The capacities needed to assure adequate local (and State) public health preparedness include the following:

1. A workforce of adequate size and with adequate training.
2. Adequate public health laboratory capacity.
3. Increased epidemiology capacity including significantly upgraded surveillance systems.
4. Information systems which are secure, continually updated, and highly accessible to local and state public health officials.

5. Communications systems which are secure, offer redundancy, and provide full time high speed access to information systems.

6. Federal, state and local laws and policies which fully support the emergency powers needed by public health officials to respond fully and quickly to bio-events and other public health emergencies.

7. Capacity to participate in community preparedness planning, which includes the testing and practice of such plans.

Financing Needed

An absolute funding level needed to assure local public health preparedness cannot be defined, especially given the many shortcomings associated with the lack of consistency among local health departments nationally. It has been estimated that between $835 million and $1.3 billion are needed annually for five years to develop a fully prepared local and state public health system. Clearly, such federal investment would need to be continued at a maintenance level once the system has been built.

Accountability

Accountability for expenditures and outcomes is an essential aspect of the large investment necessary to build a fully prepared system of local and state health departments. Financial accountability measures should assure that states maintain their current levels of support for state and local public health activities. The gains achieved in capacities and competencies should be documented and compared against a set of standards. The National Public Health Performance Standards Program has been developed and is scheduled for implementation in 2002. That program may provide a basis for mandated performance once federal funding has been assured and is in place.

PUBLIC HEALTH PRIORITIES FOR RESPONDING TO BIOTERRORISM

Ruth L. Berkelman, M.D.
Department of Epidemiology
Rollins School of Public Health, Emory University

Senator Frist challenged us at the beginning of this workshop to articulate better to the public what is meant by the term public health infrastructure; he challenged us to explain clearly and in lay language why public health infrastructure is critical to bioterrorism preparedness. We have not fully addressed this concern the past two days, and we need to accept this challenge and share with people what we mean by "public health infrastructure," in a way that everyone understands.

We need to assure our communities that a good public health infrastructure will provide them with nurses, doctors, and others who are trained in public health—experts who make it their business to know what is happening in terms of health in their community, such as whether there is an epidemic of influenza in the community, or a meningitis outbreak at a local high school or a bioterrorism attack—experts who know what to do to protect their community when public health threats such as these occur. The public needs to know that professionals in public health are in place that can investigate problems and provide guidance to individual doctors and other healthcare providers in their community. Public health professionals also work with schools, industry and the general public to protect the community. They assure that families are safe from epidemics and other threats to their health, that vaccines and antibiotics are available to the whole community and can be administered and/or distributed as necessary to protect the community's health. We need to talk to the community in concrete terms about public health infrastructure and protection of public health, just as we do with fire and police protection.

I want to turn now to six issues related to bioterrorism preparedness that have emerged in the discussions the past two days. These are not comprehensive, but they do represent priorities for public health preparedness.

First, there is a need to strengthen the local and state public health departments. In the context of this forum, there is a great deal of overlap between the public health infrastructure and bioterrorism preparedness. They are not identical, but public health infrastructure is required to assure bioterrorism preparedness. This means, in part, an infusion of resources and trained personnel. At the same time, there is a need to examine the current organization of public health departments.

Jerry Gibson described the striking disparities among local health departments. There may be a part-time nurse with 1 or 2 clinics a week for a community of 4,000 people; there may be no one available for emergencies. The state of Georgia has 19 health districts and 159 health departments. Each state is going to have to look at its needs and decide how best to proceed. Is Georgia, for example, better off working with the 19 districts or the 159 departments, or is yet another balance needed? Perhaps some of those departments should be consolidated for the purpose of preparedness for bioterrorism. As new resources are appropriated for use by local and state public health departments, a reexamination of the existing organizations may be helpful to assure that their use is efficient and effective for preparedness for bioterrorism and other major public health threats.

A second issue is disease detection and the need to strengthen surveillance. There currently is great interest in syndromic surveillance as a tool for early detection of bioterrorist events. We need to explore these systems further—systems based on pharmaceutical data, 911 calls, clinic visits, and so forth—to determine the diseases, syndromes, and conditions for which they might be important, and those for which they are not. Rigorous evaluation will be needed.

While syndromic surveillance may prove useful for some health alerts, the traditional system of having a doctor, infection control practitioner, or other health professional know what to be alert to, and who to call when they are concerned will remain fundamental. For example, it is unlikely that a system utilizing emergency department visits and based on ICD codes for fever and rash will substitute for a well-trained health care provider for the early detection of smallpox. Ed Eitzen from the Department of Defense said one of the most important defensive measures we can take is training of the healthcare provider. We need to assure education of physicians and other healthcare providers and strengthen the liaison between public health agencies and the healthcare community.

A third issue is vaccine development and procurement. The number of companies producing vaccines has declined dramatically in the past two decades, and the production of some important vaccines like the one for adenovirus was curtailed although the vaccine was still needed. There has been relatively little incentive for companies to produce new classes of antibiotics. Who is responsible for ensuring that the American public has the vaccines and antimicrobial agents that they need? What agency? There needs to be a clear mandate defining responsibilities in this realm.

A fourth issue is surge capacity. Renu Gupta raised the important need to look at the possibility of having/using reserve expertise in the private sector. We need to utilize both industry and academia during times of need both for their expertise and their surge capacity. During the anthrax crisis, they came forward and offered their help, but it was difficult for public health agencies to harness these resources. It is far more difficult to organize volunteers in the midst of a crisis if there has not been advance planning. It is likely that in the wake of the terrorist attack, private industry and academia will be even more inclined to participate in such planning today than before the terrorist attacks. One example of the need was the shortage of surge capacity in the laboratory during the recent anthrax incident. We should take steps to avoid the situation where CDC needs to conduct Level A lab analysis for anthrax, and states may be back-logged to a degree that could jeopardize public health.

A fifth issue is the need for interdisciplinary groups to work on applied research questions in the area of bioterrorism. Applied research to answer simple questions sometimes falls through the cracks. Although this may be due to lack of resources, sometimes it may be due to the lack of clarity as to who is responsible for defining and conducting needed research that does not fit with one specific field of scientific inquiry. Yet, applied research on such issues as the potential for dissemination of anthrax powder through handling of envelopes and on decontamination following release of anthrax powder can be critical. Establishing interdisciplinary groups to determine research needs and to implement the needed research may be useful. It is good to hear of CDC's plan to conduct meetings in the coming weeks to establish an applied research agenda for anthrax, meetings which will include experts from a variety of scientific fields and who represent both the public and private sector.

The final issue I want to address today is the necessity for the intelligence community and the scientific community to work together. The intelligence community needs to inform scientists, but scientists need to help the intelligence community, too. When a scientist is disaffected, the intelligence community may need to be alerted; when a scientist is discovered by other scientists to be conducting work that may inadvertently lead to adverse consequences and where the risk is deemed greater than the potential benefit, the scientific community needs to stand collectively against such work.

We also need to think about cross-training between the intelligence community and the public health community. We in public health have been hearing about the importance of documenting the chain of custody of samples for forensic purposes, and many have not understood the term "chain of custody"; public health professionals may need some training in forensic sciences to better understand the needs of that community. Also, the intelligence community may need training in public health concerns. The FBI may want to consider having some of its experts trained in the Epidemic Intelligence Service (EIS) at the CDC. The Department of Defense has had people trained in EIS for several years, and that has been quite beneficial.

6

Scientific and Policy Tools for Countering Bioterrorism

OVERVIEW

Several scientific, scientific policy, and legal tools that were presented and discussed during the workshop but have not been addressed elsewhere in this report summary are included here. These include innovative surveillance; detection and diagnostic tools and technology; scientific policy issues unique to bioterrorism response preparedness; and, bioterrorism-related legal needs and obstacles. These ideas include multiple components of the individual sessions and have cross-cutting implications for an overall response to biological threats.

Surveillance

Surveillance and rapid detection are crucial to an effective response to a bioterrorist attack. Delayed detection results in delayed prophylaxis and aggressive treatment measures. Because it is practically impossible to predict when or where a bioterrorist attack is going to happen, there are limitations to the "drop-in" terrorism surveillance systems that have been used to monitor specific places or events such as the Super Bowl or Democratic National Convention. Nor can bioterrorism surveillance be solved simply with pentium chips. Comprehensive bioterrorism surveillance will require integrating human resources, laboratory resources, and information management in innovative, legal, and acceptable ways that allow for early detection and characterization of threats.

There are several innovative surveillance systems in use or being developed. ESSENCE, for example, is an automated syndromic surveillance system that initially relied on the already extant automated health care information system across D.C., Maryland, and Virginia. Since September 11, ESSENCE has been extended to over 300 installations around the world. Every 24 hours, 30,000 ambulatory diagnoses from these various installations are downloaded, automatically analyzed, prioritized based on expected values from historic data, and visualized using geographic information systems. However, none of the systems currently being developed are likely to be adequate in and of themselves. The best solution will probably be a system of systems that is sensitive enough to detect specific conditions and even small outbreaks.

Detection and Diagnosis

In terms of environmental and clinical detection and diagnosis, although reasonably good assays are available for a limited range of specific agents, the immense diversity of microorganisms, including bioengineered pathogens, presents a major challenge. There can be considerable variability even within a strain, let alone a species. Pathogens are a natural part of the environment and can confuse detection efforts. For both environmental and clinical settings, we need rapid, standardized methods that allow for the detection of a broad spectrum of potential biological weapons in a quantitative fashion.

Rapid detection and diagnosis requires access to an extensive sequence database and high throughput laboratories. Biotechnological barriers in the public health infrastructure must be identified so that the proper tools can be appropriately distributed or accessed. Academia, industry, and government laboratories must all be brought in at appropriate levels and in appropriate ways to help build new capabilities.

Specimen collection needs to standardized and automated. For example, there is no standardized collection method for samples from the inside of a computer. Indeed, specimen collection is often the major obstacle to rapidly processing a large number of samples and the weak link in what seems to be an otherwise very promising detection and diagnosis technology.

The capability to use molecular sequences to rapidly detect and identify bioterrorist agents could serve as an important form of deterrence and might possibly prevent bioterrorist attacks from occurring in the first place. One vision is an international molecular forensics lab that would rely on a molecular fingerprint global database to identify the source of the bioterrorist agent. This capability could provide the biological equivalent of the threat of nuclear retaliation. Again, it must be emphasized that bioterrorism is a national security issue and bioterrorism preparedness efforts are a strategic defense.

Scientific Policy Issues

The fact that bioterrorism preparedness is a national security imperative raises many important and new scientific policy issues:

- It was suggested that we need a new peer review system for screening new bioterrorism defense research ideas.
- There needs to be improved communication between scientists in the clinical response laboratories and law enforcement personnel, for example with regards to resolving crime scene versus public health needs and ensuring that the physicians or other individuals who have provided samples can receive results in a timely manner.
- Law enforcement investigators need to be educated about relevant microbiological issues, and the scientific and public health communities need to be informed of how criminal investigations proceed.
- There is concern about who should have access to certain scientific materials, equipment, and information and whether access to select agents should be restricted. At the same time, it is crucial that as much information as possible be in the hands of the biomedical community so that scientists can conduct the type of research that is necessary to build a strong biodefense arsenal.
- Finally, computational modeling is an important but undervalued scientific component of bioterrorism defense preparedness. New computational capabilities can be used to model interactions between digital microbes (as opposed to actual, biological microbes) and digital immune systems. This kind of simulation approach could be used in guiding decisions about experimental design as well as in testing various policy and response scenarios.

Legal Issues

Bioterrorism preparedness as a national security imperative also raises many important legal issues. The first step toward evaluating the necessity of a legal strategy for bioterrorism is to assess the adequacy of the existing legal infrastructure for dealing with bioterrorism issues. Do provisions in the law exist that enable authorities to do what needs to be done in the context of a bioterrorism event, for example decontaminating a building or quarantining individuals? Are there any legal obstacles that would interfere with a public health response to bioterrorism?

The primary legal authority for bioterrorism preparedness and response is at the state level. Recently, the Center for the Study of Law and the Public Health at the Georgetown Law Center and Johns Hopkins University prepared a model state emergency health powers act (see Appendix G) in an effort to facilitate the analysis of public health law at the state level. The proposal is being given to states for their consideration either for adoption or simply as a tool for review of their own public health statutes in the context of bioterrorism. The proposal has stimulated much controversy. Indeed, this controversy may be a reflection of the importance of the legal infrastructure for an effective public health response to bioterrorism.

INNOVATIVE SURVEILLANCE METHODS FOR MONITORING DANGEROUS PATHOGENS

Julie A. Pavlin, M.D., M.P.H.,[1] Patrick Kelley, M.D., Dr.P.H.,[1] Farzad Mostashari, M.D., M.S.P.H.,[2] Mark G. Kortepeter, M.D., M.P.H.,[3] Noreen A. Hynes, M.D., M.P.H.,[4] Rashid A. Chotani, M.D., M.P.H.,[5] Yves B. Mikol, Ph.D.,[6] Margaret A. K. Ryan, M.D., M.P.H.,[7] James S. Neville, M.D., M.P.H.,[8] Donald T. Gantz, Ph.D.,[9] James V. Writer, M.P.H.,[1] Jared E. Florance, M.D., M.S.,[10] Randall C. Culpepper, M.D., M.P.H.,[3] Fred M. Henretig, M.D.[11]

Historically, emergent public health problems have been recognized by astute health care providers who then report their suspicions to public health authorities, rather than the reverse (Thacker, 1994; Thacker and Berkelman, 1998). Even with luck, this approach usually falls short of optimal public health care. Outbreak surveillance seeks to reduce reliance on the epidemiological insights of individual practitioners or facilities and to significantly decrease the time needed to collect and assess data, thereby allowing officials to be alerted more rapidly of an emerging threat.

A review of some recent disease outbreaks will help define the requirement for more timely, high-quality systems. Past epidemics have characteristics that can help identify the epidemiological, behavioral, and political factors that affect the detection of and response to an emerging infectious disease epidemic. Many emerging infections present as syndromes that initially do not point to a specific underlying pathogen. Potential bioterrorism events could also pose similar challenges of delayed recognition (see Table 6-1).

In addition to earlier detection of events, surveillance systems for emerging infections, including bioterrorism, are essential for focusing limited response assets and for providing evidence-based information for governmental risk communicators attempting to manage community concerns. Plans to improve

---

[1] Department of Defense Global Emerging Infections System, Silver Spring, MD

[2] New York City Department of Health, New York, NY (current affiliation OutbreakDetect, Inc., New York, NY)

[3] US Army Medical Research Institute of Infectious Diseases, Fort Detrick, MD

[4] Human Health Services Division, Food Safety and Inspection Service, Washington, DC

[5] Johns Hopkins University Applied Physics Laboratory, Laurel, MD

[6] New York City Department of Environmental Protection, Valhalla, NY

[7] Naval Health Research Center, San Diego, CA

[8] Air Force Institute for Environment, Safety and Occupational Health Risk Analysis, Brooks Air Force Base, TX

[9] George Mason University, Fairfax, VA

[10] Prince William County Health District, Manassas, VA

[11] Clinical Toxicology and Poison Control, Children's Hospital of Philadelphia, Philadelpia, PA

public health capabilities to identify and address such disease emergencies must include determining how surveillance systems can be made more timely, flexible, and sensitive without overly compromising other aspects of quality.

A recent meeting addressing innovative, responsive surveillance systems focused on three areas: 1) defining the existing functional capabilities in need of improvement, 2) examining existing prototype systems attempting to meet these needs, and 3) identifying the ideal features of a "system of surveillance systems" that would meet the need for more timely, sensitive, and flexible detection and response (Pavlin et al., 2001).

Examples of Recently Developed Innovative Surveillance Systems

In recent years, agencies and municipalities have attempted to improve public health capabilities with novel and innovative approaches to surveillance (see Table 6-2).

New York City—911 Calls

Beginning in March 1998, the New York City Department of Health, in collaboration with the Mayor's Office of Emergency Management and the Fire Department's Emergency Medical Services, began monitoring the chief clinical complaints noted in daily 911 calls as a citywide health indicator. The intent was timely detection of public health events, with particular emphasis on influenza-like illness. Several complaints, such as "difficulty breathing" and "sick"—thought to represent influenza-like illness—were selected. A review of data from 1991 to 1998 found a temporal association between the onset of annual influenza epidemics and a rise in the volume of the selected call-types. Thresholds were developed that "detected" all four annual influenza epidemics from 1994 to 1998. In addition, the system generated very few false-negative alarms—times when the expected threshold was exceeded outside of periods of peak influenza activity. This system indicated the 1999–2000 influenza epidemic approximately two weeks before recognition by traditional influenza surveillance systems.

New York City—Diarrheal Diseases Surveillance

New York City also developed three independent and complementary systems to monitor for community-wide gastrointestinal outbreaks: 1) sales of antidiarrheal medications, 2) submission of stool samples for laboratory tests, and 3) incidence of gastroenteritis in nursing home populations.

Monitoring sales of antidiarrheal medication approximates the incidence of diarrheal illness in a community. Large increases in sales of anti-diarrheal medicines have been reported during outbreaks of gastrointestinal diseases (Proctor et al., 1998; Collin et al., 1981; Rodman et al., 1997; Miller and Mikol, 1999). Volume of over-the-counter medication sales is obtained from a regional distributor and a chain of drugstores (see Figure 6-1). Sales data are received weekly and analyzed for unexpected variation.

The number of stool specimens examined for 1) bacterial culture and sensitivity (three laboratories); 2) ova and parasites (three laboratories) (see Figure 6-2); and 3) *Cryptosporidium parvum* (one laboratory) is collected daily and indicates the incidence of gastrointestinal illness in the population. This information is provided in addition to test results for giardiasis and cryptosporidiosis from active disease surveillance.

The number of new cases of gastrointestinal disease in nursing homes provides information on the incidence of illness in a population with limited exposure to the wider community. Twelve nursing homes participate in the program, with 1,850 residents. Numbers of new cases of gastrointestinal disease are provided daily.

DoD-GEIS Electronic Surveillance System for the Early Notification of Community-based Epidemics (ESSENCE)

The DoD-GEIS monitors patient data from medical facilities for changes in disease incidence in the National Capital Area (see Figure 6-3). For every outpatient encounter within the DoD medical system in the US, the provider electronically enters a code describing the reason for the patient visit. All encounters are coded at or near the time of service, even if the definitive cause of illness is not established during the visit. Most codes chosen by providers reflect this prompt diagnosis and may include syndrome-based codes such as cough and fever in addition to empiric diagnoses such as pneumonia.

All personal identifiers are removed from the data when received and the diagnoses are categorized, if applicable, into one of nine syndromic clusters. The frequency of outpatient visits in the different syndromic categories is then plotted and compared to previous years' experience. It can also be depicted geographically through geographic information systems (GIS) software.

Sandia National Laboratory—Rapid Syndrome Validation Project (RSVP)

The RSVP is an Internet-based reporting system, intended for daily routine use by physicians and epidemiologists. The system features rapid data input of clinical and demographic information via a touch sensitive monitor, automated screening of reports for signs and symptoms correlated with reportable diseases and subsequent instantaneous notification of public health officials if indicated, and rapid feedback to clinicians of the geographic and temporal distribution of recent similar syndromes in their community in recent weeks. Public health officials may use RSVP to inform users of current disease outbreaks of public health importance, specific to each of six syndromes. RSVP is easily expandable to multiple sites, requires no specialized client software other than a web browser, and may be operated as an intranet or for general data sharing.

Expectations for a Health Indicator Surveillance System

The variety of information used to monitor the health of a community can be called "health indicator" surveillance. The requirements of different users should be documented while these new systems are still in development.

National Needs

Surveillance systems can assist federal public health management of epidemics in many ways, not only in the traditional detection of outbreaks and monitoring of the effectiveness of preventive measures, but in determining how to assist and augment local health and emergency response activities. The systems should be capable of being integrated with other surveillance systems at all levels of government.

Local Needs

The ability of the local health department to identify, evaluate, and contain the effects of disease outbreaks is dependent on the timeliness and accuracy of reporting. Optimally, a surveillance system should contrast local data with data from other sources (e.g., CDC, nearby states, other jurisdictions, and military installations). Surveillance systems with greater sensitivity, completeness, and timeliness than existing systems are needed to allow the most effective identification of unexpected events across jurisdiction boundaries.

U.S. Military

The military services maintain centralized health-related surveillance systems for routine public health policy and disease control. However, more rapid and sensitive recognition of a disease outbreak wherever U.S. forces are located requires enhancements to both the level of detail available and the speed of data transmission and analysis. New health indicator surveillance systems for military communities should cooperate with civilian public health personnel since disease outbreaks do not respect military installation boundaries.

Specific Needs for Bioterrorism Surveillance

To maximize patient survival after a bioterrorist attack, we need surveillance systems that: 1) facilitate the rapid recognition of a bioterrorist event, 2) assist in determining the site of exposure, 3) maximize efficient delivery of limited medical countermeasures to the infected population, and 4) assess containment and mitigation. Expansion and improvement of surveillance systems for bioterrorism will likely have a dual benefit of strengthening the public health infrastructure for detection of naturally occurring infectious disease outbreaks and emerging diseases.

Needs for Animal and Plant Surveillance

There are many points along the farm-to-table continuum at which infectious agents can arise from or be introduced into the food supply. The increasingly centralized nature of food production in the United States, with subsequent widespread distribution, means that the impact of contaminated products would be national (and potentially international) in scope rather than regional. This argues for an integrated human-animal pathogen and antimicrobial resistance surveillance system providing rapid feedback to public health and food-regulatory agencies. Food monitoring data for microbial pathogens and food recall information are already collected by federal and state regulatory agencies and could be integrated into existing food-borne disease and outbreak surveillance systems. This system could 1) enhance the speed with which outbreaks are identified and control measures implemented, 2) identify patterns of product contamination that would lead to more rapid intervention, and 3) identify unusual illness patterns and pathogens that are dispersed but possibly related.

The use of information on disease in animals, both wild and domestic, can prove a useful tool in monitoring the health of human populations (Jaroff, 1999; Steele et al., 2000; CDC, 1999). A surveillance system will include data on animal morbidity and mortality to achieve the greatest sensitivity.

Key Issues for Developing a Surveillance System

Data Sources

Health indicator surveillance is the foundation for early recognition of an emerging infectious disease. There is a critical need for a system of systems, with the flexibility to fit the needs of each level and locality. Measurable alterations in personal behaviors within the first hours or days of illness can assist early detection of an event or epidemic. These include work or school absenteeism, changes in usage of public transportation or toll roads, and the purchase of over-the-counter remedies. Data about delivery of medical care have value not only for outbreak detection but also for the ongoing management of an epidemic

(Rodman et al., 1998). These include emergency response calls, required disease reporting, outpatient clinic and emergency room activity, inpatient and intensive care unit records, and laboratory and prescription drug requests.

An infrastructure for pre-clinical or many types of clinical data is not readily available but might use data already collected for other purposes such as billing, inventory control, or resource management. Concerns over ownership may block access to existing data deemed valuable for surveillance. Resolving these issues will require high-level leadership, commitment, and prioritization.

The following questions addressed the usefulness of health surveillance data.

- Are the data sufficiently representative of the entire population of interest? Are there important sub-populations that will not be captured by this surveillance system?
- Are the data timely? How much time will elapse between the onset of symptoms and detectable change in the data?
- Are the data available electronically? Electronic data can be transferred and analyzed more quickly than paper reports.
- Can the data be categorized as symptom clusters or syndromes? Summary data (e.g., total number of admissions or transit ridership) may indicate an event is occurring, but interpretation of the cause will be difficult in the absence of more specific information.
- Are retrospective data available? Without baseline data, it will be difficult to assess whether the data can detect new events. Furthermore, alarm thresholds using historic data obviate the need for a lengthy "run-in" period.

Improvement in Active Patient Data Collection

Although use of existing data sources can help, some situations, geographic areas, or types of medical practices may require additional data for an effective surveillance system. If a system requires new data collection, it is imperative to work closely with medical practitioners to achieve a workable solution and to provide feedback so they can benefit from their participation.

Possible features of an active data-entered, real-time surveillance system include:

- Syndrome-based reporting from a pre-determined list of signs and symptoms;
- Touch screen or personal digital assistant (PDA)-based electronic data reporting, collection, and submission;
- Graphical presentation of data based on GIS and temporal information;
- Automatic alerts to public health officials of specific signs and symptoms (e.g., fever with skin rash in young adults) suggestive of serious communicable disease; and,

- Alerts from public health officials to health care providers that can be easily updated.

The Need for a System of Systems

Ideally, a surveillance system will be sensitive enough to identify the emergence of an outbreak, categorize its nature, and identify those affected so that the outbreak can be quickly and effectively contained. Bringing together information from various health indicator data sets can allow the public health practitioner to 1) evaluate many indicators simultaneously, 2) compare variations and identify common trends, and 3) track confounding factors and reduce noise.

The compilation of information provided by independent and complementary data sources allows inter-system comparisons. Comparing the data from several indicators, some of which are more sensitive than others for different scenarios, can enable a trend observed in any single system to be confirmed by the systems. Simultaneous unexpected but concordant variations in multiple data sets may suggest actual emergence of an outbreak. Clinical reports are needed for confirmation.

The importance of collecting data through an intricate surveillance system is to use it to quickly identify and respond to an adverse event rather than to develop an archive. Ideally one organization would collect, compile, integrate, and analyze all data. Moreover, this data would need to be shared effectively and efficiently at different levels of the existing health systems. Of utmost importance, the fundamental issue of personal and organizational privacy needs to be addressed when setting up such a system.

TABLE 6-1 Selected infectious disease outbreaks characterized by delayed recognition, characterization, or response

Disease Outbreak	Characteristics
Influenza Worldwide, 1918–1919 [3,4(pp153–191)]	Rapid spread over large geographic area Overwhelms health care system Overwhelms essential services (e.g., burial of dead) Person-to-person transmission No available treatment
Legionellosis (Legionnaires' Disease) Pennsylvania, 1976[5]	Common-source Exposed population disperses from point of exposure throughout the state of Pennsylvania Unknown agent Rapid spread and demise Mimics a biological terrorist attack
Acute Respiratory Distress Syndrome (Hantavirus)[4(pp538–549),6] Southwest US	Affects small population spread over a large geographic area Cultural concerns Zoonotic
Salmonellosis Oregon, 1984, US, 1993[7]	Bioterrorism attack that mimics naturally-occurring outbreak Unrecognized as bioterrorism at time Community-wide outbreak Common agent
Encephalitis (West Nile virus) New York City, 1999[8,9]	Zoonotic (birds are first victims) Initial diagnosis wrong Limited geographic area affected (humans) Specific population group affected (elderly humans) New agent to New York City Suspicion of bioterrorism

TABLE 6-2 Possible sources of health indicator surveillance data

Data Source	Pros	Cons and Confounders
Outpatient and ER visits	Reflects incidence of disease in general population	Nonspecific- May be difficult to document definitive information
ICU diagnoses	Best indicator of rare events like West Nile virus or Hantavirus pulmonary syndrome	Will not capture milder cases
OTC pharmacy sales	Reflects symptomatology most broadly	Subject to promotions/sales Nonspecific
Clinical lab submissions	Ordered by clinicians	May not be ordered for all (most) patients
Medicare or Medicaid claims	Ease of data capture	Problems with timeliness and accuracy Not broadly representative
Nursing homes	Reported by medical personnel Immobile population with limited exposure possibilities	Immobility reduces exposure potential Not broadly representative
Systematic testing for specific disease agents of specimens submitted to public health lab	Specificity of diagnoses	Broad screening not likely to capture meaningful data Difficulty getting information on positive samples Not timely
School and work absenteeism	May occur earlier than visits to clinician	May be absent for nonmedical reasons Delays in obtaining data
Ambulance call chief complaints	Many communities with timely access to data	Non-specific
Poison info calls	Ability to access real-time	May not be related to infectious diseases
HMO/Nurse hotline calls	Occur very early in disease outbreak	May be difficult to categorize

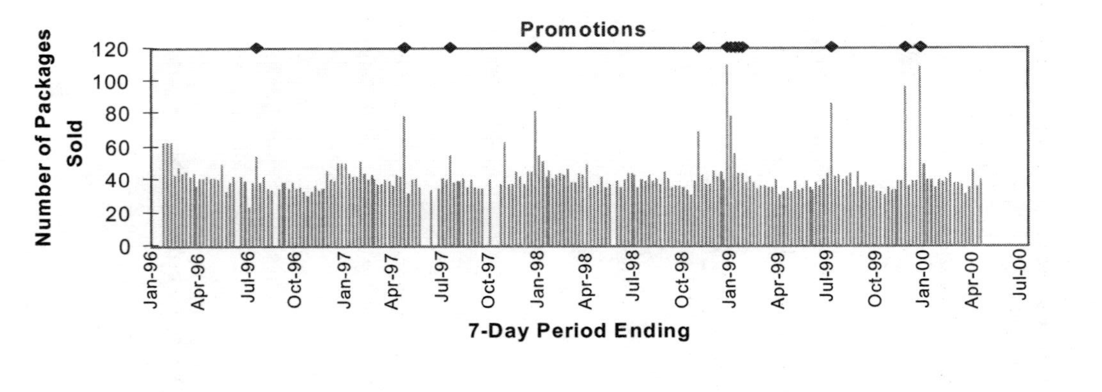

FIGURE 6-1 Average daily sales of anti-diarrheal medicine from a chain of 38 drugstores (Feb 96–Apr 00)

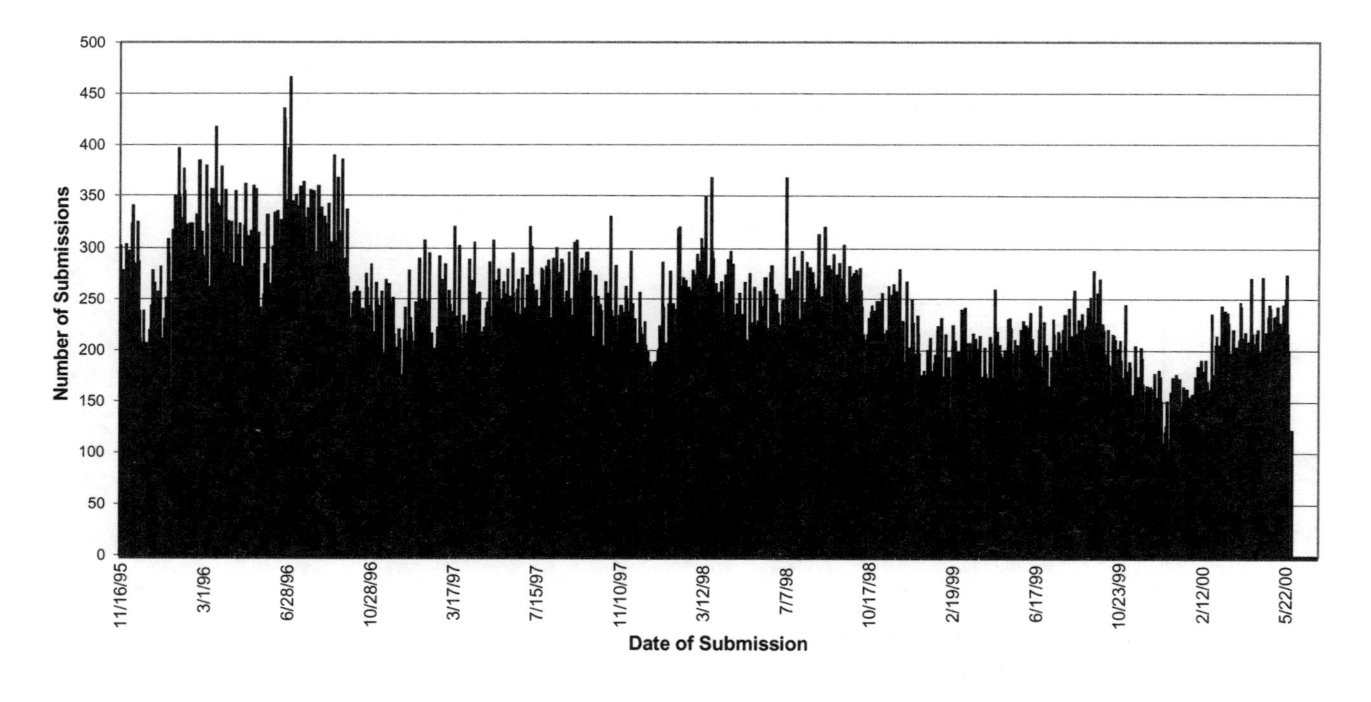

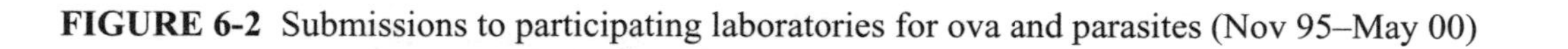
FIGURE 6-2 Submissions to participating laboratories for ova and parasites (Nov 95–May 00)

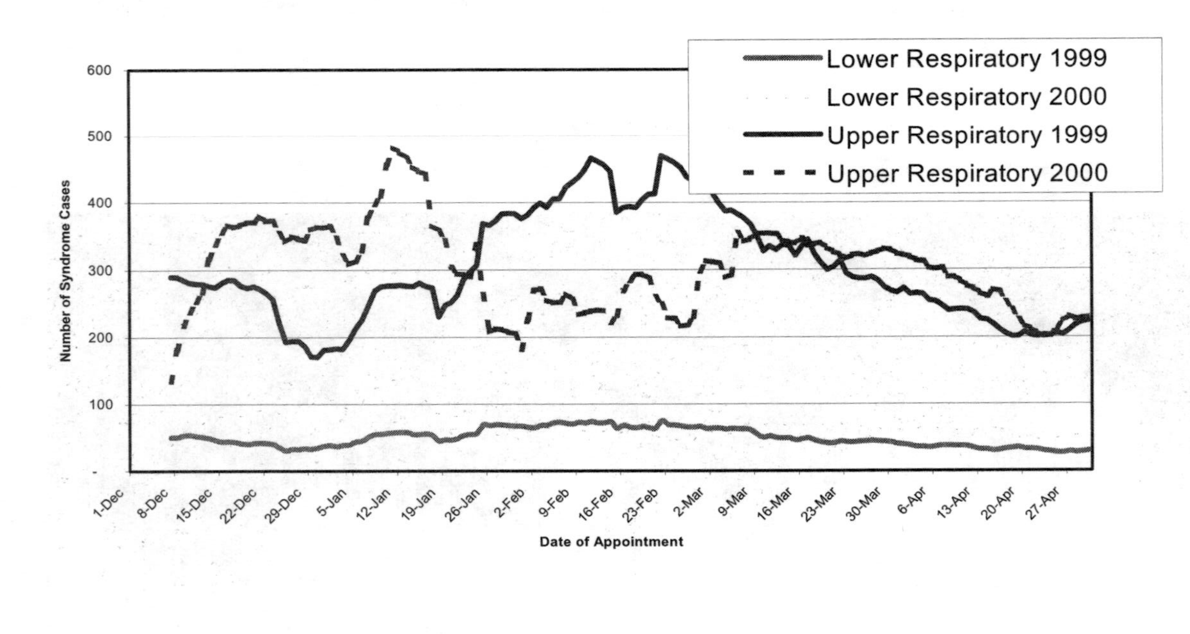

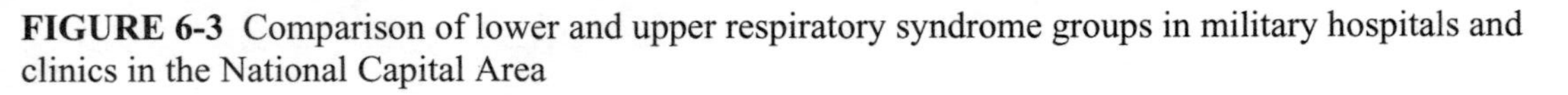

FIGURE 6-3 Comparison of lower and upper respiratory syndrome groups in military hospitals and clinics in the National Capital Area

THE USE OF COMPUTATIONAL MODELING IN RESPONDING TO BIOLOGICAL THREATS

Donald S. Burke, M.D.
Professor, Department of International Health
Director, Center for Immunization Research, Johns Hopkins Bloomberg School of Public Health

Computational modeling and simulation is an important but undervalued component of the scientific strategy to defend against emerging infectious diseases and bioterrorism. The extraordinary sustained growth in computational power has already given rise to ambitious modeling efforts in a wide variety of other scientific disciplines including nuclear physics, astronomy, economics, and other fields. My own experience in collaborations with computer scientists at the Navy Center for Applied Research in Artificial Intelligence has lead me to realize that computational approaches can productively be used to analyze and conceptualize the behavior of epidemic microbes. I suggest that we are fast approaching an era in which we will use computational modeling and simulation to guide public health policy decisions with regards to emerging infectious diseases and bioterrorism. We may be able to predict and prevent, rather than just respond to, epidemic infectious diseases.

A variety of computational approaches can be used. Using a "top/down" approach analytic techniques borrowed from physics can be used to study epidemiologic data. Fourier transforms and wavelet decomposition—standard tools in electrical engineering—can be used to decompose a temporal or spatio-temporal epidemiological data set into its various components. Decomposition in effect reduces noise and permits careful analysis of individual temporal and spatial components. Such techniques will probably prove useful in understanding environmental forcing of epidemics, such as the cyclical influence of weather and climate on disease transmission. Evolutionary and genetic algorithms, used routinely by computer scientists, can be used to model rapidly evolving organisms such as RNA viruses. All RNA viruses (HIV, influenza, Ebola) have very small genomes (total length approximately 10,000 nucleotides), so that full length genomic sequencing of viral isolates has become routine. We can now study in complete detail, reconstruct—and using evolutionary algorithms, simulate—the evolutionary patterns of these and other future threats. Analysis of the evolutionary tactics used by small "digital organisms" may permit discovery of yet unknown rules that govern real-world patterns of viral evolution. Another powerful computational approach comes from the field of socio-economics, known as "agent-based" modeling. This involves computational creation of populations of digital entities ("agents"), each of which can be thought of as a person. Sets of rules are written that govern the behaviors of each agent, either individually or as part of a group, such as movement patterns, likes and dislikes, lifespan, or even resistance and susceptibility to disease. Agents move about in two dimensional space and interact with other agents. Such models can be used to simulate and study the efficacy of public health interventions such as targeted immunization or quarantine measures. Finally, combining agent-based modeling with genetic algorithms may provide a novel strategy to understand—and some day predict—how microbes evolve and emerge as epidemic diseases.

My view is that we who work in the field of epidemiology are far behind other scientific disciplines in exploitation of computational techniques. Some of the current large computational modeling and simulation efforts in other scientific fields include:

- The Department of Transportation, in collaboration with the Department of Energy, is using agent-based modeling to model automobile traffic flows in major cities.
- The Virgo Consortium has used supercomputers to simulate the structure of the universe. In the simulation, each particle represents a galaxy. One can easily imagine that instead of a galaxy, each particle represents an RNA virus sequence navigating (evolving) through sequence space.
- Nuclear weapons are simulated at the Los Alamos National Laboratory's Strategic Computing Complex. At the complex, there are more than 300 nuclear weapons scientists, modelers, and designers working in a 300,000 square foot facility. Their simulations have a 30 trillion per second calculation capability, which results in 30 million pixel simulation displays. Such computational power could permit simulation of infectious disease epidemics with realistic features.

Computational modeling and simulation should be an important new area for initiatives in our efforts to confront emerging infectious diseases and bioterrorism. Clearly there are some excellent opportunities for developing some very creative new approaches. Arguably a new generation of models and simulations may prove to be the only way that we can predict—and hope to prevent—the emergence and epidemic spread of future infectious disease threats. At the very least, this type of modeling and simulation should have heuristic value—it should be extremely useful in teaching about infectious disease epidemiology, as well as in guiding the design and conduct of field research.

DIAGNOSTICS AND DETECTION METHODS: IMPROVING RAPID RESPONSE CAPABILITIES

David A. Relman, M.D.
Departments of Microbiology & Immunology, and Medicine
Stanford University

Ideally, one would like to be able to detect a bioterrorist attack early on, after an agent has been released but before there is any clinical evidence of pathology or overt damage. At this time point, there are opportunities for environmental detection, as well as opportunities for diagnosis of the exposed and infected but pre-symptomatic host. Intervention during this early preclinical incubation phase provides the greatest opportunity for benefit. However, it also poses the greatest challenges: current detection and diagnostic methods are not very useful at this point. It is usually not until a person becomes acutely or severely ill that available methods readily identify and characterize the nature of the attack. Thus, there are several issues related to diagnostics and detection that must be addressed in order to improve rapid response capabilities.

What needs to be done to improve rapid diagnostics and detection?

There are two aspects of diagnostics and detection—environmental and clinical—but their needs are very similar. Environmental detection requires rapid, semi-automated methods that enable detection of a broad spectrum of potential biological agents. These methods need to be standardized, rigorous, reproducible, and based on a thorough understanding of the natural background. Natural background refers to an agent's genetic variability, antigenic variability, geographic distribution, and how each of these changes over time. It also refers to non-biological aspects of the environment that could complicate and obfuscate detection efforts.

In the clinical setting, diagnostic methods should be rapid enough to reveal causative agents early in the course of disease. They should provide quantitative results, because sometimes the only difference between health and disease with respect to a particular agent is its relative abundance. These methods should be standardized and available at the point of care. As with environmental detection, accurate clinical diagnosis is based on a sound understanding of the natural background of the agent, including related agents, of the normal flora, and of non- or abiotic parts of clinical specimens that could potentially complicate these approaches.

What is the current status of available diagnostic and detection methods?

For environmental detection, there are reasonably good assays for a limited range of specific agents. These assays are based on cultivation, immunologic detection, and nucleic acid-based detection (e.g., PCR). However, current environmental detection systems are generally only capable of intermittent monitoring, plus they require an unacceptably high degree of hands-on maintenance. They are, by and large, designed for idealized conditions and settings that are very simple and free of the extraneous variables that characterize true natural backgrounds. Available information on natural backgrounds is spotty at best. Current protocols are nonstandardized or poorly standardized.

For clinical diagnosis, specific assays are available for most high priority agents. They are based on the same technologies as the environmental detection assays and work reasonably well in fairly idealized conditions. Their use is based on clinical suspicion; they are not generally used for automated implementation based upon broader or less specific information. As with available environmental detection methods, the collection protocols are largely nonstandardized or poorly standardized. And again, they are based on only spotty information about the natural background. In many cases they are too slow to be clinically useful.

Examples of current enabling procedures and technologies

There are some reasonably good collectors available that can sample large volumes of air over short periods of time by concentrating and impacting air content onto either a water filter or solid substrate; the collected material is then introduced into a detection scheme. The swab is another common collection device.

As has been learned from recent anthrax surveillance, swabbing and collecting in a standardized fashion from environmental surfaces and from the internal features of complicated three-dimensional objects is an extremely difficult problem. The most successful currently available collection devices are best suited for sampling air, which one would expect to be one of the simplest entities in the natural world to sample. But air is actually very complex. For example, black rubber particles in the air that come from car tires are a major PCR inhibitor. Still, air sampling methods are better than what is currently available for collecting and concentrating target in water, food, soil, and other materials.

Detection procedures still rely heavily on cultivation. When available, cultivated organisms allow for a more comprehensive evaluation of important phenotypes, than does molecular information generated by newer technologies. Although some phenotypes are not reliable for identification purposes, the availability of a pure culture facilitates acquisition of more reliable molecular information. Thus we should continue to consider ways of optimizing and enhancing our ability to recover organisms in culture in addition to other options that we choose to pursue. In theory, this method should allow for the detection of a single organism, but the process is slow. And such a high sensitivity can be problematic if the organism is part of the biological background.

There are many available immunology-based detection platforms, a number of which have received considerable attention in the recent press. Although the so-called "Smart Tickets" offer some true utility, they have major limitations. For example, they rely on a few standard, well-used, and well-tested antibodies. This dependency creates a potential critical vulnerability for detection and diagnostic methods in terms of specificity and reliability. Naturally-occurring epitope variability is only the first problem, as we begin to face the possibility of engineered organisms with deliberately modified epitopes. There are only a few developers venturing beyond a small set of well-characterized antibodies in their explorations of new immunologic platforms. Also, immunology-based assays typically have fairly high lower limits of detection, so sensitivity is another potential problem. The analysis time, however, is reasonably good.

There are a large number of nucleic acid detection technologies that work reasonably well in idealized settings. As with the immunological detection platforms, the fact that these assays generally rely on the sequences of only a few typed strains is a potential vulnerability. The sensitivity of nucleic acid-based assays, however, is much greater than that of immunology-based methods. The level of detection can be as low as a few sequence copies, bacteria, or viral particles.

What are some of the stumbling blocks and problems in the quest for better diagnostics and detection methods?

- The diversity of potential bioterrorist agents present a major challenge. This includes not only all of the naturally occurring pathogens scattered throughout the bacterial, viral, and eukaryotic worlds, but also all of the bioengineered, chimeric organisms that may not even yet exist. The immense variability within strains, let alone species, as well as varying degrees of relative abundance or evenness in nature make it very difficult to distinguish causal agents from other agents in the environment or host. With increasing use of the more sensitive PCR-based technologies, it is likely that the natural, variable biological background will show up more clearly. Additionally, the likelihood that even healthy individuals have a quantifiable amount of bacterial and other microbial DNA or RNA circulating through their blood and other sequestered anatomic compartments further complicates the issue.
- The causal nature of the association between detected agent and disease in an experimental subject is not easily established; nor are the methods and criteria needed for such an effort well-defined. Likewise, the correlation

between environmental detection of an agent and the risk of exposure and disease in a nearby host is a difficult proposition.

- The field is cluttered with unsubstantiated claims and ill-defined validation standards. Tests and methods are often validated for analytical, but not clinical performance characteristics. There are a number of eager vendors on the market who are selling detection and diagnostic kits that have not been well-validated.
- There are few standardized collection methods for complex, real-world specimens, such as the insides of computers.
- Laboratory surge capacity is inadequate, specimen analysis throughput is low, and turnaround times are slow.
- Delivery and implementation of state-of-the-art technologies is poor. For example, although rapid, real-time PCR offers major benefits, it is not available at the point of care and in places where validation needs to occur, such as in the environmental or clinical workplace.
- Because of limited sensitivity and inadequate attention to optimized specimen selection and processing, diagnostic methods are limited to the late stages of disease.

What can be done, and where can we be in five years?

Near-term goals include the following:

- We need a library of high affinity binding reagents for detection of a wide spectrum of biological agents, their variants and components. These reagents should include not just antibodies but also other high affinity ligands such as aptamers and peptide nucleic acids.
- We need an extensive sequence database effort. This is already underway, but we still need to consider its breadth, depth, and what kind of information we need to mine. For example, we still need broad range oligonucleotide sequences as primers or probes for a number of families of pathogens, viral and otherwise, for which we do not have good reagents currently.
- We need high throughput laboratories with much greater surge capacity. These laboratories might also be involved in methods and technology development. These could be dual purpose labs that are also used for other routine diagnostic purposes, e.g., influenza virus typing.
- We need to focus more effort on standardization and automation of specimen collection and processing in order to analyze more efficiently and accurately large numbers of samples. Devices that are currently being developed need to be validated with real world problems.
- We need to consider how to best apply some of the new biotechnologies, such as nanotechnology, microfluidics, and microarrays. Microarrays

in particular will likely have a major impact on strain and host typing; for example, a number of organism-specific whole genome microarrays have illustrated the power of genome-wide hybridization profiles for strain classification and potentially forensic strain typing. Recent developments in genomics suggest that we may now have the capability to perform genomics on a single bacterial cell, such that it will no longer be necessary to cultivate an organism from a clinical specimen or from the environment. DNA microarrays will also facilitate the identification and detection of diagnostic and prognostic host signatures. Expression analysis using DNA microarrays may prove valuable in diagnosing clinical disease and classifying infectious agents by comparing an individual's gene expression pattern to known pathogen associated patterns. But there are many unresolved issues, such as knowing which cells serve as the best source for signature data, how well and on what basis the host discriminates between different pathogens, and what role host specificity plays. In general, one hopes and expects that variability between healthy individuals is limited in comparison to the differences between healthy and diseased hosts, and that it will not obscure the recognition of a common infectious agent. But, the sources and nature of human host-specific "intrinsic-ness" must be defined.

- We need to consider how certain physics applications, such as hyperspectral analysis, may be relevant to infectious disease detection and diagnosis.

NATIONAL SECURITY AND OPENNESS OF SCIENTIFIC RESEARCH

Ronald M. Atlas, Ph.D.
Professor of Biology, University of Louisville
President Elect—American Society for Microbiology

The post September 11 anthrax attack has forced the recognition that bioterrorism is a reality of the 21st century. Infectious diseases pose grave threats to U.S. national security. Our response to these threats requires new and very focused research efforts along with enhancements to the public health infrastructure. Investments aimed at protecting against bioterrorism are best harmonized with the overall efforts to combat infectious diseases. Outbreaks of infectious diseases, whether naturally occurring or intentionally initiated, can represent threats to national and global security. Sustainable efforts that diminish the threat of bioterrorism and lessen the natural occurrences of infectious diseases (a dual function approach) will have the greatest long-term benefits.

Efforts to enhance detection and surveillance systems will be particularly valuable in improving our daily health, and that of our children, while at the same time protecting against acts of bioterrorism. Similarly, the discovery and development of new vaccines and antimicrobial drugs will be especially valuable in combating infectious diseases, especially with the continued emergence of antibiotic resistance, while helping protect against the potential use of antibiotic-resistant and genetically engineered agents by bioterrorists. Both antibiotic resistance and immune modulation must be considered given that a sophisticated bioterrorist could employ modern molecular approaches to design especially deadly biothreat agents. Broad spectrum antibacterials and antivirals could offer protection against a wide variety of infectious agents and could offer major protection against bioterrorism—this would offer broad protection against unknown biothreat agents while specific narrower spectrum drugs would be most appropriate when the exact nature of an agent had been determined.

Ensuring the vigor of research and development efforts to combat infectious diseases, with an appropriate focus on biothreats, will require an influx of new investments and the strategic redirection of some ongoing efforts through reallocation. If properly directed, the investments in bioterrorism response will be sustainable and will help diminish the overall threat to national and global security posed by infectious diseases.

As we move forward we need to ensure the health of the biomedical research enterprise. While protecting against inappropriate use, we need to ensure that legitimate scientists can access the materials, equipment, and information to move this biomedical agenda forward. We cannot let terrorists undermine our efforts to find new vaccines, drugs, and diagnostics in the battle against infectious disease.

Never before has the biomedical community faced greater challenges about protecting public health from the spread of infectious agents while facing increased scrutiny about the misuse of science by terrorists. *Are new mechanisms needed to govern scientific research so as to lessen the probability of the development of advanced biological weapons?* If so, what should be done? The National Academy of Sciences and the American Society for Microbiology (ASM) must assume leadership roles in fostering a public discourse that will ensure the advancement of science for the betterment of humankind and enhanced protection against bioterrorism. We must ensure that the biomedical community is assisted and not deterred from finding the diagnostic tools, vaccines, and medicinals needed to combat bioterrorism.

The scientific, biomedical, and public health communities must work with law enforcement to combat bioterrorism. But as Gerald Epstein says in his article entitled *Controlling Biological Warfare Threats: Resolving Potential Tensions Among the Research Community, Industry, and the National Security Community*, which will appear in the December issue of *Critical Reviews in Microbiology*, the research and national security communities have different objectives, cultures, and norms, and are likely to weigh the costs and benefits of proposed policy measures differently. It is not surprising, therefore, that these differences would sharpen and that there would be a greater need to seek resolution following the events of September 11 and the ensuing anthrax attacks. Epstein, who had worked in the Office of Science and Technology during the Clinton Administration raises the following issues, which I will discuss individually:

- Tightening restrictions on access to dangerous pathogens;
- Restricting access and dissemination of "relevant information," i.e., classifying research reports and censoring research publications; and,
- Imposing restrictions on the conduct of "contentious research," i.e., limiting fundamental biological or biomedical investigations that produce organisms or knowledge that could have immediate weapons implications.

> – *How can we better control access to potential biothreat agents?*
> – *Are there individuals that should not be permitted to conduct certain categories of research, or that should not be given access to dangerous pathogens?*
> – *What should be done about physical security at institutions that maintain cultures of potentially dangerous biological agents to help prevent unauthorized individuals from obtaining such agents?*
> – *Are locks enough or should armed guards be required to secure laboratories possessing select agents?*

ASM has supported imposing reasonable restrictions on access to select agents that pose high risks as potential biological weapons. It has supported legislation and regulation that control the exchange of certain dangerous pathogens including the CDC Laboratory Registration/Select Agent Transfer Program

These regulations, which place shipping and handling requirements on laboratory facilities that transfer or receive select agents capable of causing substantial harm to human health, are designed to ensure that select agents are not shipped to parties who are not equipped to handle them appropriately or who lack proper authorization for their requests. Under the regulations which have been in effect since April 15, 1997, the CDC regulates the shipment of 36 select agents. The list of agents was developed in consultation with the ASM. It includes agents that are especially dangerous, such as the agents that cause smallpox, anthrax, plague, and other deadly diseases. The regulations require adherence to CDC biosafety manual that includes various biosecurity measures. Thus, it begins to limit access. The ASM publicized Select Agent shipping regulations to the scientific community and has repeatedly exhorted microbiologists to adhere to all the regulatory requirements it imposes. But lest there be any excessive sense of security, it must be realized that these regulations apply only to U.S. facilities and that with the exception of smallpox, all the other select agents occur in nature.

ASM also supported the USA Patriot Act which was signed into law on October 26, 2001. This Act imposes restrictions on who may possess select agents, specifically restricting possession of select agents for aliens from countries designated as supporting terrorism and from individuals who are not permitted to purchase handguns, including some individuals with a history of mental illness or a criminal record. ASM felt that these were reasonable protective measures that would not have a significant adverse impact on biomedical research and that might increase national security by making it more difficult to obtain select agents by individuals who might misuse them. Although ASM sought provisions for exemptions Congress decided otherwise. Thus, the law contains no provision for exemptions under any circumstances

While prohibiting the possession of select agents for purposes that are not for bona fide research and other beneficial purposes, the USA Patriot act does not impose registration requirements for the possession of select agents for legitimate purposes ASM has supported such registration since 1999. On December 4, 2001, the Senate Appropriations Committee approved HR 3338, the DoD Appropriations Bill for FY 2002, which Senator Diane Feinstein (D-CA) and Senator Judd Gregg (R-NH) amended to include Section 8134 Regulation of Biological Agents and Toxins. The amendment they introduced was the same as Section 216 of S 1765, the Bioterrorism Preparedness Act of 2001, which was reintroduced by Senator Frist and Senator Kennedy on December 4, 2001. It also is similar to a House resolution introduced by Representative Tauzin. These Bills restate the CDC select agent transfer rules and require safeguards to prevent access to such agents and toxins for use in domestic or international terrorism or for any other criminal purpose. They mandate biennial updating of the select agent list which seems important given the pace of science. They mandate the imposition of regulations and standards for possession of select agents that ensure exclusion of individuals restricted by the USA Patriot Act and require registration for anyone possessing a select agent. Background checks would have to be conducted and steps might also be mandated to ensure that law enforcement could prevent suspected terrorists from gaining access to select agents. Appropriate security requirements for persons possessing, using, or transferring biological agents and toxins would be imposed and information would have to be provided, if available, that would facilitate the traceability of select agents if those agents were ever misused.

Beyond the laws and regulations that limit access to select agents, lies the question of blocking the dissemination of select information that might be useful to bioterrorists, what Epstein terms opacity and what others might call secrecy and censorship. Should more research be declared classified? Should there be criteria that would warrant restrictions on publication or other dissemination of research results? Should we stop revealing genomes? Should some aspects of research be withheld from publication, e.g., methods or selective results? Should there be review boards to consider the national security implications of all publications?

ASM has had to grapple with many of these questions, both before and after September 11. For example, among the information posted at ASM's website in an effort to provide relevant information were the abstracts of the 4th International Conference that was organized by scientists from the U.S. Army Medical Research Institute, the British Defense Research Agency, NIH, and the Pasteur Institute. One of the abstracts, In Vitro Selection and Characterization of High-Level Fluoroquinolone Resistance in *Bacillus anthracis*. By L. Price, A. G. Vogler, S. James, and P. Keim of Northern Arizona State University, described a study that showed increasing exposure to ciprofloxacin resulted in the evolution of fluoroquinolone resistance in *Bacillus anthracis*. This meant that antibiotic resistant *Bacillus anthracis* could be intentionally produced. Did it represent a roadmap for a bioterrorist? It also meant multiple antibiotic treatment was warranted in cases of inhalational anthrax. Thus did it present information useful to the biomedical community. Should the abstract have been published? Should it have been removed after September 11?

After some discussion we decided to leave this and other information about bioterrorism and anthrax on the ASM website for the education of the scientific community. My view was in favor of the benefits accrued from openness in science—if someone wished to publish legitimate research I did not want ASM to act as the censor. This position in favor of openness of science drew some concern as reported by Eric Lichtblau in his article *Response to Terror: Rising Fears That What We Do Know Can Hurt Us* that appeared in the Los Angeles Times on November 18, 2001. The article quoted University of Pennsylvania ethicist Arthur Caplan as saying, "We have to get away from the ethos that knowledge is good, knowledge should be publicly available, that information will liberate us...Information will kill us in the techno-terrorist age, and I think it's nuts to put that stuff on Websites."

The question of openness of science was considered by ASM years ago when it considered whether the smallpox virus genome should be classified or whether it should be published in the open scientific literature. ASM supported open publication and considered that the sequence data would be especially valuable to understand the virulence of smallpox virus and to provide targets for potential therapeutic drug design. Indeed analysis of the smallpox viral genome has revealed the basis for its virulence including the basis for immunomodulation. The genome analysis revealed targets for vaccine, drug, and detection development, information that seems to be of far more biomedical value than as an aid to bioterrorists.

The same sort of questions have been raised about the genome of *Bacillus anthracis*. In this case, the nucleotide sequence of plasmid pXO1 was published showing that it contains a "pathogenicity island," with the three toxin genes (cya, lef, and pagA), regulatory elements controlling the toxin genes, three germination response genes, 19 additional ORFs and 3 sequences that may encode enzymes responsible for the synthesis of a polysaccharide capsule usually associated with serotype-specific virulent streptococci. The conclusion was that major virulence elements of *Bacillus anthracis* are plasmid encoded. Despite the fact that the critical pathogenicity data was already published the question was raised about publishing the full genome which recently was completed. In fact some have questioned whether the genomic information already published should somehow be expunged from the open literature? So far the decision has been to continue to release genomic data although there continue to be expressions of concern. Given that the genomic data is viewed as relevant for the identification of targets for therapeutic drugs and vaccines, then it also can be viewed as relevant for identifying targets for increased virulence and the avoidance of current therapies, vaccines, and detection protocols. There is no doubt that there is a duality of potential good and evil encoded within genomes.

The same questions can be raised about other scientific findings, for example the recent demonstration by reverse genetics that a single mutation at position 627 in the PB2 protein of an H5N1 influenza A virus influenced the outcome of infection in mice, i.e., one mutation can greatly increase virulence. Thus, it may be much more simple to create more virulent biothreat agents than previously thought again raising the question of whether such information should have been revealed? Should journals censor such information in such articles? Should meetings remove abstracts and presentations of such information? Should sponsoring agencies require advance review of any presentations and publications so as to restrict release of such information? What would this do to the future of biomedical research in the United States, especially if we were the only country to begin to restrict the communication of scientific results that are of clear biomedical importance.

In an effort to respond to these questions I asked the editors of the 11 journals published by ASM to consider under what conditions they would consider rejecting a paper based upon ethical and national security issues. They responded that they did not want to publish papers that violated the ASM code of ethics nor those that violated other guidelines, including the NIH guidelines for recombinant organisms. They also were very sensitive to the national security implications but were not prepared to restrict the flow of legitimate scientific communications that were clearly aimed at our understanding of microbiology and that held potential for advancing biomedical science. After due deliberation they drafted for me the following statement: "The ASM recognizes that there are valid concerns regarding the publication of information in scientific journals that could be put to inappropriate use. The ASM hopes to participate in the public debate on these issues. Until a national consensus is reached, the rare manuscript that might raise such issues will be reviewed by the ASM Publications Board prior to the Society proceeding to publication." This statement with an accompanying introduction will be sent to all Editors of all ASM journals in order that they be alerted as to their responsibilities in this matter. Thus, the editors of ASM journals are trying to be responsible stewards of scientific information and communication by carefully balancing national security with the value of advancing science for the benefit of humanity.

Besides questions about communication of scientific information Gerald Epstein also discusses the possibility of constraining research, i.e., restricting researchers from conducting certain types of research. Are there areas of research or types of experimentation that should not be conducted at all? Are there others that should require advance approval? Is molecular biology a threat—will recombinant DNA technology be used to create horrific biothreat agents? Should certain molecular biology experiments and methodologies be prohibited?

Much of this concern emanates from experiments in which IL-4 genes were inserted into mousepox viruses. The result was suppression of the immune response to a much greater extent than anyone had predicted. Virus-encoded IL-4 not only suppresses primary antiviral cell-mediated immune responses but also can inhibit the expression of immune memory responses. A poxvirus can be simply genetically engineered for which immunization will be totally ineffective. The implications for possible genetic engineering of a horrific strain of smallpox virus are enormous. In hindsight some have asked whether this research should have been permitted? Shouldn't we have known in advance how dangerous the results might have been. Others who clearly were surprised by the results feel that this study alerts us to the need for more research on the immune response and antiviral drugs.

Yet other concerns have been raised about DNA shuffling because of its potential power to create new biothreat agents. Some point to the fact that this methodology potentially allows the rapid production of numerous biothreat agents with enhanced virulence. They raise the fear that DNA shuffling increases threat of being able to create a deadly new pathogen—intentionally or accidentally. They ask whether this methodology is too powerful and hence whether we should prohibit its use. But the potential benefits regarding new drug discoveries seem to outweigh these risks. In my view, the scientific community must move forward as quickly as possible in eliminating the threat of bioterrorism by finding effective preventative measures and cures so that infectious diseases are not a credible threat to humanity.

Beyond the obvious need to further biomedical research and to strengthen the public health infrastructure, one can ask about the appropriate role of the scientific community in identifying misconduct. What obligation do members of the research community have to identify, call attention to, or clarify activities of others that may appear suspicious?

There may be areas of research or types of experiments that pose such sensitivity regarding potential bioweapons application that merit extraordinary obligations for transparency and openness. There are clear aspects of bioethics that require scientists to be whistle blowers when public health is threatened.

Concerning the area of bioethics, the Council Policy Committee of ASM passed a resolution following September 11 affirming the longstanding position of the Society that microbiologists will work for the proper and beneficent application of science and will call to the attention of the public or the appropriate authorities misuses of microbiology or of information derived from microbiology. ASM members are obligated to discourage any use of microbiology contrary to the welfare of humankind, including the use of microbes as biological weapons. Bioterrorism violates the fundamental principles expressed in the Code of Ethics of the Society and is abhorrent to ASM and its members.

In conclusion I want to share some thoughts from Abigail Salyers, the current President of ASM: "Terrorism feeds on fear, and fear feeds on ignorance. The best defense against anthrax or any other infectious disease is information—information in a form that can be used by scientists and by members of the public to guide rational and effective actions to ensure public safety. Placing major new barriers in the path of the flow of information between scientists and between scientists and the public more likely may ultimately contribute to terrorism by interfering with our ability to prepare and to respond to the threat of the misuse of science by bioterrorists."

COORDINATING THE INTELLIGENCE, PUBLIC HEALTH, AND RESEARCH COMMUNITIES

Craig Watz*
Federal Bureau of Investigation

An intentional biological terrorism event requires a law enforcement response. Regardless of whether it was for political, social, or other reasons, the responsible individuals inflicted terror, committed an act of terrorism, and need to be aggressively pursued, investigated and prosecuted. Otherwise, there may be a repeat incident. Thus the role of the FBI in the bioterrorism arena.

The FBI's capability to apprehend bioterrorists is based on effective laws, federal statutes, and the ability to enforce these laws. Recent legal initiatives include an expansion of the Biological Weapons Anti-Terrorism act, which now applies to the possession of a biological agent that is beyond reasonable means for peaceful prophylactic protective or bona fide research. And, as of November 1, 2001, sentencing guidelines became effective such that anyone who does violate the WMD statute enters into a matrix to determine the sentence received. Prior to that, sentencing was at the discretion of the judge. The new guidelines were established in hope that more structured sentencing would serve as a stronger deterrence factor. During recent events, it has become clear that more consideration must be given to educating prosecuting attorneys and investigators in microbiology. It is also very important that we utilize available resources, including experts within the scientific community.

Equally important is educating the public health community, including public health and state epidemiologists, in how the FBI conducts their investiga-

* This statement reflects the professional view of the author and should not be construed as an official position of the Federal Bureau of Investigation.

tions. For example, it is important that the FBI collects environmental swabs and maintains a strict chain of custody in order to ensure that that evidence is in the same or similar condition at the time of trial.

The recent anthrax case has illustrated both a covert and an overt release which the FBI is still investigating. The D.C. incident is an example of an incident in which there existed a known crime scene so the FBI knew where to go and respond. The D.C. incident grew exponentially and also spread to the post offices and the Senate Building. Both types of releases have required the FBI to work very closely with public health and have illustrated the importance of communication, coordination, and the sharing of information—including intelligence information—between the FBI and public health.

Finally, recent events have stressed the need to minimize the overlap between federal, state and local agencies; and the need to set aside personal or agency agendas in order to work together to protect the public and hopefully prevent repeat incidents.

VIRTUALLY ASSURED DETECTION AND RESPONSE: UTILIZING SCIENCE, TECHNOLOGY, AND POLICY AGAINST BIOTERRORISM

Scott P. Layne, M.D., Ph.D.
Associate Professor, School of Public Health
University of California, Los Angeles

Homeland Security and the Biological Weapons Convention

The United States must control bioweapons threats on two major fronts. Domestically, it must seek new ways to boost homeland security and respond to terrorists attacks in several American cities. Internationally, it must seek new ways to overhaul the long stalemated Biological Weapons Convention (BWC) Protocol or propose an alternative way to establish legally binding verification and compliance procedures. The challenges are enormous and demand rapid, reliable, and complete information on which to make decisions.

The development of bioweapons requires three key elements: knowledge, equipment, and infectious agents. These elements have "dual uses" and thereby pose serious challenges to verification, compliance, and security. The general scientific knowledge required to develop bioweapons is conveyed in many microbiologic texts and is not feasible to remove. The United States seeks measures that thwart the migration of technical expertise and first-hand knowledge from past and present bioweapons programs. Likewise, the small-scale laboratory equipment required to create bioweapons is all but impossible to restrict. The United States supports regulations that block the export of industrial-scale laboratory equipment to potential proliferators. With the exception of *variola major* (smallpox), the various infectious agents required to create bioweapons are found in nature. The United States seeks regulations that constrains the sale of weaponizable seed stocks to qualified researchers and institutions.

Yet the United States is only one of many countries that supply such knowledge, equipment, and infectious agents. For example, *Bacillus anthracis* is the subject of research in many countries and conventional forensic methods may not be able to identify the source of *B. anthracis* used in any particular biological weapon. However, science and technology have opened up an extremely powerful means to address this problem. Infectious disease agents from specific origins exhibit unique molecular fingerprints that are all but impossible to erase (Jackson et al., 1998; Keim et al., 2000). These fingerprints are inherent to many, if not all, bioweapons agents on the A-List, including bacteria and viruses against humans and animals. It is therefore feasible to sequence the genes of such agents, organize that information in large databases, and use this molecular information to strengthen future BWC agreements and homeland security efforts. The elements of the plan are as follows.

Molecular Forensics

The United States, the world's leader in biotechnology, is in a position to create a new kind of high-throughput molecular forensics laboratory against bioweapons agents. Optimally, there would be two such facilities. The first would be domestically based, used to enhance homeland security, and serve as a model to states that are parties to the BWC. The second would be internationally based and offer improved verification and compliance capabilities to future BWC agreements. These two facilities could generate complementary and corroborative information.

A dedicated high-throughput laboratory against bioweapons agents would offer several important capabilities. First, it would enable exhaustive molecular fingerprinting and taxonomic positioning for a broad spectrum of known threat agents. Second, it would perform such analyses in a consistent and chain-of-custody manner. Third, it would produce high-resolution information within hours to days after sample receipt. In addition, the domestic facility could operate with a "closed" compartment, offering capabilities to the national and homeland security communities, and a separate "open" compartment, offering capabilities to the scientific community. The international facility could operate with capabilities and compartments established by future BWC agreements. Such arrangements would enable the United States to maintain its own molecular forensics and database capability yet share powerful testing methods and technologies with states that are parties to the BWC.

In 1992, the Australia Group identified nearly 100 bacteria, viruses, fungi, and toxins against people, animals, and plants with potentials for weaponization.

To date, however, only about 20 infectious agents have been used to produce biological weapons. A realistic goal would be therefore to fingerprint and catalog this "low hanging fruit." From a technical, economic, and political standpoint, the result would be to make it more difficult to mount and maintain a secret offensive bioweapons program.

Available Technologies

All the necessary technologies are available to build and operate a high-throughput molecular forensics laboratory and database system against bioweapons agents (Layne et al., 2001). More than a hundred companies manufacture the necessary equipment, which generally consist of flexible "plug-and-work" modules, and such technologies are often integrated into one of two kinds of system designs. The first are portable devices offering relatively simple and rapid tests. The second are high-throughput automation and robotic systems offering highly definitive tests. These larger systems must be housed in a semi-tractor trailer or suitable building, where samples must be brought to them. As outlined below, the optimal system would integrate both designs.

Several portable laboratory devices are available that fit into a suitcase and perform simple (yes/no) detection tests on the spot. The tests are based on polymerase chain reaction (PCR) methods and utilize tailored molecular primers against specific biothreat agents, such as *B. anthracis*. A larger set of primers is capable of screening for a larger list of biothreat agents. Such portable devices are able to detect very small traces of organisms but cannot actually sequence their genes. They often incorporate a personal computer to control and monitor tests, an Internet link to enable real-time data acquisition, and a global positioning device to automatically track locations. With such technologies, a trained individual can screen about two dozen samples per hour. To increase testing capacity, multiple devices can be deployed.

More definitive molecular forensics tests require more steps. A large assortment of automated and robotic equipment is available for this kind of work. Such industrial-scale technologies (e.g., robotic arms/conveyers, bar code readers, liquid handlers, incubators, genomic sequencers, flow cytometers, and image analyzers) are capable of performing all the procedures required by the proposed high-throughput molecular forensics laboratory. From a design standpoint, the various plug-and-work modules would be integrated into a flexible working system that could be upgraded with the latest commercial technologies. Incoming samples would follow an orderly flow, with different mass-analysis lines focusing on different biothreat agents. Because of automation and miniaturization, the entire facility (which permits the growing, extracting, sequencing, and archiving of samples) would fit into a surprisingly compact space that contains biohazardous materials and safeguards workers.

More sequence information is always better for molecular forensics, yet there are tradeoffs between laboratory productivity and definitive identifications. Complete viral genomes range in size from 10,000 to 300,000 DNA or RNA bases, whereas complete bacterial genomes range in size from 1,000,000 to 6,000,000 DNA bases. (In comparison, the human genome is composed of about 4,000,000,000 bases.) To fingerprint and taxonomically position biothreat virus, the molecular forensics laboratory would have to sequence and analyze 50 percent to 100 percent of each isolate's genome. On the other hand, to do this for biothreat bacteria, the laboratory would have to sequence only 5 percent to 10 percent of each isolate's genome. Current technologies would enable a high-throughput molecular forensics laboratory to sequencing about 10,000,000 bases per day. This would correspond roughly to fingerprinting and positioning about 500 viruses or 50 bacteria per day. Such procedures could be completed within hours or days after receiving samples.

The high-throughput laboratory would also be able to perform the simpler (yes/no) PCR-based tests described above. A surge capacity of 10,000 samples per day would be feasible with current technologies. At such rates, however, the limiting factors would be sample collection and transportation rather than rapid testing.

The high-throughput molecular forensics laboratory would generate a sizeable database within a few years. In addition to cataloguing molecular fingerprints, the laboratory would also be able to analyze the taxonomic position and natural genetic history of threat agents (genealogies). In reach-back and attribution scenarios, genealogies could prove to be more powerful than fingerprints alone. The most recent generation of teraflop computers, which can achieve speeds of 30×10^{12} calculations per second, would be well suited to analyze the threat agent database. Domestically, the goal would be to support decision-making processes and offer surge capacity for public health, emergency medical, agricultural, and law enforcement efforts. Internationally, the goal would be to support United States national security and intelligence operations as well as future BWC agreements. The toolbox for such undertakings includes currently available tracking, mapping, and modeling technologies.

Virtually Assured Detection And Response (VADAR)

The United States has mature policies to deter nuclear attacks, set forth as mutual assured destruction (MAD). It also has established policies to deter conventional attacks, set forth by the ability to fight on one or two major fronts and several minor fronts at once. But the United States has few well-developed policies to deter biological attacks. A high-throughput molecular forensics laboratory and database facility would help to fill this gap by enabling a new policy of virtually assured detection and response (VADAR) regarding biological attacks. The framework is as follows.

The collapse of the system of two opposing superpowers has led to an uncertain world order characterized by one global ultrapower, a majority of responsible governments, several rogue states, multiple religious fringe groups, and some shadowy international syndicates that are forming new networks and posing new challenges to global security. Today, at least 17 countries are known to be developing or producing bioweapons and the list may be expanding.

The scale of global trade also poses a major challenge. For example, more than 14,000 loaded 40-foot marine containers enter the United States each day (Flynn, 2000). Containers routinely travel through the country before reaching a port of entry and the system tracking their intended course and location is rudimentary. Furthermore, few containers undergo any form of inspection and, even when this occurs, specialized inspection technologies are rarely used. The ease of smuggling bioweapons constitutes a significant threat to homeland security, in part because the problems associated with marine containers represents only the "tip of the iceberg" in our leaky border controls.

The foot-and-mouth disease outbreak in Great Britain and Europe, where the economic loss is estimated above £25 billion, reflects another aspect of the problem. Current methods of disease control, which rely on veterinarians inspecting animals for signs of infection, collecting mucosal and blood samples, and analyzing them with manual laboratories, have cycle times of three to five days. Foot-and-mouth disease can spread from one location to another, however, in far less time. Consequently, the current system with manual laboratories cannot support science-based decisions on quarantine zones, animal destruction, and resource allocation. At the heart of the problem is a lack of rapid, accurate, and complete information on which to make dependable decisions. A quantum leap in threat agent surveillance and data analysis is needed.

In a bioattack on the United States, as few as 50 sickened people in one major city could stretch public health, emergency medical, and law enforcement services beyond local capabilities. Larger attacks involving major metropolitan areas would be overwhelming and require the delivery of tons of antibiotics to exposed persons within days, challenging national capabilities. A coherent program that strengthens homeland security thus requires sizeable laboratory and informatic resources that can be organized in terms of four overall phases.

First, in preventing attacks, the United States would rely on the ability to fingerprint and catalogue bioweapons agents with high-throughput technologies. An extensive database of molecular fingerprints and associated origins would offer a new means of rapid attribution and therefore deterrence. It would put rogue states, religious fringe groups, and international syndicates on notice that there is little chance to evade blame for bioattacks.

Second, in the unfortunate event of an attack, public health laboratories would be overwhelmed simply because there would be too many samples to analyze quickly. Manual laboratories would be unable to answer even the simplest questions: Is the agent present? How many different infectious agents were released? How do they differ? What are the best initial therapies to treat those afflicted and exposed? Information from high-throughput laboratories would reduce confusion and save lives by offering rapid testing in acute situations.

Third, in the aftermath of an attack, public health, agricultural, and law enforcement officials would need accurate answers to another set of questions. What are the geographic boundaries of each infectious agent? What are their stabilities? What are the effects on animals and plants? Information from high-throughput laboratory and mapping systems would speed the recovery process by offering testing for cleanup and investigatory operations.

Fourth, in response to the attack, law enforcement officials must collect evidence in accordance with chain of custody procedures. Intelligence agencies and military services must make accurate attributions and take swift actions to protect national security. Information from high-throughput laboratories and their associated databases could prevent further attacks by rapidly pinpointing suspected sources.

The relatively small anthrax attacks in a few American cities flooded the bioterrorism response network. Thousands of samples were sent to a patchwork of state and federal laboratories which, at best, were equipped to handle about 100 samples per day (Kahn et al., 2000). Even with many laboratories working around-the-clock, they could not keep pace with emergency testing demands.

Implementation

Strengthening homeland security against bioterrorism needs enhanced public health and emergency medical preparedness at home and expanded human intelligence capabilities abroad. Moving beyond the BWC Protocol stalemate requires reliable disclosure of dual use facilities, timely inspection of suspicious programs, and systematic testing for certain (i.e., a short A-List) weaponizable agents. The common element among such undertakings is rapid, complete, and reliable information on which to make assessments and decisions.

A high-throughput molecular forensics laboratory and database facility would cost several hundred million dollars to build and operate over the first five years. Since the needed technologies already are available, it could be operational within two years.

Such a facility could be operated under the newly created Homeland Security Council. The mission of this new national medical forensics and intelligence support laboratory would be to complement and cooperate with existing government agencies such as health, agriculture, emergency management, justice, defense, intelligence, and the national laboratories. It would support public health, law enforcement, and homeland security programs without usurping their long-established missions. It would provide needed surge capacity in the acute and cleanup phases of terrorist bioattacks. It would also have mechanisms to support certain scientific and technical research.

In building the first molecular forensic laboratory against bioweapons agents, the overall testing methods and high-throughput capabilities would be shared with the scientific community. The design of certain molecular primers against specific biothreat agents and resulting fingerprint and genealogies, however, would be available to the national and homeland security communities only. Such open architectures would facilitate the development a second internationally-based laboratory that parallels the initial design.

Conclusion

In the aftermath of the terrorist attacks on the World Trade Center and Pentagon and organized anthrax attacks in several American cities, there has been renewed debate on the risks of further biological attacks. At present, the risk remains unclear. Yet it is clear that terrorist attacks have become more spectacular and lethal and have now reached our homeland soil. The question is: When will the shift to more devastating forms of bioterrorism take place? The United States now has the opportunity to organize effective prevention, deterrence, and response measures.

The United States must also act on domestic and international fronts. In mitigating bioterrorism, is VADAR a perfect solution? *No*. Is it an improvement over existing methods and policies? *Yes*. Is it possible to circumvent? *Yes*. But with secret offensive bioweapons programs possibly assisting organized terrorism, can we afford to wait?

RESEARCH AND THE PUBLIC HEALTH RESPONSE

Eric Eisenstadt,[*] Ph.D.
Defense Advanced Research Projects Agency

Technology could help public health enormously; but to help focus the development of technology for public health (as well as for the FBI and other law enforcement agencies who cope with forensic issues that resemble the diagnostic ones faced by public health), the public health community needs to articulate its technology needs. Once these needs are defined, then the science and technology communities, including funding organizations such as (DARPA), can begin to define the science and technology programs required to develop the desired capability. In this way, bridges can be built between public health and the technical community. Indeed, agencies like DARPA are very good at assembling the kind of interdisciplinary scientific and technical efforts—involving academia, industry, and government laboratories—that are required to develop new capabilities. Suppose, for example, the case could be made for a routine molecular

[*] This statement reflects the professional view of the author and should not be construed as an official position of the Defense Advanced Research Projects Agency.

diagnostic capability that would provide a point-of-care physician with the information needed to make a decision about which treatment to prescribe within 30 minutes of taking a blood sample from a patient. A research and development effort might then be mounted to develop this new diagnostic capability by assembling researchers from the appropriate technical and user communities—e.g., molecular biology, materials science, signal processing, and clinical microbiology—to work together to create a new technology. The challenge would be enormous but the magnitude of the development effort will be a strong function of how strongly the case had been made for doing it in the first place.

Genomics-based technologies, for example, have great potential for improving public health. Fulfillment of this potential would be accelerated if the public health community participated in developing a vision of how the application of genomics information could enhance health care. Such a vision might serve to rally the nation to develop technological capabilities that enhance our ability to cope with many of the bioterrorism response and preparedness issues that have been identified in our discussions.

During World War II, for example, it was recognized that radar had tremendous potential for identifying U boats. The proof of principle had been done, but the technology still needed to be developed. A vision of what radar might be capable of doing for the military led to the initiation of the radar program at MIT from which great science and technology emerged including the foundations of the microelectronics industry.

Finally, it is very difficult to bound all of the bioterrorism response capabilities that have been discussed during this workshop. There are simply too many imaginable bioterrorist scenarios (multiple agents and multiple ways to create mischief with them). We do not have sufficient resources to address an unbounded set of problems. So we must try in some rational way to bound bioterrorism and define the set of bioterrorism issues that need to be addressed. We must focus and develop a big vision that the country can respond to. For example, why not identify as a national goal the removal of infectious disease as a public health threat? This does not mean that we need to define how to eliminate infectious disease. When, a few hundred years ago, the British parliament recognized the need "to find longitude" they didn't know how it was going to be done. But by crisply stating the problem and offering a prize to the one who solved it, some fantastic science and technology emerged. Could we not rally the country behind a campaign to eliminate the infectious disease threat?

REFERENCES

P. Kelly:

Centers for Disease Control and Prevention (CDC). 1999. Update: West Nile-like viral encephalitis – New York, 1999. *Mortality and Morbidity Weekly Record* 48:890–892.

Collin JF, Melet JJ, Mortot M, Foliquet JM. 1981. Eau d'adduction publique et gastro-enterites en Meuthe-et-Moselle. *Journal of Français d'Hydrobiologie* 35:155–174.

Jarroff L. 1999. Of mosquitoes, dead birds and epidemics. *Time* 154(15):74–75.

Miller JR, Mikol Y. 1999. Surveillance for diarrheal disease in New York City. *Journal of Urban Health* 76:388–390.

Pavlin JA, Mostashari F, Kortepeter MG, Hynes NA, Chotani RA, Mikol YB, Ryan MAK, Neville JS, Gantz DT, Writer JV, Florance JE, Culpepper RC, Henretig FM, Kelley PW. 2001. Innovative Surveillance Methods for Rapid Detection of Emerging Infections—The Results of an Inter-Agency Workshop on Health Indicator Surveillance.

Proctor ME, Blair KA, Davis JP. 1998. Surveillance data for waterborne illness detection: an assessment following a massive waterborne outbreak of *Cryptosporidium* infection. *Epidemiology and Infection* 120:43–54.

Rodman JS, Frost F, Davis-Burchat L, Fraser D, Langer J, Jakubowski W. 1997. Pharmaceutical sales a method of disease surveillance. *Journal of Environmental Health* 60:8–14.

Rodman JS, Frost F, Jakubowski W. 1998. Using nurse hot line calls for disease surveillance. *Emerging Infectious Diseases* Apr–Jun;4(2):329–32.

Steele KE, Linn MJ, Schoepp RJ, Steele KE, Linn MJ, Schoepp RJ, Komar N, Geisbert TW, Manduca RM, Calle PP, Raphael BL, Clippinger TL, Larsen T, Smith J, Lanciotti RS, Panella NA, McNamara TS. 2000. Pathology of fatal West Nile virus infections in native and exotic birds during the 1999 outbreak in New York City, New York. *Vet Pathol* 37:208–224.

Thacker SB. 1994. Historical development. In: Teutsch SM and Churchill RE eds. *Principles and Practice of Public Health Surveillance*. New York, NY: Oxford University Press 8–9.

Thacker SB, Berkelman RL. 1988. Public health surveillance in the United States. *Epidemiology Review* 10:164–90.

S. Layne:

Flynn SE. 2000. Beyond Border Control. *Foreign Affairs* 79: 57–68.

Jackson PJ, Hugh-Jones, ME, Adair DM, Green G, Hill KK, Kuske CR, Grinberg, LM, Abramova, FA, and Keim P. 1988. PCR analysis of tissue samples from the 1979 Sverdlovsk anthrax victims: The presence of multiple *Bacillus anthracis* strains in different victims. *Proceedings of the National Academy of Sciences, USA* 95: 1224–1229.

Kahn AS, Morse S, and Lillibridge S. 2000. Public health preparedness for biological terrorism in the USA. *Lancet* 356: 1179–1182.

Keim P, Price LB, Klevytska AM, Smith KL, Schupp JM, Okinaka R, Jackson, PJ, and Hugh-Jones ME. 2000. Multiple-locus variable-number tandem repeat analysis reveals genetic relationships within *Bacillus anthracis. Journal of Bacteriology* 182: 2928–2936.

Layne SP, Beugelsdijk TJ, and Patel CKN (editors). 2001. Firepower in the Lab: Automation in the Fight Against Infectious Diseases and Bioterrorism. Washington, D.C.: Joseph Henry Press.

Appendix A

Biological Threats and Terrorism: How Prepared Are We? Assessing the Science and Our Response Capabilities

November 27–29, 2001

Lecture Room
National Academy of Sciences
2101 Constitution Avenue, NW
Washington, DC 20418

AGENDA

TUESDAY, NOVEMBER 27, 2001

8:30 am Continental Breakfast

9:00 **Welcome and Opening Remarks**
Adel Mahmoud, Chair, Forum on Emerging Infections
President, Merck Vaccines
Stanley Lemon, Vice-Chair, Forum on Emerging Infections
Dean of Medicine, The University of Texas Branch at Galveston

9:15 **Framing the Debate: Real-Time Considerations for Addressing Bioterrorism**
Hon. William Frist, United States Senate
Anthony Fauci, National Institute of Allergy and Infectious Diseases
Edward Eitzen, U.S. Army Medical Research Institute of Infectious Diseases
James Hughes, Centers for Disease Control and Prevention
Michael Osterholm, University of Minnesota
Margaret Hamburg, Nuclear Threat Initiative

Session I: Assessing Our Understanding of the Threats

Moderator: Joshua Lederberg, The Rockefeller University

10:15 **Anthrax**
Arthur Friedlander, U.S. Army Medical Research Institute of Infectious Diseases

10:45 **Smallpox**
Peter Jahrling, U.S. Army Medical Research Institute of Infectious Diseases

11:15 **Tularemia and Plague**
David Dennis, NCID, Centers for Disease Control and Prevention

11:45 **Botulinum Toxin**
Stephen Arnon, Infant Botulism Treatment and Prevention Program, California Department of Health Services

12:15 pm **Innovative Surveillance Methods for Monitoring Dangerous Pathogens**
Patrick Kelley, Walter Reed Army Institute for Research

1:00 **Lunch**

Session II: Vaccines: Development, Production, Supply, and Procurement Issues

Moderator: Carole Heilman, National Institute of Allergy and Infectious Diseases

2:00 **Vaccines for Threatening Agents: Ensuring the Availability of Countermeasures to Bioterrorism**
Philip Russell, Office of the Secretary, Department of Health and Human Services

2:30 **The Department of Defense and the Development and Procurement of Vaccines Against Dangerous Pathogens: A Role in the Military and Civilian Sector?**
Anna Johnson-Winegar, Office of the Secretary of Defense, Chemical and Biological Matters

3:00 **Applications of Modern Technology to Emerging Infections and Vaccine Development**
Gary Nabel, National Institute of Allergy and Infectious Diseases, Vaccine Research Center

3:30 **Meeting the Regulatory and Product Development Challenges for Vaccines and Other Biologics to Address Terrorism**
Jesse Goodman, Center for Biologics Evaluation and Research, Food and Drug Administration

Session III: Discussion Panel of Biological Threats and the Research Implications

4:30 **Moderator: Adel Mahmoud,** Merck Vaccines
Donald Burke, Johns Hopkins University
Stanley Plotkin, Aventis Pasteur
Ken Alibek, Hadron, Inc.

6:30 Adjournment of the first day

WEDNESDAY, NOVEMBER 28, 2001

8:00 am Continental Breakfast

8:30 **Opening Remarks / Summary of Day 1**
Stanley Lemon, Vice Chair, Forum on Emerging Infections

Session IV: The Research Agenda and Emerging Technologies

Moderator: Gail Cassell, Eli Lilly and Company

9:30 **The Role of Antivirals**
C.J. Peters, University of Texas-Galveston

10:00 **New Research in Antitoxins**
John Collier, Harvard Medical School

10:30 **Recombinant Human Antibody: Immediate Immunity for Botulinum Neurotoxin and Other Class A Agents**
James Marks, University of California, San Francisco

11:00 **Diagnostics and Detection Methods: Improving Rapid Response Capabilities**
David Relman, Stanford University

11:30 **Meeting the Regulatory and Product Development Challenges for Drugs to Address Terrorism**
Andrea Meyerhoff, Office of the Commissioner, Food and Drug Administration

12:30 pm **Q & A Session/Working Lunch**

Session V: The Response Infrastructure

Moderator: Michael Osterholm, University of Minnesota

1:30 **Lessons Being Learned: The Challenges and Opportunities**
Scott Lillibridge, Office of the Secretary, Department of Health and Human Services
Julie Gerberding, Centers for Disease Control and Prevention
Bradley Perkins, Centers for Disease Control and Prevention
Kevin Yeskey, Centers for Disease Control and Prevention

3:00 **The Progress, Priorities, and Concerns of Public Health Laboratories**
Mary Gilchrist, University Hygienic Laboratory, Iowa

3:30 **Centers for Public Health Preparedness**
Stephen S. Morse, Mailman School of Public Health, Columbia University

4:00 **The Role of Coordinated Information Dissemination: The CASCADE program in the United Kingdom**
John Simpson, Head of Emergency Planning Co-ordination Unit, Department of Health, UK

4:30 **The Legal Infrastructure for an Effective Public Health Response**
David Fidler, Indiana University School of Law

Session VI: Discussion Panel of the Spectrum of Research and Public Health Responses

5:00 **Moderator: James Hughes,** NCID, Centers for Disease Control and Prevention
Michael Ascher, Office of the Secretary, Department of Health and Human Services
Craig Watz, Federal Bureau of Investigation
Donald Wetter, U.S. Public Health Service
David Shlaes, Wyeth-Ayerst Research

Jerry Gibson, South Carolina Department of Health and Environmental Control
Eric Eisenstadt, Defense Advanced Research Projects Agency
Renu Gupta, Novartis

6:30 **Closing Remarks /Adjournment**
Stanley Lemon, Vice-Chair, Forum on Emerging Infections

THURSDAY, NOVEMBER 29, 2001

8:30 am **Continental Breakfast**

9:00 **Opening Remarks**
Adel Mahmoud, Chair
Stanley Lemon, Vice-Chair

Priorities for the Next Steps in Countering Bioterrorism

9:15 am **Panel Discussion**
Moderator: Fred Sparling, UNC-Chapel Hill
Panelists:
D.A. Henderson, Office of the Secretary, Department of Health and Human Services
Ruth Berkelman, Emory University
Scott Layne, UCLA
Susan Maslanka, Centers for Disease Control and Prevention
Kristi Koenig, Department of Veterans Affairs
Tom Milne, National Association of County and City Health Officials
Ronald Atlas, University of Louisville, President-Elect, American Society of Microbiology

11:15 **Round-the-Table Discussion (Lunch will be served)**

2:00 pm **Closing Remarks/Adjournment**

Appendix B

Information Resources

PUBLISHED LITERATURE

Overview

Carus WS. 2001. The Illicit Use of Biological Agents since 1990. Working Paper: Bioterrorism and Biocrimes. Center for Counterproliferation Research. Washington, DC: National Defense University.

Christopher GW, Cieslak TJ, Pavlin JA, Eitzen EM Jr. 1997. Biological warfare. A historical perspective. JAMA 278(5):412–417. Available at: http://jama.ama-assn.org/issues/v278n5/ffull/jsc7044.html.

Fauci AS. 2001. Infectious diseases: Considerations for the 21st century. Clinical Infectious Diseases 32(5):675–685. Available at: http://www.journals.uchicago.edu/CID/journal/contents/v32n5.html.

Franz DR, Zajtchuk R. 2000. Biological terrorism: Understanding the threat, preparation, and medical response. Disease-a-Month 46(2):125–190.

Hamburg MA. 2000. Bioterrorism: A challenge to public health and medicine. Journal of Public Health Management and Practice 6(4):38–44.

Hawley RJ, Eitzen EM. 2001. Biological weapons: A primer for microbiologists. Annual Review of Microbiology 55:235–253. Available at: http://micro.annualreviews.org/cgi/content/full/55/1/235.

Henderson DA. Bioterrorism. International Journal of Clinical Practice Supplement 115:32–36.

Kortepeter MG, Cieslak TJ, Eitzen EM. 2001. Bioterrorism. Journal of Environmental Health 63(6):21–24.

Lane HC, Fauci AS. 2001. Bioterrorism on the home front: A new challenge for American medicine. JAMA 286(20):2595–2597. Available at: http://jama.ama-assn.org/issues/v286n20/fpdf/jed10079.pdf.

Anthrax

Bradley KA, Mogridge J, Mourez M, Collier RJ, Young JA. 2001. Identification of the cellular receptor for anthrax toxin. Nature 414(6860):225–229. Available at: http://www.nature.com/nature/anthrax/.

Inglesby TV, Henderson DA, Bartlett JG, Ascher MS, Eitzen E, Friedlander AM, Hauer J, McDade J, Osterholm MT, O'Toole T, Parker G, Perl TM, Russell PK, Tonat K. Working Group on Civilian Biodefense. 1999. Anthrax as a biological weapon: Medical and public health management. JAMA 281(18):1735–1745. Available at: http://jama.ama-assn.org/issues/v281n18/ffull/jst80027.html.

Pannifer AD, Wong TY, Schwarzenbacher R, Renatus M, Petosa C, Bienkowska J, Lacy DB, Collier RJ, Park S, Leppla SH, Hanna P, Liddington RC. 2001. Crystal structure of the anthrax lethal factor. Nature 414(6860):229–233. Available at: http://www.nature.com/nature/anthrax/.

Pearson, H. October 24, 2001. Anthrax action shapes up. Nature. Online. Available at: www.nature.com/nsu/011025/011025-9.html. Accessed October 26, 2001.

Smallpox

Cohen J. 2001. Bioterrorism. Smallpox vaccinations: How much protection remains? Science 294(5544):985. Available at: http://www.sciencemag.org/cgi/content/full/294/5544/985.

Henderson DA, Inglesby TV, Bartlett JG, Ascher MS, Eitzen E, Jahrling PB, Hauer J, Layton M, McDade J, Osterholm MT, O'Toole T, Parker G, Perl T, Russell PK, Tonat K. Working Group on Civilian Biodefense. 1999. Smallpox as a biological weapon: Medical and public health management. JAMA 281(22):2127–2137. Available at: http://jama.ama-assn.org/issues/v281n22/ffull/jst90000.html.

Other Dangerous Pathogens

Arnon SS, Schechter R, Inglesby TV, Henderson DA, Bartlett JG, Ascher MS, Eitzen E, Fine AD, Hauer J, Layton M, Lillibridge S, Osterholm MT, O'Toole T, Parker G, Perl TM, Russell PK, Swerdlow DL, Tonat K. Working Group on Civilian Biodefense. 2001. Botulinum toxin as a biological weapon: Medical and public health management. JAMA 285(8):1059–1070. Available at: http://jama.ama-assn.org/issues/v285n8/ffull/jst00017.html.

Centers for Disease Control and Prevention. 2001. National Report on Human Exposure to Environmental Chemicals. Atlanta: Centers for Disease Control and Prevention. Available at: http://www.cdc.gov/nceh/dls/report/PDF/CompleteReport.pdf.

Dennis DT, Inglesby TV, Henderson DA, Bartlett JG, Ascher MS, Eitzen E, Fine AD, Friedlander AM, Hauer J, Layton M, Lillibridge SR, McDade JE, Osterholm MT, O'Toole T, Parker G, Perl TM, Russell PK, Tonat K. Working Group on Civilian Biodefense. 2001. Tularemia as a biological weapon: Medical and public health management. JAMA 285(21):2763–2773. Available at: http://jama.ama-assn.org/issues/v285n21/ffull/jst10001.html.

Inglesby TV, Dennis DT, Henderson DA, Bartlett JG, Ascher MS, Eitzen E, Fine AD, Friedlander AM, Hauer J, Koerner JF, Layton M, McDade J, Osterholm MT, O'Toole T, Parker G, Perl TM, Russell PK, Schoch-Spana M, Tonat K. Working Group on Civilian Biodefense. 2000. Plague as a biological weapon: Medical and public health management. JAMA 283(17):2281–2290. Available at: http://jama.ama-assn.org/issues/v283n17/ffull/jst90013.html.

Vaccines

Cohen J, Marshall E. 2001. Bioterrorism: Vaccines for biodefense: A system in distress. Science 294(5542):498–501. Available at: http://www.sciencemag.org/cgi/content/full/294/5542/498.

National Vaccine Advisory Committee. 1997. United States vaccine research: A delicate fabric of public and private collaboration. Pediatrics 100(6):1015–1020. Available at: http://www.pediatrics.org/cgi/content/full/100/6/1015.

Russell PK. 1999. Vaccines in civilian defense against bioterrorism. Emerging Infectious Diseases 5(4):531–533. Available at: http://www.cdc.gov/ncidod/EID/vol5no4/russell.htm.

U.S. Department of Defense. July 2001. Report on Biological Warfare Defense Vaccine Research and Development Programs. Available at: http://www.defenselink.mil/pubs/ReportonBiologicalWarfareDefenseVaccineRDPrgras- July2001.pdf.

Vandersmissen W. 1992. Availability of quality vaccines: The industrial point of view. Vaccine 10(13):955–957.

Widdus R. 2001. Public-private partnerships for health: Their main targets, their diversity, and their future directions. Bulletin of the World Health Organization 79(8):713–720. Available at: http://www.who.int/bulletin/pdf/2001/issue8/vol79.no.8.713-720.pdf.

Antimicrobials

Barrett, A. November 5, 2001. How to get pharma's big guns aimed at microbes. Business Week, pp. 40–41.

Cassell GH, Mekalanos J. 2001. Development of antimicrobial agents in the era of new and reemerging infectious diseases and increasing antibiotic resistance. JAMA 285(5):601–605. Available at: http://jama.ama-assn.org/issues/v285n5/ffull/jsc00411.html.

Wheeler C, Berkley S. 2001. Initial lessons from public-private partnerships in drug and vaccine development. Bulletin of the World Health Organization 79(8):728–734. Available at: http://www.who.int/bulletin/pdf/2001/issue8/vol79.no.8.728-734.pdf.

Emerging Discovery and Technologies

Dubensky TW Jr, Liu MA, Ulmer JB. 2000. Delivery systems for gene-based vaccines. Molecular Medicine 6(9):723–732.

Gu ML, Leppla SH, Klinman DM. 1999. Protection against anthrax toxin by vaccination with a DNA plasmid encoding anthrax protective antigen. Vaccine 17(4):340–344.

Klinman DM, Verthelyi D, Takeshita F, Ishii KJ. 1999. Immune recognition of foreign DNA: A cure for bioterrorism? Immunity 11(2):123–129.
Liu MA, McClements W, Ulmer JB, Shiver J, Donnelly J. 1997. Immunization of non-human primates with DNA vaccines. Vaccine 15(8):909–912.
Mourez M, Kane RS, Mogridge J, Metallo S, Deschatelets P, Sellman BR, Whitesides GM, Collier RJ. 2001. Designing a polyvalent inhibitor of anthrax toxin. Nature Biotechnology 19(10):958–961. Available at: http://www.nature.com/nature/anthrax/.

Safety and Regulatory Challenges

Kolata G. November 13, 2001. Bioterror drugs stall over rules and logistics. New York Times. Online. Available at: www.nytimes.com. Accessed November 15, 2001.
LEADER: When drugs can be too safe. November 7, 2001. Financial Times. Online. Available at: www.news.ft.com. Accessed November 15, 2001.
Michaels A, Dyer G. November 7, 2001. U.S. may tighten guidelines on drug approvals. Financial Times. Online. Available at: www.news.ft.com. Accessed November 8, 2001.
Pollack A. November 13, 2001. Antibiotics business is again popular. New York Times. Online. Available at: www.nytimes.com. Accessed November 15, 2001.
Smith HA, Klinman DM. 2001. The regulation of DNA vaccines. Current Opinion in Biotechnology 12(3):299–303.
Zoon KC. 1999. Vaccines, pharmaceutical products, and bioterrorism: Challenges for the U.S. Food and Drug Administration. Emerging Infectious Diseases 5(4):534–536. Available at: http://www.cdc.gov/ncidod/EID/vol5no4/zoon.htm.

Proliferation of Dangerous Pathogens

Breithaupt H. 2000. Toxins for terrorists. Do scientists act illegally when sending out potentially dangerous material? European Molecular Biology Organization Reports 1(4):298–301. Available at: http://embo-reports.oupjournals.org/cgi/content/full/1/4/298?.
Fraser CM, Dando MR. 2001. Genomics and future biological weapons: The need for preventive action by the biomedical community. Nature Genetics 29(3):253–256. Available at: http://www.nature.com/nature/anthrax/.
Malakoff D, Enserink M. 2001. Bioterrorism. New law may force labs to screen workers. Science 294(5544):971–973. Available at: http://www.sciencemag.org/cgi/content/full/294/5544/971.
Tansey B, McCormick E. November 12, 2001. 22,000 U.S. labs handle deadly germs: Feinstein backs bill for government tracking system. San Francisco Chronicle. Online. Available at: www.sfgate.com. Accessed November 14, 2001.
Zelicoff A. May 2001. Arms control today: An impractical protocol. Arms Control Association. Online. Available at: www.armscontrol.org. Accessed October 30, 2001.

Preparedness and Emergency Response

Benjamin GC. 2001. Public health infrastructure: Creating a solid foundation. Physician Executive 27(2):86–87.

Caruso JT. November 6, 2001. Bioterrorism. Statement of Deputy Assistant Director of the Counterterrorism Division of the Federal Bureau of Investigation to the Senate Judiciary Subcommittee on Technology, Terrorism and Government Information.

Centers for Disease Control and Prevention. 2001. The public health response to biological and chemical terrorism: Interim planning guidance for state public health officials. Atlanta: Centers for Disease Control and Prevention. Available at: http://www.bt.cdc.gov/Documents/Planning/PlanningGuidance.PDF.

Fidler DP. 2001. The malevolent use of microbes and the rule of law: Legal challenges presented by bioterrorism. Clinical Infectious Diseases 33(5):686–689. Available at: http://www.journals.uchicago.edu/CID/journal/contents/v33n5.html.

Fine A, Layton M. 2001. Lessons from the West Nile viral encephalitis outbreak in New York City, 1999: Implications for bioterrorism preparedness. Clinical Infectious Diseases 32(2):277–282. Available at: http://www.journals.uchicago.edu/CID/journal/contents/v32n2.html.

Fraser MR, Fisher VS. 2001. Elements of effective bioterrorism preparedness: A planning primer for local public health agencies. Washington, DC: National Association of County and City Health Officials. Available at: http://www.naccho.org/files/documents/Final_Effective_Bioterrism.pdf.

Gallo RJ, Campbell D. 2000. Bioterrorism: Challenges and opportunities for local health departments. Journal of Public Health Management and Practice 6(4):57–62.

Heymann, D. September 5, 2001. Strengthening global preparedness for defense against infectious disease threats. Statement for the Committee on Foreign Relations, United States Senate. Hearing on the Threat of Bioterrorism and the Spread of Infectious Diseases. Available at: http://www.who.int/emc/pdfs/Senate_hearing.pdf.

Hughes J. April 20, 1999. Statement of Director, National Center for Infectious Diseases to the Senate Judiciary Subcommittee on Technology, Terrorism and Government Information. Available at: http://www.cdc.gov/ncidod/diseases/biojud.htm.

Illinois Department of Public Health. 2001. Surviving Disasters: A Citizen's Emergency Handbook. Available at: http://www.idph.state.il.us/pdf/SurvivingDisasters.pdf.

Inglesby T, Grossman R, O'Toole T. 2001. A plague on your city: Observations from TOPOFF. Clinical Infectious Diseases 32(3):436–445. Available at: http://www.journals.uchicago.edu/CID/journal/contents/v32n3.html.

Inglesby TV, O'Toole T, Henderson DA. 2000. Preventing the use of biological weapons: Improving response should prevention fail. Clinical Infectious Diseases 30(6):926–929. Available at: http://www.journals.uchicago.edu/CID/journal/contents/v30n6.html.

Keim M, Kaufmann AF. 1999. Principles for emergency response to bioterrorism. Annals of Emergency Medicine 34(2):177–182.

Khan AS, Ashford DA. 2001. Ready or not—Preparedness for bioterrorism. New England Journal of Medicine 345(4):287–289. Available at: http://content.nejm.org/cgi/content/full/345/4/287.

Moser R Jr, White GL, Lewis-Younger CR, Garrett LC. 2001. Preparing for expected bioterrorism attacks. Military Medicine 166(5):369–374.
O'Toole T, Inglesby T. June 2001. Shining light on Dark Winter. Online. Available at: www.hopkins-biodefense.org. Accessed October 30, 2001.
Rotz LD, Koo D, O'Carroll PW, Kellogg RB, Sage MJ, Lillibridge SR. 2000. Bioterrorism preparedness: Planning for the future. Journal of Public Health Management and Practice 6(4):45–49.
Schoch-Spana M. 2000. Implications of pandemic influenza for bioterrorism response. Clinical Infectious Diseases 31(6):1409–1413. Available at: http://www.journals.uchicago.edu/CID/journal/contents/v31n6.html.
Teeter DS, Koenig KL. 2000. VA's role in bioterrorism preparations. American Journal of Infection Control 28(4):321.
Terriff CM, Tee AM. 2001. Citywide pharmaceutical preparation for bioterrorism. American Journal of Health-System Pharmacists 58(3):233–237.
World Health Organization. 2001. Public health response to biological and chemical weapons: WHO Guidance. Second edition. Geneva: World Health Organization. Available at: http://www.who.int/emc/pdfs/BIOWEAPONS_exec_sum2.pdf.

Monitoring and Surveillance Tools

ESSENCE: Electronic Surveillance System for the Early Notification of Community-Based Epidemics. Information available at: www.geis.ha.osd.mil.
Morse SS, Rosenberg BH, Woodall J. 1996. ProMED global monitoring of emerging diseases: Design for a demonstration program. Health Policy 38(3):135–153.

Detection and Diagnostic Capabilities

Krenzelok EP. 2001. The critical role of the Poison Center in the recognition, mitigation and management of biological and chemical terrorism. (Abstract only) Przegl Lek 58(4):177–181.
Morse SS. 1996. Importance of molecular diagnostics in the identification and control of emerging infections. Molecular Diagnostics 1(3):201–206.

Laboratory Capacity

Gilchrist MJ. 2000. A national laboratory network for bioterrorism: Evolution from a prototype network of laboratories performing routine surveillance. Military Medicine 165(7 Supplement 2):28–31.
Peterson LR, Hamilton JD, Baron EJ, Tompkins LS, Miller JM, Wilfert CM, Tenover FC, Thomson Jr RB. 2001. Role of clinical microbiology laboratories in the management and control of infectious diseases and the delivery of health care. Clinical Infectious Diseases 32(4):605–611. Available at: http://www.journals.uchicago.edu/CID/journal/contents/v32n4.html.

Training Capacity

American Public Health Association. 2001. Effective public health assessment, prevention, response, and training for emerging and re-emerging infectious

diseases, including bioterrorism. American Journal of Public Health 91(3):500–501. Available at: http://www.ajph.org/content/vol91/issue3/.
Franz DR, Jahrling PB, Friedlander AM, McClain DJ, Hoover DL, Bryne WR, Pavlin JA, Christopher GW, Eitzen EM Jr. 1997. Clinical recognition and management of patients exposed to biological warfare agents. JAMA 278(5):399–411. Available at: http://jama.ama-assn.org/issues/v278n5/ffull/jsc71014.html.
Pesik N, Keim M, Sampson TR. 1999. Do U.S. emergency medicine residency programs provide adequate training for bioterrorism? Annals of Emerging Medicine 34(2):173–176.
Waeckerle JF, Seamans S, Whiteside M, Pons PT, White S, Burstein JL, Murray R, Task Force of Health Care and Emergency Services Professionals on Preparedness for Nuclear, Biological, and Chemical Incidents. 2001. Executive summary: Developing objectives, content, and competencies for the training of emergency medical technicians, emergency physicians, and emergency nurses to care for casualties resulting from nuclear, biological, or chemical incidents. Annals of Emerging Medicine 37(6):587–601.

Hospital Capacity

Wetter DC, Daniell WE, Treser CD. 2001. Hospital preparedness for victims of chemical or biological terrorism. American Journal of Public Health 91(5):710–716. Available at: http://www.ajph.org/content/vol91/issue5/.

Protecting Food and Water Supplies

Kaferstein FK, Motarjemi Y, Bettcher DW. 1997. Foodborne disease control: A transnational challenge. Emerging Infectious Diseases 3(4):503–510. Available at: http://www.cdc.gov/ncidod/eid/vol3no4/kaferste.htm.
Khan AS, Swerdlow DL, Juranek DD. 2001. Precautions against biological and chemical terrorism directed at food and water supplies. Public Health Reports 116(1):3–14.
Logan-Henfrey L. 2000. Mitigation of bioterrorist threats in the 21st century. Annals of the New York Academy of Sciences 916:121–133.
Nara PL. 1999. The status and role of vaccines in the U.S. food animal industry. Implications for biological terrorism. Annals of the New York Academy of Sciences 894:206–217.
Neher NJ. 1999. The need for a coordinated response to food terrorism. The Wisconsin experience. Annals of the New York Academy of Sciences 894:181–183.
Pellerin C. 2000. The next target of bioterrorism: Your food. Environmental Health Perspectives 108(3):A126–129. Available at: http://ehpnet1.niehs.nih.gov/members/2000/108-3/spheres.html.
Sequeira R. 1999. Safeguarding production agriculture and natural ecosystems against biological terrorism. A U.S. Department of Agriculture emergency response framework. Annals of the New York Academy of Sciences 894:48–67.
Torok TJ, Tauxe RV, Wise RP, Livengood JR, Sokolow R, Mauvais S, Birkness KA, Skeels MR, Horan JM, Foster LR. 1997. A large community outbreak of salmonellosis caused by intentional contamination of restaurant salad

bars. JAMA 278(5):389–395. Available at: http://jama.ama-assn.org/issues/v278n5/ffull/joc71206.html.
Williams JL, Sheesley D. 2000. Response to bio-terrorism directed against animals. Annals of the New York Academy of Sciences. 916:117–120.

INTERNET RESOURCES

Federal Government

Centers for Disease Control and Prevention:

Public Health Emergency Preparedness and Response: http://www.bt.cdc.gov
Centers for Public Health Preparedness: http://www.phppo.cdc.gov/owpp/centersforPHP.asp
Health Alert Network: http://www.phppo.cdc.gov/han
Bioterrorism Preparedness and Response Core Capacity Project 2001 Draft Report:
http://www.bt.cdc.gov/RegionalMeetings/2001/2001RMSummary.asp
Facts about Anthrax: http://www.bt.cdc.gov/DocumentsAPP/facts_about.pdf
Use of Anthrax Vaccine in the United States: Recommendations of the Advisory Committee on Immunization Practices (ACIP). MMWR. Dec.15, 2000;49:RR-15: http://www.cdc.gov/mmwr/PDF/rr/rr4915.pdf
Anthrax Vaccine: What You Need to Know: http://www.cdc.gov/nip/publications/VIS/vis-anthrax.pdf
Biological and Chemical Terrorism: Strategic Plan for Preparedness and Response. MMWR. April 21, 2000;49:(RR04);1–14. Available at: http://www.cdc.gov/mmwr/preview/mmwrhtml/rr4904a1.htm
Bioterrorism Alleging Use of Anthrax—and Interim Guidelines for Management—United States 1998. MMWR. Feb. 5, 1999;48(4):69–74. Available at: http://www.cdc.gov/mmwr/PDF/wk/mm4804.pdf
Updated Recommendations for Handling Suspicious Packages or Envelopes (An Official CDC Health Advisory)
http://www.bt.cdc.gov/DocumentsApp/Anthrax/10272001AM/han47.asp
What Every Physician Should Know About Anthrax, Part I
http://www.sph.unc.edu/about/webcasts/bioter_10-18_stream1.htm
What Every Physician Should Know About Anthrax, Part II
http://www.phf.org/anthrax2HTML.htm
PulseNet--Foodborne Disease Surveillance
http://www.cdc.gov/ncidod/dbmd/pulsenet/pulsenet.htm

Central Intelligence Agency: http://www.cia.gov/terrorism/index.html
Department of Defense: http://www.anthrax.osd.mil/
Department of Health and Human Services:
http://www.hhs.gov/hottopics/healing/biological.html
Department of Justice: http://www.usdoj.gov/
Department of Veterans Affairs: http://www.va.gov/emshg
Federal Bureau of Investigation:
http://www.fbi.gov/pressrel/attack/attacks.htm
Federal Emergency Management Agency: http://www.fema.gov/
Food and Drug Administration: http://www.fda.gov/cber/faq/cntrbfaq.htm
FirstGov:
http://firstgov.gov/featured/usgresponse.html?ssid=1004388686307_172

National Domestic Preparedness Office: http://www.ndpo.gov
Federal Weapons of Mass Destruction Training Compendium: http://www.ndpo.gov/compenium.pdf
National Library of Medicine: MEDLINEplus Health Information
Anthrax: http://www.nlm.nih.gov/medlineplus/anthrax.html
Biological and Chemical Weapons:
http://www.nlm.nih.gov/medlineplus/biologicalandchemicalweapons.html
Office of Homeland Security: http://www.whitehouse.gov/homeland/
Sandia National Laboratories: http://www.sandia.gov/NERA/extdocs.htm
Surgeon General: http://www.surgeongeneral.gov/sgoffice.htm
Anthrax/Bioterrorism:
http://www.surgeongeneral.gov/topics/bioterrorism.htm
U.S. Department of Agriculture:
http://www.usda.gov/special/biosecurity/safeguard.htm
U.S. Postal Service: http://www.usps.com/news/2001/press/serviceupdates.htm

State and Local Governments

California: http://www.dhs.ca.gov/bioterrorism/
Colorado: http://www.cdphe.state.co.us/dc/bioterror/bioterrorismhom.asp
District of Columbia:
http://dchealth.dc.gov/news_room/health_alert.asp?id=6&mon=200110
Florida: http://www.doh.state.fl.us/terrorism/index.htm
Georgia: http://www.ph.dhr.state.ga.us/programs/emerprep/bioterrorism.shtml
DeKalb County Board of Health Bioterrorism Response Plan prepared by the Center for Public Health Preparedness. Information available at: www.dekalbhealth.net.
Illinois: http://www.idph.state.il.us/Bioterrorism/bioterrorismfaqs.htm
Indiana: http://www.state.in.us/isdh/healthinfo/bioterrorism.htm
Iowa: http://www.idph.state.ia.us/Terrorism/default.htm
Kansas: http://www.kdhe.state.ks.us/han/bioterror.html
Maryland: http://www.dhmh.state.md.us/phdsec/html/phalert.htm
Massachusetts: http://www.state.ma.us/dph/topics/bioterrorism/BT.htm
Minnesota: http://www.health.state.mn.us/bioterrorism/
New Jersey: http://www.state.nj.us./health/er/biofs.htm
New York: http://www.health.state.ny.us/nysdoh/bt/bt.htm
New York City: http://www.ci.nyc.ny.us/html/doh/html/cd/wtc8.html
Oregon: http://www.ohd.hr.state.or.us/acd/bioterr/facts.htm
Tennessee: http://www.state.tn.us/health/CEDS/bioterrorism.htm
Texas: http://www.tdh.state.tx.us/bioterrorism/default.htm
Virginia: http://www.vdh.state.va.us/bt/index.htm
Wisconsin: http://www.dhfs.state.wi.us/dph_bcd/Bioterrorism/

Other state and local health departments: http://www.cdc.gov/other.htm

Educational and Research Institutions

Center for Nonproliferation Studies: http://cns.miis.edu/research/cbw/

Columbia University: Center for Public Health Preparedness: http://cpmcnet.columbia.edu/dept/sph/CPHP/index.html

Harvard University John F. Kennedy School of Government: Belfer Center for Science and International Affairs: http://www.ksg.harvard.edu/bcsia/

Humanitarian Resource Institute: http://www.humanitarian.net/biodefense

Johns Hopkins University: Center for Civilian Biodefense Studies: http://www.hopkins-biodefense.org/

Massachusetts Institute of Technology: Center for International Studies: http://web.mit.edu/cis/

National Academy of Sciences: http://www.nap.edu/terror/

University of Maryland: Center for International and Security Studies at Maryland: http://www.puaf.umd.edu/CISSM

University of Minnesota: Center for Infectious Disease Research and Policy: http://www.cidrap.umn.edu

Domestic and International NGOs

CBACI Report: Bioterrorism in the United States: Threat, Preparedness, and Response: http://www.cbaci.org/CDCSectionLinksMain.htm

Center for Strategic and International Studies: http://www.csis.org/homeland/index.html

Chemical and Biological Arms Control Institute (CBACI): http://www.cbaci.org

Chemical and Biological Weapons Nonproliferation Project: http://www.stimson.org/cwc/index.html

Henry L. Stimson Center: http://stimson.org

Institute for Homeland Security: http://www.homelandsecurity.org/index.cfm

RAND Corporation: http://www.rand.org/hot/newslinks.html#terror

TrainingFinder.org: www.TrainingFinder.org. Provides information on over 30 distance learning courses for public health professionals on bioterrorism and emergency preparedness.

World Health Organization: http://www.who.int/home-page/
Communicable Disease Surveillance and Response: http://www.who.int/emc/diseases/index.html

Appendix C

Testimony of Joshua Lederberg, Ph.D.

Professor Emeritus of Molecular Genetics and Informatics and Sackler Foundation Scholar
The Rockefeller University, New York, N.Y.
For a Hearing on
The Threat of Bioterrorism and the Spread of Infectious Diseases
Before the
Committee on Foreign Relations, United States Senate
August 24, 2001

I am honored to address the committee on a matter of transcendent importance to U.S. security and global human welfare. I define biological warfare as use of agents of disease for hostile purposes. This definition encompasses attacks on human health and survival and extends to plant and animal crops. Biological warfare was the focus of billion-dollar investments by the United States and the former Soviet Union until President Nixon's unilateral abjuration in 1969. This declaration was followed by the negotiation, ratification, and coming into force (in 1975) of the Biological Weapons Convention, a categorical ban on the development, production, and use of biological weapons.

Biological weapons are characterized by low cost and ease of access; difficulty of detection, even after use, until disease has advanced; unreliable but open-ended scale of predictable casualties; and clandestine stockpiles and delivery systems. Per kilogram of weapon, the potential lives lost approach those of nuclear weapons, but less costly and sophisticated technology is required.

Intelligence estimates indicate that up to a dozen countries may have developed biological weapons. Considerable harm (on the scale of 1,000 casualties) could be inflicted by rank amateurs. Terrorist groups, privately or state-sponsored, with funds up to $1 million, could mount massive attacks of 10 or 100 times that scale. For each 1,000 persons on the casualty roster, 100,000 or 1,000,000 are at risk and in need of prophylactic attention, which in turn necessitates a massive triage. Studies of hypothetical scenarios document the complexity of managing bioterrorist incidents and the stress that control of such incidents would impose on civil order.

While powerful nations maintain a degree of equilibrium through mutual deterrence and shared interests, less powerful elements may find in biological warfare opportunities to harm their enemies. Under current levels of preparedness (e.g., physical facilities and organization and operational doctrines), biological warfare is probably the most perplexing and gravest security challenge we face.

President Nixon's abjuration of biological warfare as a U.S. military weapon in 1969 set in motion the most important diplomatic and legal steps towards its eradication globally, laying the groundwork for the Biological Weapons Convention treaty. The treaty lacks robust verification mechanisms, mainly for reasons intrinsic to the technology. However, verification is not the foundation of the U.S. stance; the United States has long since abandoned the idea that it would respond in kind to such an attack. Were it not for the Biological Weapons Convention, a gradually escalating technology race would have amplified even further this threat to human existence. The treaty does set a consensually agreed-upon standard of behavior: it has become institutionalized into international law, and infractions open the door to enforcement.

Although further provisions for verification would do little to enhance our knowledge of those infractions, they would nevertheless have important symbolic value in reaffirming international commitment to the principles of the treaty. Creative leadership is needed to develop other ways to strengthen that reaffirmation. The real problem with the Biological Weapons Convention is enforcement, not verification. We have all-but-certain knowledge that Saddam Hussein has continued Iraq's biological weapons development program. To convince our allies, much less neutral nations and potential adversaries, of what is at stake, we may have to elevate the priority we give to this threat. We must also become more knowledgeable about the local political and cultural terrain and more ingenious in designing sanctions that will not impose undue hardship on the Iraqi population. Our public diplomacy is predicated on the stated proposition that use of biological weapons is an offense to civilization. This major accomplishment of the Biological Weapons Convention needs to be reaffirmed both in the attention we give to our own defense and in our stern responses to substantial infractions from any quarter.

Unlike the aftermath of nuclear or high-explosive bombardment, attack with biological weapons is amenable to interventions for some hours or days after the event, depending on the agent used. With the most publicized agent, anthrax, administration of appropriate antibiotics can protect the majority of those exposed. The other side of the coin is recognizing the syndrome within hours of the earliest symptoms. Biosensors are being developed to confirm suspicions of anthrax. We will have to rely on early diagnosis of the first human (or animal) cases to provide the basis for focusing those sensors. Because a wide list of diseases must be considered, this surveillance entails reinvigorating our overall public health infrastructure. In contrast to the explosive rise of health-care expenditures, public health funding has been allowed to languish, boosted only very recently by public arousal about emerging infections and bioterrorism. That boost entails personnel and organizational structures, but improvement also depends on funding for new as well as established programs.

In addition to diagnostic capability, we need organizational and operational doctrines that can confront unprecedented emergencies, we need trained personnel on call, and we need physical facilities for isolation, decontamination, and care. We also need stockpiles of antibiotics and vaccines appropriate to the risk, preceded by careful analysis of what kinds and how much. We need research on treatment methods (e.g., how should inhalational anthrax be managed with possibly limited supplies of antibiotics). Still more fundamental, research could give us sharper tools for diagnosis and more usable ranges of antibacterial and antiviral remedies.

Organizing the government to deal with mass contingencies is a goal that is vexing and still poorly addressed. It entails coordination of local, state, and federal assets and jurisdictions and the intersection of law enforcement, national security, and public health. A time of crisis is not ideal for debates over responsibility, authority, and funding.

Our main bulwark against direct large-scale attack is the combination of civic harmony and firm retaliation. Better intelligence is key to retaliation, apprehension, and penal containment and sanctions. This territory is technically unfamiliar to most of the intelligence community, which has taken many positive steps but has a long way to go. Resources for managing biological threats are fewer than those allocated to other, more familiar threats.

I have already alluded to public diplomacy (starting with firm conviction at home) about the level of priority to be given to the biological weapons threat if a successful attack is to be averted. A dilemma is how to study the threats of biowarfare in detail and develop vaccines and other countermeasures, while maintaining the policy of abhorrence at the idea of using disease as a weapon. The central premise of the Biological Weapons Convention is that infectious disease is the common enemy of all humans and that joining with that enemy is an act of treason against humanity. This premise clearly inspired adherence to the Convention, even by countries that might otherwise exploit biological weapons to level the playing field against a superpower. Having set aside biological weapons as of small advantage to U.S. military power, we are fortunate that we share the treaty's interests and conclusions. They can only be strengthened if we internalize them and participate ever more fully in global campaigns for health. Current levels of funding for AIDS, malaria, and tuberculosis are small but are certainly steps in the right direction. We should assume leadership among nations cooperating with the World Health Organization to bolster global systems of surveillance and outbreak investigation of diseases that could threaten us all.

SOURCE: Lederberg, J. November/December 2001. Biological warfare. Emerging Infectious Diseases. Online. Available at: www.cdc.gov/ncidod/eid/vol7no6/lederberg.htm.

Appendix D

Summary of the Frist-Kennedy "Bioterrorism Preparedness Act of 2001"

December 4, 2001

The Frist-Kennedy "Bioterrorism Preparedness Act of 2001" is designed to address gaps in our nation's biodefense and surveillance system and our public health infrastructure. This new legislation builds on the foundation laid by the "Public Health Threats and Emergencies Act of 2000" by authorizing additional measures to improve our health system's capacity to respond to bioterrorism, protect the nation's food supply, speed the development and production of vaccines and other countermeasures, enhance coordination of federal activities on bioterrorism, and increase our investment in fighting bioterrorism at the local, state, and national levels. The legislation would authorize approximately $3.2 billion in additional funding for Fiscal Year 2002 (and such sums in years thereafter) toward these activities.

Title I—National Goals for Bioterrorism Preparedness

Title I of the "Bioterrorism Preparedness Act" states that "the United States should further develop and implement a coordinated strategy to prevent and, if necessary, to respond to biological threats or attacks." It further states that it is the goal of Congress that this strategy should: (1) provide federal assistance to state and local governments in the event of a biological attack; (2) improve public health, hospital, laboratory, communications, and emergency response preparedness and responsiveness at the state and local levels; (3) rapidly develop and manufacture needed therapies, vaccines, and medical supplies; and (4) enhance the safety of the nation's food supply and protect its agriculture from biological threats and attacks.

Title II—Improving the Federal Response to Bioterrorism

Title II requires the Secretary of Health and Human Services (HHS) to report to Congress within one year of enactment, and biennially thereafter, on progress made toward meeting the objectives of the Act. It provides statutory authorization for the strategic national pharmaceutical stockpile, provides additional resources to the Centers for Disease Control and Prevention (CDC) to carry out education and training initiatives and to improve the nation's federal laboratory capacity, and establishes a National Disaster Medical Response System of volunteers to respond, at the Secretary's direction, to national public health emergencies (with full liability protection, re-employment rights, and other worker protections for such volunteers similar to those currently provided to those who join the National Guard).

The bill further amends and clarifies the procedures for declaring a national public health emergency and expands the authority of the Secretary during the emergency period. In declaring such an emergency, the Secretary must notify Congress within 48 hours. Such emergency period may not be longer than 180 days, unless the Secretary determines otherwise and notifies Congress of such determination. During that emergency period, the Secretary may waive certain data submittal and reporting deadlines.

A recent report by the General Accounting Office raised concerns about the lack of coordination of federal anti-bioterrorism efforts. Therefore, the bill contains a number of measures to enhance coordination and cooperation among various federal agencies. Title II establishes an Assistant Secretary for Emergency Preparedness at HHS to coordinate all functions within the Department relating to emergency preparedness, including preparing for and responding to biological threats and attacks.

Title II also creates an interdepartmental Working Group on Bioterrorism that includes the secretaries of HHS, Defense, Veteran's Affairs, Labor, and Agriculture, the Director of the Federal Emergency Management Agency, the Attorney General of the United States, and other appropriate federal officials. The Working Group consolidates and streamlines the functions of two existing working groups first established under the "Public Health Threats and Emergencies Act of 2000." It is responsible for coordinating the development of bioterrorism countermeasures, research on pathogens likely to be used in a biological attack, shared standards for equipment to detect and protect against infection from biological pathogens, national preparedness and response for biological threats or attacks, and other matters.

Title II also establishes two advisory committees to the Secretary. The National Task Force on Children and Terrorism will report on measures necessary to ensure that the health needs of children are met in preparing for and responding to any potential biological attack or event. The Emergency Public Information and Communications Task Force will report on appropriate ways to communicate to the public information regarding bioterrorism. Both of these committees sunset after one year.

The title also contains a congressional recommendation that there be established an official federal Internet website on bioterrorism to provide information to the public, health professionals, and others on matters relevant to bioterrorism. The title further requires that states have a coordinated plan for providing information relevant to bioterrorism to the public.

Additionally, Title II helps the federal government better track and control biological agents and toxins. The Secretary of HHS is required to review and update a list of biological agents and toxins that could pose a severe threat to public health and safety and to enhance regulations regarding the possession, use, and transfer of such agents or toxins. Violations of these regulations could trigger civil penalties of up to $500,000, and criminal sanctions may be imposed. Existing law already regulates the transfer of these pathogens.

Title III—Improving State and Local Preparedness Capabilities

Numerous reports in recent years have found the nation's public health infrastructure lacking in its ability to respond to biological threats or other emergencies. For example, nearly 20 percent of local public health departments have no e-mail capability, and fewer than half have high-speed Internet or broadcast facsimile transmission capabilities. Before September 11, only one in five U.S. hospitals had bioterrorism preparedness plans in place.

Title III addresses this situation by including several enhanced grant programs to improve state and local public health preparedness. In addition to converting the current public health core capacity grants established under the "Public Health Threats and Emergencies Act of 2000" to non-competitive grants, the bill replaces the current 319F competitive bioterrorism grant with a new state bioterrorism emergency program that provides resources to states based on population and that would guarantee each state a minimum level of funding for preparedness activities. States must develop bioterrorism preparedness plans to be eligible for such funding. Activities funded under this grant include conducting an assessment of core public health capacities, achieving the core public health capacities, and fulfilling the bioterrorism preparedness plan. This program would only be authorized for two years.

The bill also establishes a new grant program for hospitals that are part of consortia with public health agencies, and counties or cities. To be eligible for the grant, the hospital's grant proposal must be consistent with its state's bioterrorism preparedness plan. Using these grants, hospitals will acquire the capacity to serve as regional resources during a bioterrorist attack. This program is authorized for five years.

Title IV—Developing New Countermeasures Against Bioterrorism

To better respond to bioterrorism, Title IV expands our nation's stockpile of smallpox vaccine and critical pharmaceuticals and devices. The bill also expands research on biological agents and toxins, as well as new treatments and vaccines for such agents and toxins.

Since the effectiveness of vaccines, drugs, and therapeutics for many biological agents and toxins often may not ethically be tested in humans, Title IV ensures that the Food and Drug Administration (FDA) will finalize by a date certain its rule regarding the approval of new priority countermeasures on the basis of animal data. Priority countermeasures will also be given expedited review by the FDA.

Because of the limitations on a market for vaccines for these agents and toxins, Title IV gives the Secretary of HHS authority to enter into long-term contracts with sponsors to "guarantee" that the government will purchase a certain quantity of a vaccine at a certain price. The government has the authority, through an existing Executive Order, to ensure that sponsors through these contracts will be indemnified by the government for the development, manufacture, and use of the product as prescribed in the contract.

Title IV also provides a limited antitrust exemption to allow potential sponsors to discuss and agree upon how to develop, manufacture, and produce new priority countermeasures, including vaccines and drugs. Federal Trade Commission and Department of Justice approval of such agreements is required to ensure they are not anti-competitive.

Title V—Protecting Our Nation's Food Supply

With 57,000 establishments under its jurisdiction and only 700–800 food inspectors, including 175 import inspectors for more than 300 ports of entry, FDA needs increased resources for inspections of imported food. The President's emergency relief budget included a request for $61 million to enable FDA to hire 410 new inspectors, lab specialists, and other experts, as well as invest in new technology and equipment to monitor food imports.

Title V grants FDA needed authorities to ensure the safety of domestic and imported food. It allows FDA to use qualified employees from other agencies and departments to help conduct food inspections. Any domestic or foreign facility that manufactures or processes food for use in the United States must register with FDA. Importers must provide at least four hours notice of the food, the country of origin, and the amount of food to be imported. FDA's authority is made more explicit to prevent "port-shopping" by marking food shipments denied entry at one U.S. port to ensure such shipments do not reappear at another U.S. port.

The bill gives additional tools to FDA to ensure proper records are maintained by those who manufacture, process, pack, transport, distribute, receive, hold or import food. The FDA's ability to inspect such records will strengthen its ability to trace the source and chain of distribution of food and determine the scope and cause of the adulteration or misbranding that presents a threat of serious adverse health consequences or death to humans or animals. Importantly, the bill also enables FDA to detain food for a limited period of time while FDA seeks a seizure order if such food is believed to present a threat of serious adverse health consequences or death to humans or animals. The FDA may also debar a person who engages in a pattern of seeking to import such food.

Title V also includes several measures to help safeguard the nation's agriculture industry from the threats of bioterrorism. Toward this end, it contains a series of grants and incentives to help encourage the development of vaccines and antidotes to protect the nation's food supply, livestock, or crops, as well as preventing crop and livestock diseases from finding their way to our fields and feedlots.

It also authorizes emergency funding to update and modernize USDA research facilities at the Plum Island Animal Disease Laboratory in New York, the National Animal Disease Center in Iowa, the Southwest Poultry Research Laboratory in Georgia, and the Animal Disease Research Laboratory in Wyoming. Also, it funds training and implements a rapid response strategy through a consortium of universities, the USDA, and agricultural industry groups.

SOURCE: http://www.senate.gov/~frist/Issues/Issues-National_Defense/FightingTerrorism /Bioterrorism/biobilllsum/biobilllsum.html

Appendix E

Department Of Health And Human Services: Food and Drug Administration [Docket No. 01N–0494]

Prescription Drug Products; Doxycycline and Penicillin G Procaine Administration for Inhalational Anthrax (Post-Exposure)

SUMMARY: The Food and Drug Administration (FDA) is clarifying that the currently approved indications for doxycycline and penicillin G procaine drug products include use in cases of inhalational exposure to *Bacillus anthracis* (the bacterium that causes anthrax). We also are providing dosing regimens that we have determined are appropriate for these products for this use. We encourage the submission of supplemental new drug applications (labeling supplements) to add the dosage information to the labeling of currently marketed drug products.

ADDRESSES: Submit labeling supplements to the Center for Drug Evaluation and Research, Food and Drug Administration, Central Document Room, 12229 Wilkins Ave., Rockville, MD 20852.

FOR FURTHER INFORMATION CONTACT:

Dianne Murphy, Center for Drug Evaluation and Research (HFD–950), Food and Drug Administration, 5600 Fishers Lane, Rockville, MD 20857, 301-827-2350.

SUPPLEMENTARY INFORMATION:

I. Anthrax

Anthrax is caused by the spore-forming bacterium *Bacillus anthracis*. There are three types of anthrax infection in humans: cutaneous, gastrointestinal, and inhalational. Until recently, most human experience with anthrax was associated with exposure to infected animals or animal products. Anthrax is reported annually among livestock. In areas where these animal cases occur, most human cases are the cutaneous form. Such cases occur among workers who have handled infected hoofed animals or products from these animals. Gastrointestinal anthrax has been reported following the ingestion of undercooked or raw meat from infected animals. Inhalational anthrax, resulting from inhalation of aerosolized spores, was associated with industrial processing of infected wool, hair, or hides in the United States in the past. Before October 2001, no case of inhalational anthrax had been reported in the United States since 1978. In 1979, at least 64 people died in Sverdlovsk (currently Ekaterinburg), Russia, of inhalational anthrax after *Bacillus anthracis* spores were accidentally released from a Soviet military laboratory. Administration of certain antimicrobial agents may prevent or reduce the incidence of disease following inhalational exposure to *Bacillus anthracis*.

II. Approved Drug Products

Drug products containing doxycycline, doxycycline calcium, doxycycline hyclate,[1] and penicillin G procaine are currently approved with indications for anthrax.[2] The approved labeling for the doxycycline products states that the drugs are indicated in infections caused by *Bacillus anthracis*. The approved labeling for penicillin G procaine drug products states that the drugs are indicated for anthrax. Presently, the labeling for these drug products do not specify a dosing regimen for inhalational exposure to *Bacillus anthracis*. The indication sections of approved labeling for these drug products does not specify cutaneous, gastrointestinal, or inhalational anthrax. We have determined that the language in the labeling of drug products containing doxycycline, doxycycline calcium, doxycycline hyclate, and penicillin G procaine is intended to, and does, cover all forms of anthrax, including inhalational anthrax (post-exposure): to reduce the incidence or progression of disease following exposure to aerosolized *Bacillus anthracis*. On August 30, 2000, we approved supplements to provide an indication for inhalational anthrax (post-exposure) for ciprofloxacin hydrochloride tablets and ciprofloxacin intravenous (IV) solution, IV in 5 percent dextrose, IV in 0.9 percent saline, and oral suspension. The approved labeling for these ciprofloxacin products provides for a 60-day dosing regimen. Because ciprofloxacin drug products are already specifically indicated for inhalational anthrax (post-exposure) and their approved labeling provides a regimen for inhalational anthrax (post-exposure), we do not discuss ciprofloxacin any further

---

[1]Doxycycline hyclate tablets, equivalent to 20 milligrams (mg) base, and doxycycline hyclate 10 percent for controlled release in subgingival application are not subjects of this notice because they have periodontal indications and do not have indications for anthrax or infections caused by *Bacillus anthracis*.

[2]Other drug products are currently approved with indications for anthrax or infections caused by *Bacillus anthracis*, i.e., minocycline, tetracycline, oxytetracycline, demeclocycline, and penicillin G potassium. We have not completed a review on these other drugs. We will not discuss these other drugs further in this notice.

in this notice. It is relevant, however, that the rhesus monkey study supporting the approval of ciprofloxacin for inhalational anthrax also included separate doxycycline and penicillin G procaine treatment arms. Each of these arms showed a survival advantage over placebo.[3] No other antimicrobial drugs were tested in this study.

III. Doxycycline Drug Products

We have determined that 100 mg of doxycycline, taken orally twice daily for 60 days, is an appropriate dosing regimen for administration to adults who have inhalational exposure to *Bacillus anthracis*. The corresponding oral dosing regimen for children under 100 pounds (lb) is 1 mg per (/) lb of body weight (2.2 mg/kilogram (kg)), given twice daily for 60 days. We have determined that IV doxycycline can be administered to adults in a 100 mg dose twice daily for inhalational anthrax (post-exposure). The corresponding IV dosing regimen for children under 100 lb is 1 mg /lb of body weight (2.2 mg/kg), twice daily. Intravenous therapy is indicated only when oral therapy is not indicated. Intravenous therapy should not be given over a prolonged period of time. Patients should be switched to oral doxycycline, or another antimicrobial drug product, as soon as possible, to complete a 60-day course of therapy.

A. Safety

Doxycycline drug products have been used for over 30 years, and the literature on the products is voluminous. We have reviewed the literature dealing with the long-term administration of doxycycline for treatment of diseases other than anthrax. Several articles report the results of studies involving the administration of doxycycline in amounts comparable to the doses recommended in this notice. They also involve administration of doxycycline for 60 days and periods approaching and exceeding 60 days. We have also reviewed data from our Adverse Event Reporting System (AERS). Analysis of these articles and data indicates no pattern of unlabeled adverse events has been associated with the long-term use of doxycycline. Doxycycline and other members of the tetracycline class of antibiotics are not generally indicated for the treatment of any patients under the age of 8 years. Tetracyclines are known to be associated with teeth discoloration and enamel hypoplasia in children and delays in bone development in premature infants after prolonged use. We have balanced the nature of the effect on teeth and the fact that this delay in bone development is apparently reversible against the lethality of inhalational anthrax, and concluded that doxycycline drug products can be labeled with a pediatric dosing regimen for inhalational anthrax (post-exposure). We are not recommending that IV doxycycline be administered for

---

[3]Friedlander AM et al., "Postexposure Prophylaxis Against Experimental Inhalation Anthrax," Journal of Infectious Diseases, 167:1239–1243, 1993.

prolonged periods because of the possibility of thrombophlebitis and other complications of IV therapy. Thrombophlebitis as a possible adverse reaction is already described in the approved labeling for IV doxycycline drug products. Patients administered IV doxycycline for inhalational anthrax (post-exposure) should be switched to oral doxycycline or another antimicrobial drug product as soon as possible to complete a 60-day course of therapy.

B. Effectiveness

We have reviewed minimal inhibitory concentration (MIC) data for the tetracycline class and *Bacillus anthracis*, pharmacokinetic data, data from the Sverdlovsk incident, and the outcome data from a study of inhalational exposure to *Bacillus anthracis* in rhesus monkeys.[4] We have concluded that 100 mg of doxycycline, administered twice a day for 60 days, is an effective dosing regimen for adults who have inhalational exposure to *Bacillus anthracis*. The corresponding dosing regimen for children under 100 lb of 1 mg/lb of body weight (2.2 mg/kg), given twice daily for 60 days, is also effective.

C. Labeling for Oral Doxycycline

We encourage the submission of labeling supplements for orally administered doxycycline, doxycycline calcium, and doxycycline hyclate drug products. The revised labeling should contain a specific indication for inhalational anthrax (post-exposure), the recommended dosing regimen, safety information relevant to use in children, and other information described below. The following specific changes to the current approved labeling are recommended:

•*Indications and Usage*. The indication for anthrax should be revised from "Anthrax due to *Bacillus anthracis*" to "Anthrax due to *Bacillus anthracis*, including inhalational anthrax (post-exposure): to reduce the incidence or progression of disease following exposure to aerosolized *Bacillus anthracis*." This indication should be removed from the paragraph of the "Indications and Usage" section that begins "When penicillin is contraindicated, doxycycline is an alternative drug in the treatment of the following infections:" and inserted at the end of the preceding paragraph that begins "Doxycycline is indicated for the treatment of infections caused by the following gram-positive microorganisms when bacteriologic testing indicates appropriate susceptibility to the drug:."

•*Warnings*. The last sentence in the first paragraph of the "Warnings" section should be revised to read as follows: "TETRACYCLINE DRUGS, THEREFORE, SHOULD NOT BE USED IN THIS AGE GROUP, EXCEPT FOR ANTHRAX, INCLUDING INHALATIONAL ANTHRAX (POST-

[4] Friedlander.

EXPOSURE), UNLESS OTHER DRUGS ARE NOT LIKELY TO BE EFFECTIVE OR ARE CONTRAINDICATED."

•*Dosage and Administration*. The following text should be inserted as the last item of the "Dosage and Administration" section: "Inhalational anthrax (post-exposure): ADULTS: 100 mg of doxycycline, by mouth, twice a day for 60 days. CHILDREN: weighing less than 100 lb (45 kg); 1 mg/lb (2.2 mg/kg) of body weight, by mouth, twice a day for 60 days. Children weighing 100 lb or more should receive the adult dose."

D. Labeling for IV Doxycycline

We encourage the submission of labeling supplements for doxycycline hyclate injectable drug products. The revised labeling should contain a specific indication for inhalational anthrax (post-exposure), the recommended dosing regimen, safety information relevant to use in children and prolonged use, and other information described below. We recommend that labeling supplements for doxycycline hyclate injectable drug products include the following specific changes:

•*Indications*. The indication for anthrax should be revised from "*Bacillus anthracis*" to "Anthrax due to *Bacillus anthracis*, including inhalational anthrax (post-exposure): to reduce the incidence or progression of disease following exposure to aerosolized *Bacillus anthracis*." This indication should be removed from the paragraph of the "Indications" section that begins "When penicillin is contraindicated, doxycycline is an alternative drug in the treatment of infections due to:" and inserted at the end of the preceding paragraph that begins "Doxycycline is indicated for the treatment of infections caused by the following gram-positive microorganisms when bacteriologic testing indicates appropriate susceptibility to the drug:."

•*Warnings*. The last sentence in the first paragraph of the "Warnings" section should be revised to read as follows: "TETRACYCLINE DRUGS, THEREFORE, SHOULD NOT BE USED IN THIS AGE GROUP, EXCEPT FOR ANTHRAX, INCLUDING INHALATIONAL ANTHRAX (POST-EXPOSURE), UNLESS OTHER DRUGS ARE NOT LIKELY TO BE EFFECTIVE OR ARE CONTRAINDICATED."

•*Dosage and Administration*. The following paragraph should be inserted in the "Dosage and Administration" section after the paragraph describing the treatment for syphilis: "In the treatment of inhalational anthrax (post-exposure) the recommended dose is 100 mg of doxycycline, twice a day. Parenteral therapy is only indicated when oral therapy is not indicated and should not be continued over a prolonged period of time. Oral therapy should be instituted as soon as possible. Therapy must continue for a total of 60 days." The following paragraph should be inserted in the "Dosage and Administration" section after the paragraph describing the dosages for children above 8 years of age: "In the treatment of inhalational anthrax (post-exposure) the recommended dose is 1 mg/lb (2.2 mg/kg) of body weight, twice a day in children weighing less than 100 lb (45 kg). Parenteral therapy is only indicated when oral therapy is not indicated and should not be continued over a prolonged period of time. Oral therapy should be instituted as soon as possible. Therapy must continue for a total of 60 days."

IV. Penicillin G Procaine Drug Products

We have determined that 1,200,000 units of penicillin G procaine, administered every 12 hours, is an appropriate dosing regimen for adults who have inhalational exposure to *Bacillus anthracis*. The corresponding dosing regimen for children is 25,000 units/kg of body weight (maximum 1,200,000 units) every 12 hours.

A. Safety

Penicillin drug products have been used for over 50 years. The amount of literature on penicillin is correspondingly large. We have reviewed published literature on the safety of penicillin G procaine. We have also reviewed data from AERS. Analysis of these articles and data indicates that no pattern of unexpected adverse events is associated with the use of penicillin G procaine as described in the recommended dosing regimen. All adverse events that we have identified are described in the approved labeling. We note that there may be an increased risk of neutropenia and an increased incidence of serum sickness-like reactions associated with use of penicillin for more than 2 weeks. Because prescribing health care professionals should take those factors into consideration when continuing administration of penicillin G procaine for longer than 2 weeks for inhalational anthrax (post-exposure), we are suggesting that the labeling for the drug products reflect these concerns about neutropenia and serum sickness-like reactions.

B. Effectiveness

We have reviewed MIC data for penicillin G and *Bacillus anthracis*, pharmacokinetic data, data from the Sverdlovsk incident, clinical data regarding the use of penicillins in treatment of primarily cutaneous anthrax, and the outcome data from a study of inhalational exposure to *Bacillus anthracis* in rhesus monkeys.[5] We have concluded that the recommended dosing regimens are effective for adults and children who have inhalational exposure to *Bacillus anthracis*.

---

[5]Friedlander.

C. Labeling

We encourage the submission of labeling supplements for penicillin G procaine injectable drug products. The revised labeling should contain a specific indication for inhalational anthrax (post-exposure), the recommended dosing regimen, safety information relevant to prolonged use and use in children, and other information described below. The following specific changes to the current approved labeling are recommended:

•*Indications*. In the "Indications" section, the indication for anthrax should be revised from "Anthrax" to "Anthrax due to *Bacillus anthracis*, including inhalational anthrax (post-exposure): to reduce the incidence or progression of the disease following exposure to aerosolized *Bacillus anthracis*."

•*Precautions*. In the "Precautions" section, at the end of the paragraph that begins "In prolonged therapy with penicillin, and particularly with high-dosage schedules, periodic evaluation of the renal and hematopoietic systems is recommended," the following text should be added: "In such situations, use of penicillin for more than 2 weeks may be associated with an increased risk of neutropenia and an increased incidence of serum sickness-like reactions."

•*Dosage and Administratio*n. In the "Dosage and Administration" section, immediately following "Anthrax— cutaneous: 600,000 to 1,000,000 units/ day," the following text should be inserted: "Anthrax— inhalational (post-exposure): 1,200,000 units every 12 hours in adults, 25,000 units per kilogram of body weight (maximum 1,200,000 unit) every 12 hours in children. The available safety data for penicillin G procaine at this dose would best support a duration of therapy of 2 weeks or less. Treatment for inhalational anthrax (post-exposure) must be continued for a total of 60 days. Physicians must consider the risks and benefits of continuing administration of penicillin G procaine for more than 2 weeks or switching to an effective alternative treatment."

V. Conclusions

Drug products containing the following active ingredients are currently approved for administration in cases of inhalational anthrax:

- Doxycycline
- Doxycycline calcium
- Doxycycline hyclate
- Penicillin G procaine

We encourage the submission of labeling supplements for these drug products. The revised labeling should specifically mention inhalational anthrax (post-exposure), the recommended dosing regimen, safety information relevant to prolonged exposure (60 days or longer), and other information described in this notice. The requirement for data to support these labeling changes may be met by citing the published literature we relied on in publishing this notice. A list of the published literature and reprints of the reports will be available for public inspection in the Dockets Management Branch (HFA–305), Food and Drug Administration, 5630 Fishers Lane, Rm. 1061, Rockville, MD 20852. It is unnecessary to submit copies and reprints of the reports from the listed published literature. We invite applicants to submit any other pertinent studies and literature of which they are aware.

VI. Published Literature

The published literature we have relied on in making our recommendations will be placed on display in the Dockets Management Branch (address above) and may be seen by interested persons between 9 a.m. and 4 p.m., Monday through Friday. A list of this published literature will be on display in the Dockets Management Branch and on the Internet at

www.fda.gov/cder/drug/infopage/penG...doxy/bibliolist.htm.

Dated: October 26, 2001

Bernard A. Schwetz, *Acting Principal Deputy Commissioner*

[FR Doc. 01–27493 Filed 10–29–01; 4:35 pm]

SOURCE: Federal Register/Vol. 66, No. 213/Friday, November 2, 2001/Notices

Appendix F

Veterans Affairs/Department of Defense Contingency

The Department of Veterans Affairs (VA) serves as the primary medical backup to the military health care system during and immediately following an outbreak of war or national emergency. The VA/Department of Defense (DoD) Contingency Hospital System Plan outlines how the Veterans Health Administration (VHA) supports that effort.

BACKGROUND

The VA/DoD Health Resources Sharing and Emergency Operation Act (Public Law 97–174) was enacted on May 4, 1982. This law gave VA a new mission: to serve as the principal health care backup to DoD in the event of war or national emergency that involves armed conflict. In addition to the contingency mission, this public law amended Title 38, United States Code (U.S.C.), to promote greater peacetime sharing of health care resources between VA and DoD.

In response to Public Law 97–174, a Memorandum of Understanding (MOU) was executed between the Secretary of Defense and the Administrator of Veterans Administration (presently the Secretary of Veterans Affairs), specifying each agency's responsibilities under the law.

DoD maintains medical operations plans that would coordinate the receipt, distribution, and treatment of returning military casualties. The VA/DoD Contingency Hospital System Plan describes how VA hospital beds would be made available to treat returning military casualties.

VA/DOD CONTINGENCY ANNUAL REPORT

Annually, VA medical centers estimate the number of beds that could potentially be made available to receive returning military casualties. These reported bed estimates take into account the impact on local operations of VA employees subject to military call up. This annual report includes Estimated VA Contingency Beds and VHA Employees Subject to Mobilization. An annual report is provided by the Secretary of Veterans Affairs to select members of Congress.

BED AVAILABILITY REPORTING EXERCISES

Quarterly estimates of VA/DoD contingency beds are gathered from VA medical centers nationally. These exercises are conducted quarterly in order to maintain VA's awareness and readiness to respond in a timely fashion should the VA/DoD Contingency Hospital System be activated.

POINT OF CONTACTS FOR VA/DOD CONTINGENCY

VA/DoD Annual Report:	Philip Wooten Director, Plans Phone: 304-264-4837 Fax: 304-264-4499
VA/DoD Quarterly Bed Report:	Michael Vojtasko Director, Operations Phone: 304-264-4801 Fax: 304-264-4810

SOURCE: www.va.gov/emshg

Appendix G

The Model State Emergency Health Powers Act

as of December 21, 2001

a Draft for Discussion Prepared by:

The Center for Law and the Public's Health at Georgetown and Johns Hopkins Universities

For the Centers for Disease Control and Prevention [CDC]

to Assist:
National Governors Association [NGA],
National Conference of State Legislatures [NCSL],
Association of State and Territorial Health Officials [ASTHO], and
National Association of County and City Health Officials [NACCHO]

Contact Information:
Lawrence O. Gostin, J.D., LL.D. (Hon.), Professor and Director,
Center for Law and the Public's Health, Georgetown University Law Center
600 New Jersey Avenue, N.W., Washington, D.C. 20001
(202) 662-9373
gostin@law.georgetown.edu
Full text available at: www.publichealthlaw.net

PREAMBLE

In the wake of the tragic events of September 11, 2001, our nation realizes that the government's foremost responsibility is to protect the health, safety, and well being of its citizens. New and emerging dangers—including emergent and resurgent infectious diseases and incidents of civilian mass casualties—pose serious and immediate threats to the population. A renewed focus on the prevention, detection, management, and containment of public health emergencies is thus called for.

Emergency health threats, including those caused by bioterrorism and epidemics, require the exercise of essential government functions. Because each state is responsible for safeguarding the health, security, and well being of its people, state and local governments must be able to respond, rapidly and effectively, to public health emergencies. The Model State Emergency Health Powers Act (the "Act") therefore grants specific emergency powers to state governors and public health authorities.

The Act requires the development of a comprehensive plan to provide a coordinated, appropriate response in the event of a public health emergency. It facilitates the early detection of a health emergency by authorizing the reporting and collection of data and records, and allows for immediate investigation by granting access to individuals' health information under specified circumstances. During a public health emergency, state and local officials are authorized to use and appropriate property as necessary for the care, treatment, and housing of patients, and to destroy contaminated facilities or materials. They are also empowered to provide care, testing and treatment, and vaccination to persons who are ill or who have been exposed to a contagious disease, and to separate affected individuals from the population at large to interrupt disease transmission.

At the same time, the Act recognizes that a state's ability to respond to a public health emergency must respect the dignity and rights of persons. The exercise of emergency health powers is designed to promote the common good. Emergency powers must be grounded in a thorough scientific understanding of public health threats and disease transmission. Guided by principles of justice, state and local governments have a duty to act with fairness and tolerance towards individuals and groups. The Act thus provides that, in the event of the exercise of emergency powers, the civil rights, liberties, and needs of infected or exposed persons will be protected to the fullest extent possible consistent with the primary goal of controlling serious health threats.

Public health laws and our courts have traditionally balanced the common good with individual civil liberties. As Justice Harlan wrote in the seminal United States Supreme Court case of *Jacobson* v. *Massachusetts*, "the whole people covenants with each citizen, and each citizen with the whole people, that all shall be governed by certain laws for the 'common good.'" The Act strikes such a balance. It provides state and local officials with the ability to prevent, detect, manage, and contain emergency health threats without unduly interfering with civil rights and liberties. The Act seeks to ensures a strong, effective, and timely response to public health emergencies, while fostering respect for individuals from all groups and backgrounds.

Although modernizing public health law is an important part of protecting the population during public health emergencies, the public health system itself needs improvement. Preparing for a public health emergency requires a well-trained public health workforce, efficient data systems, and sufficient laboratory capacity.

Appendix H

NACCHO Research Brief: Assessment of Local Bioterrorism and Emergency Preparedness

INTRODUCTION

"Is the United States ready for a bioterrorist attack?" This is a question continually posed by concerned citizens and the media. Local public health agencies (LPHAs) serve on the frontlines in responding to bioterrorism threats and other public health emergencies. LPHAs play a key role in preparing jurisdictions for bioterrorism, including rapid detection of unusual health events, coordination with response partners and healthcare facilities, providing treatment recommendations and protocols to prevent spread of infection and disease, doing "contact tracing" to assure that all individuals exposed to bioterrorism agents are reached for testing and treatment, and providing health information and resources to the public and the media.

The threat from bioterrorism has become more real since the recent September 11, 2001, terrorist attacks on the United States and recent anthrax incidents. These incidents have led many LPHAs to examine their capacity to prepare for, detect, and respond to emerging health threats. The National Association of County and City Health Officials (NACCHO) conducted a survey in October 2001 to assess local preparedness for bioterrorism and found that only 20% of LPHAs have a comprehensive response plan in place. LPHAs have made progress and learned important lessons about the challenges of bioterrorism preparedness in the last few years, but have a long way to go to achieve the capacities needed to detect and respond to an act of bioterrorism as quickly as possible, to prevent the spread of disease and save lives.

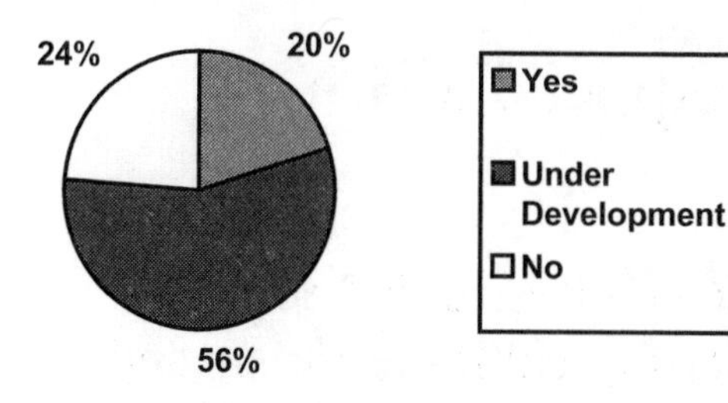

FIGURE 1: Percent of LPHAs with a Comprehensive Response Plan.

METHODOLOGY

The data presented in this research brief were obtained from an assessment of LPHA bioterrorism and emergency preparedness conducted by NACCHO. In response to the September 11, 2001, attacks and subsequent anthrax incidents, NACCHO conducted this survey to better understand how these events impacted LPHAs and how prepared they were to respond. A 9-question survey was developed by NACCHO, and was faxed and e-mailed to 999 NACCHO members and state associations of local public health agencies. In a short turnaround time of one week, 530 responses were received, a response rate of 53%.

Once survey responses were received, NACCHO staff conducted the data analysis. Open-ended questions were coded for ease of analysis, and data analysis was conducted using the statistical software package Stata®.

LPHAS IN THE AFTERMATH OF SEPTEMBER 11

Local health officials played a variety of roles in response to the September 11 terrorist events. Most of these roles revolved around communicating with various partners within their communities. A number of respondents indicated that much of the time was spent fielding questions from concerned community members and staff. Other roles included working with response partners to develop, update, and review plans and protocols to respond to emergencies. Developing fact sheets and providing information to the public and media were also mentioned.

Other roles mentioned by respondents included placing LPHA staff on alert, activating or supporting emergency operations centers (EOC) community systems, activating emergency response plans, and increasing disease surveillance. Some respondents indicated that they had to rely on the news media to be alerted and receive updates, not local disaster response agencies, state health departments, or federal agencies.

When asked how well prepared the LPHA was to play these roles, 15% indicated they were well prepared, while a large majority, 75%, said they were only fairly or somewhat prepared. Nine percent (9%) of respondents indicated they were not prepared at all.

TYPES OF INQUIRIES RECEIVED AS A RESULT OF THE ATTACKS

Sixty-six (66%) of LPHAs surveyed indicated that the health department received a significant number of questions regarding the September 11 event or terrorism/bioterrorism in general.

By far, most inquiries received were questions concerning vaccination and medication availability. Other frequently asked questions dealt with the level of local preparedness and the existence of emergency response plans. Many citizens wanted to know how prepared their community was, what the LPHA was doing to prepare the community, and how they could obtain copies of the local plan. There were also numerous questions regarding the threat of bioterrorism, such as: "What is the likelihood that a bioterrorist attack would occur?" and "Are we in danger of crop sprayers spreading bio-agents?" Lastly, there were questions about where the community would receive necessary resources and where citizens could donate blood or help in any other way.

When asked how well prepared they felt to respond to these inquiries they were receiving, 38% of the LPHAs who responded indicated they were "pretty well prepared" to respond, while another 50% said they were only "somewhat prepared," and 12% felt they were "not prepared at all."

FRUSTRATIONS AND LESSONS LEARNED

Many LPHAs indicated they faced several problems and frustrations in this time of crisis. The chief frustration voiced by respondents was the lack of resources and equipment failure. Specifically, many discussed the malfunctioning of necessary communication tools such as pagers, cell phones, e-mails, and faxes. This frustration was related to the second most frequently reported frustration: lack of or poor communication from state and federal agencies. Many local health officials interpreted this lack of information as poor leadership from federal and state health agencies.

Another common frustration was about insufficient local preparedness. Many LPHAs indicated they lacked a fully developed response plan. Others had no plan at all. Other frustrations included lack of consistent and standard information regarding bioterrorism, poor coordination between public health and emergency management, and the need for increased training among LPHA staff and the need for more specialized staff. Problems dealing with the media, such

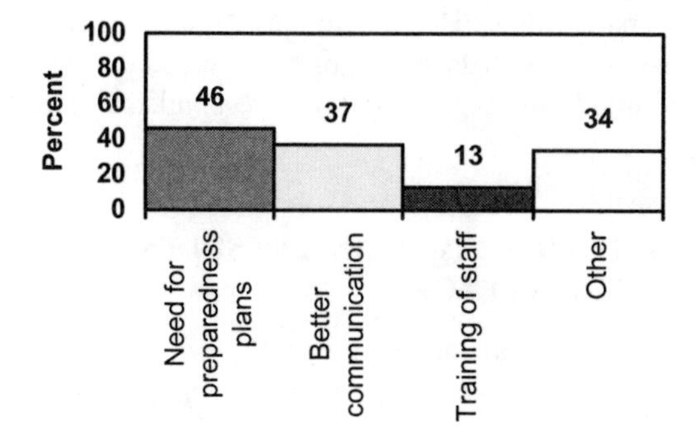

FIGURE 2: Percent of LPHAs: Lessons Learned

as countering incorrect media information and being overwhelmed with questions from the media, were also cited as common frustrations.

Sixty-four percent (64%) of the LPHAs surveyed indicated that the events of September 11 contributed some lessons learned for their health department (see Figure 2). The primary lesson was that they are not adequately prepared for a bioterrorist attack and that they need to have a response plan in place with clearly defined roles and chain of command. There is also a need for enhanced collaboration among federal, state, and local agencies, including development of up-to-date contact lists. Furthermore, many realized the importance of reliable methods of communication, including pagers, cell phones, email. The need to train LPHA staff was also noted. Other lessons learned dealt with improving syndromic surveillance capacity, addressing mental health issues, and providing information to physicians, the public, and the media.

CONCLUSIONS

Gaps in local public health preparedness were clearly realized in the aftermath of September 11. Our nation's local public health system is lacking in its preparedness to protect its communities if faced with a biological attack. Bioterrorism preparedness plans, effective communication systems, and reliable and timely information are key to a prepared public health workforce, yet these pieces are currently incomplete.

Additional resources to enhance our nation's public health system are critical at this time. LPHAs play a vital role in the response to bioterrorism and other emergency situations. The capacities needed by LPHAs to effectively respond to an act of bioterrorism allow for the development of a dual-use infrastructure that improves the capacity to respond to all local public health emergencies and hazards. NACCHO will continue its efforts to build local response capacity to benefit and improve the health of communities nationwide.

NACCHO is the national organization representing local public health agencies (including city, county, metro, district, and tribal agencies). NACCHO works to support efforts which protect and improve the health of all people and all communities by promoting national policy, developing resources and programs, and supporting effective local public health practice and systems.

This report was prepared by Anjum Hajat, MPH, Zarnaaz Rauf, MPH, and Carol Brown, MS.

SOURCE: http://www.naccho.org/files/documents/BT_brief1.pdf

Appendix I

Statement on Vaccine Development

Council of the Institute of Medicine
November 5, 2001

The events following the tragedies of September 11, 2001, have reemphasized a serious defect in America's capacity to deal with biological agents used in terrorist attacks. The capacity to develop, produce, and store vaccines to deal with these agents are inadequate to meet the nation's needs. In 1993 the Institute of Medicine published *The Children's Vaccine Initiative: Achieving the Vision.* In assessing the national and international situation, the committee said, "because the private sector alone cannot sustain the costs and risks associated with the development of most CVI vaccines, and because the successful development of vaccines requires an integrated process, the committee recommends that an entity, tentatively called the National Vaccine Authority (NVA), be organized to advance the development, production, and procurement of new and improved vaccines of limited commercial potential but of global public health need" [1].

In a 1992 report, *Emerging Infections: Microbial Threats to Health in the United States*, another IOM committee recommended the development of an integrated management structure within the federal government for acquiring vaccines, as well as a facility for developing and producing vaccines with government support [2].

Evidence for the inability of the private sector to meet the country's needs for vaccines has accumulated substantially since the 1993 report. Fewer private companies are manufacturing vaccines. Continually needed vaccines such as the tetanus and influenza vaccines are in increasingly short supply. The availability of influenza vaccines has been delayed over the past several years and in 2000, one company stopped production. Pneumococcal conjugate vaccine is unavailable in several states because of the sole source manufacturer's inability to meet demand. Only one source is currently available for meningococcal varicella and measles-mumps-rubella vaccines.

There are just four major vaccine manufacturers in the world today, and only two in the United States [3]. There were four times that number only 20 years ago. There are many small new research and development companies backed by venture capital and devoted to vaccine development. Many are working on anticancer vaccines for which market forces may be enough to keep them in production. However, good products developed by these startups to combat infectious diseases often do not come to market because of the very large costs of testing in pilot studies and in manufacturing. Currently, the United States has a single licensed anthrax vaccine product, manufactured by a single plant. Because the Food and Drug Administration (FDA) had identified problems in the manufacturing process during regular inspections, the plant was closed for renovations in 1998, and to date, no new lots of anthrax vaccine have yet been cleared for release.

Prior to the events of September 11, the delays and problems faced by both the Department of Health and Human Services and Department of Defense in developing and procuring a cell-culture smallpox vaccine provide convincing evidence that major changes are needed at the national level. With the government guaranteeing payment in this time of national need, several potential manufacturers have come forward. This is an ad hoc example of a larger national need for mechanisms to obtain other public-good vaccines on an ongoing basis, and not just under extenuating circumstances when there is a great deal of public awareness of the need for vaccines.

The Children's Vaccine Initiative committee listed the functions of a National Vaccine Authority as shown in Appendix 1. While these activities focused on the Children's Vaccine Initiative, they now have a broader importance to America, as the potential need for vaccines required to meet biological threats increases. The IOM Council believes the Authority should focus its attention upon vaccines that will not be adequately produced by existing public or private entities. Important functions of the Authority would include: conducting in-house vaccine-related research and development, assisting companies in the production of pilot lots of vaccines; and arranging and contributing to the procurement of National Vaccine Authority vaccines. An especially important function would be to provide opportunities for the production of pilot lots of vaccines developed by small biotechnology firms, and to produce vaccines when market forces are not sufficient to facilitate large-scale production.

The IOM Council further believes the Authority should facilitate communications among relevant contributors to vaccine research and development, including academic research efforts, manufacturers, regulatory agencies, and the public. The Authority should not interfere in any way with public or private research or development efforts to create new vaccines. It should be available to assist such efforts when opportunities arise. It should interact with other public and private entities to assure a timely and effective system for storage and distribution of appropriate vaccines. It should identify mechanisms to expand current forms of liability protection for the adverse effects of vaccines, including expansion of federal efforts for indemnification of manufacturers. The Authority could become a source of appropriate reliable information to the media health care personnel, policy-makers, and the public. The FDA could work closely with such an Authority to oversee vaccine development and production as well as facilitate their oversight processes and reduce regulatory complexities. In some cases, it might find mechanisms to guarantee a price for vaccines to stimulate private sector production, as has occurred with smallpox vaccine in the current situation.

Recently, proposals have been made for the creation of a government-owned, contractor-operated national vaccine facility. The IOM Council believes this is one in a spectrum of public-private ventures by which a NVA could facilitate development and production of needed vaccines. The conduct of research, development, production, and distribution of vaccines in such a facility should be the responsibility of a private contractor selected by a competitive bidding process. This effort should not preclude other collaborations with private contractors in other public-private projects. Funding for such a facility will initially require a substantial financial investment [4]. While a major priority for this facility would be to develop vaccines necessary to protect American troops and for use against bioterrorism, the facility also should be charged with production of other vaccines that are in scarce supply and would not otherwise be provided in the public or private sectors. In some cases in which there are few private sector uses, the facility would become the principal source of such vaccines. In other cases, a variety of public and private partnerships could be undertaken to produce needed vaccines [5].

The Council of the Institute of Medicine of the National Academies believes that the development of a National Vaccine Authority is long overdue. It could be created within the Department of Health and Human Services, in collaboration with the Department of Defense or as a joint effort of the two departments. Moreover, the Council believes that establishment of a government-owned, contractor-operated facility for research, development, and production of vaccines is essential to meeting the country's public health needs, particularly those related to bioterrorism and protection of our armed forces. This facility also should play a role in development and production of other vaccines required for the public health that are not currently available on the open market. The Council encourages the president of the United States, the secretary of health and human services, secretary of defense, and the director of the Office of Homeland Security to evaluate these recommendations as critical elements for maintaining the country's health.

COUNCIL OF THE INSTITUTE OF MEDICINE

Kenneth I. Shine, M.D. (Chairman)
President, Institute of Medicine, Washington, D.C.

Nancy C. Andreasen, M.D., Ph.D.
Andrew H. Woods Chair of Psychiatry, and Director, Mental Health Clinical Research Center, University of Iowa Hospitals and Clinics, Iowa City

Enriqueta C. Bond, Ph.D.
President, Burroughs Wellcome Fund, Research Triangle Park, North Carolina

Jo Ivey Boufford, M.D.
Dean, Robert F. Wagner Graduate School of Public Service, New York University, New York City

Harvey R. Colten, M.D.
President and Chief Medical Officer, iMetrikus, Inc., Carlsbad, California

Johanna T. Dwyer, D.Sc.
Director, Frances Stern Nutrition Center, New England Medical Center, Boston, Massachusetts

Anthony S. Fauci, M.D.
Director, National Institute of Allergy and Infectious Diseases, National Institutes of Health, Bethesda, Maryland

James R. Gavin III, M.D., Ph.D.
Senior Scientific Officer, Howard Hughes Medical Institute, Bethesda, Maryland

Ada Sue Hinshaw, Ph.D.
Dean, School of Nursing, University of Michigan, Ann Arbor

Brigid L.M. Hogan, Ph.D.
Investigator, Howard Hughes Medical Institute, and Hortense B. Ingram Professor, Department of Cell Biology, Vanderbilt University School of Medicine, Nashville, Tennessee

Thomas S. Inui, M.D.
President and Chief Executive Officer, The Fetzer Institute, Kalamazoo, Michigan

Michael M.E. Johns, M.D.
Executive Vice President for Health Affairs, Director, Robert W. Woodruff Health Science Center, Emory University, Atlanta, Georgia

Mary-Claire King, Ph.D.
American Cancer Society Professor of Medicine and Genetics, University of Washington, Seattle

Lawrence S. Lewin, M.B.A.
Executive Consultant, Washington, D.C.

Joseph B. Martin, M.D., Ph.D.
Dean of the Faculty of Medicine, Harvard Medical School, Boston, Massachusetts

William L. Roper, M.D.
Dean, School of Public Health, University of North Carolina, Chapel Hill

Stephen M. Shortell, Ph.D.
Blue Cross of California Distinguished Professor of Health Policy and Management, Professor of Organization Behavior, School of Public Health, University of California, Berkeley

Edward H. Shortliffe, M.D., Ph.D.
Professor and Chair, Department of Medical Informatics
Columbia Presbyterian Medical Center, Columbia University, New York City

Kenneth E. Warner, Ph.D.
Avedis Donabedian Distinguished University Professor of Public Health
Department of Health Management & Policy
School of Public Health, University of Michigan, Ann Arbor

Gail R. Wilensky, Ph.D.
John M. Olin Senior Fellow, Project HOPE, Bethesda, Maryland

Ex Officio Members (non-voting):

David R. Challoner, M.D.
IOM Foreign Secretary
Director, Institute for Science and Health Policy, University of Florida, Gainesville

Harold J. Fallon, M.D.
IOM Home Secretary
Dean Emeritus, School of Medicine, University of Alabama, Birmingham

REFERENCES

1. Mitchell, V.S., Philipose, N.M., and Sanford, J.P., eds. The Children's Vaccine Initiative: Achieving the Vision. Washington, D.C.: National Academy Press, 1993.
2. Lederberg, J., Shope, R.E., and Oaks, S.C., Jr., eds. Emerging Infections: Microbial Threats to Health in the United States. Washington, D.C.: National Academy Press, 1992.
3. Merck Vaccine Division (parent company is Merck Pharmaceuticals) and Wyeth-Lederle Vaccines (parent company is American Home Products Corporation) are U.S.-based companies. Aventis Pasteur and GlaxoSmithKline operate within the United States and have products licensed by the FDA for use in the United States, but they are companies based in other countries.
4. Department of Defense. Report on Biological Warfare Defense Vaccine Research and Development Programs. July 2001.
5. Pearson, G.W. The Children's Vaccine Initiative: Continuing Activities. Washington, D.C.: National Academy Press, 1995.

APPENDIX 1: Function of a National Vaccine Authority

- Define the need
- Assess the market
- Establish priorities for U.S. CVI vaccine development in conjunction with the global CVI
- Characterize desired vaccine products
- Assemble intellectual property rights
- Advance CVI product development through the private sector
- Conduct in-house vaccine-related research and development
- Assist companies in the production of pilot lots of vaccine
- Support clinical testing and field trials of candidate vaccines
- Transfer CVI-related vaccine technology to developing country manufacturers
- Train U.S. and overseas nationals in the principles of vaccine development, pilot manufacture, and quality control
- Arrange and contribute to the procurement of NVA vaccines
- Evaluate and redefine needs
- Represent the United States in international CVI forums, such as the Consultative Group

NOTE: Mitchell, V.S., Philipose, N.M., and Sanford, J.P., eds. The Children's Vaccine Initiative: Achieving the Vision. Washington, D.C.: National Academy Press, 1993, p. 133.

In addition to these functions, the need for vaccines to fulfill anti-terrorist and military requirements should be included.

SOURCE:
http://www.iom.edu/IOM/IOMHome.nsf/Pages/Vaccine+Development

Appendix J

Testimony of
Kenneth I. Shine, M.D.

President, Institute of Medicine, The National Academies

For a Hearing on Risk Communication:
National Security and Public Health

Before the Congress of the United States House of Representatives
Subcommittee on National Security, Veterans Affairs,
and International Relations

Committee on Government Reform, U.S. House of Representatives
November 29, 2001

I am Kenneth I. Shine, President of the Institute of Medicine of the National Academy of Sciences. For the last three years I have also served as a member of the congressionally mandated Commission on Weapons of Mass Destruction, chaired by Governor James Gilmore of Virginia, otherwise known as the Gilmore Commission. My comments reflect the opinions of the National Academies, as represented by Bruce Alberts, President of the National Academy of Sciences, and William Wulf, President of the National Academy of Engineering, whom I joined in signing a statement on October 3, 2001. In this statement, we advised the American public and health professionals to seek authoritative information on anthrax from three websites, those at the Centers for Disease Control and Prevention, The National Library of Medicine, and the Johns Hopkins University. We made the statement because of our concern about the amount of misinformation being conveyed about the anthrax incidents and the confusion that had resulted from multiple sources of analysis, commentary, and advice.

Mr. Chairman, in 1988 the Institute of Medicine issued a report called *The Future of Public Health*. The report described the state of infrastructure for public health in America as in "disarray." The report recommended renewed national attention to the infrastructure, human resource needs, educational capacity, and programs in public health in America. In 1992, the Institute of Medicine issued a report on *Emerging Infections: Microbial Threats to Health in the United States,* from a committee chaired by Nobel Laureate, Joshua Lederberg, and Dr. Robert Shope. In that report, additional recommendations were made for strengthening the capacity of public health and medicine to deal with new and emerging infections including those presented by terrorism. Although some additional resources were provided to the Centers for Disease Control and Prevention in response to these reports, these were limited. Over the past decade the overall condition of the public health system in America has continued to erode. Many of these weaknesses were graphically displayed during the anthrax episodes. Laboratory capabilities, adequate staff for investigations, the relationship and responsibilities of public health to law enforcement and, especially for purposes of this hearing, the effectiveness of communications to the public and to health professionals about the anthrax terrorism were found wanting.

Key to the role of public health is education and information for the public and for health professionals. Whether an epidemic is a naturally occurring one such as that involving West Nile virus, or whether produced by a terrorist, public health professionals and public health departments around the United States need timely, accurate, and reliable information.

Every epidemic results in new knowledge as it is studied and understood. In the case of anthrax, information about the inhalation form of the disease was limited to a very small number of cases over an extended period of time. Medical practitioners and public health officials in the United States never had direct experience with inhalation anthrax. Not only is it important to learn in an ongoing way as such an epidemic develops, but it is also important to rapidly translate that knowledge into reliable guidance to health professionals and to the community.

In this testimony I will focus on two critical methods of communication about these issues in the 21st century: verbal communication—particularly via television—and the Internet. I begin with remarks concerning verbal communication.

Within the Department of Health and Human Services, there must be a single credible medical/public health expert spokesperson who reports regularly, most likely daily, to the American people in regard to any outbreak with national significance. This is analogous to the situation in local communities where there is a need for such an individual to communicate on behalf of the local health department. Several months before the anthrax outbreak, uninformed statements on local television in a community with two cases of meningococcal meningitis resulted in thousands of individuals taking antibiotics or seeking immunizations that were not indicated. Local stores of antibiotics were depleted and many people were subjected to risk from unnecessary treatment. This episode emphasizes the need for credible medical/public health information during natural events, as well as during those that are produced by terrorism.

In the case of the anthrax episodes, the media responded by interviewing countless numbers of individuals. Among them was a self-professed pundit who announced he was an expert on the "anthrax virus." Anthrax is a bacterium, not a virus. In many cases, well-intentioned infectious disease specialists who knew a good deal about the literature on anthrax could provide accurate retrospective information, but when pressed about the current events, they were not privy to information about the cases that had occurred. They were then forced to either acknowledge their limitations, which the responsible experts did, or in the case of others less responsible, to speculate based on news reports, rumors, and a variety of other kinds of incomplete or false information.

In a national emergency, such as that experienced with anthrax, the regular appearance on television of a credible medical/public health expert spokesperson who has up-to-date knowledge of the outbreak is important. Such a responsible individual can of course consult with law enforcement agencies with regard to information that might be important in an ongoing criminal investigation. However, the goal of the terrorist is to produce terror. Terror arises from fear magnified by an exaggerated sense of risk, and perpetuated by misinformation and rumors. In these episodes, the balance should be biased in favor of providing good information to protect the public health.

In addition to the Department of Health and Human Services, the other major stakeholder that must provide public information in the case of terrorism is the Office of Homeland Security. The Gilmore Commission has urged that one of the associate directors of that office be an Associate Director for Health. We know far too little about the availability of hospital beds, burn units, decontamination capability, and a variety of other parameters required by the health system to deal with terrorism. Moreover, the necessity for dramatically improved communications among the public health system, the medical care system, and law enforcement all require a high level of coordination and communication. If this individual is also to be a spokesperson on such episodes as the anthrax outbreak, it is critical that his or her statements should be carefully coordinated with the principal credible medical/public health spokesperson within the Department of Health and Human Services. These messages must be well thought out and consistent to avoid confusion and misdirection. And clearly both individuals must be kept completely up to date with the most recent information, including the complete results of scientific and forensic analyses.

It is understandable that political leaders and Administration officials wish to be the spokespersons for their departments or agencies in the face of a threat to the national security or to the nation's health. It is important that they do so. But the impact of their communications are not diminished when they are joined by a credible medical/public health expert spokesperson who is knowledgeable about the nature of the disease and is also privy to up-to-date information about the outbreak. Turning to such an individual when technical questions are raised does not diminish, but rather enhances, the authority of the non-medical leader in addressing the public's concerns. For example, the presence of Dr. Anthony Fauci at hearings and press conferences came late in the sequence of events but his appearance was extremely valuable. Furthermore, his interviews by the media were paradigms of clarity, accuracy, and relevancy. It is noteworthy that, in one of his appearances with a number of so-called experts, he was forced to correct inaccurate statements made by others during the program.

The other major issue that was identified by the October 3 statement from my colleagues and me is the importance of authoritative, well-presented, up-to-date websites where health professionals, the public, and others can quickly obtain good information. The Internet has been flooded with multiple websites concerning anthrax. Many are reliable. But, as noted in our message, many are incorrect, inaccurate, misleading, and in some cases downright scams. Identifying the most reliable during an emergency is important for those who seek such information.

The CDC maintains a website for this purpose. Ultimately, as we indicated in our statement, excellent information appeared at that website, though it was not as well organized as it might have been. The capacity of the CDC website to respond to inquiries was, for a period of time, limited. Access was limited by the large number of inquiries. In view of the importance of credible and accurate information, accurate resources should be made available so that the CDC can provide information using the most modern technologies and the most professional presentations, and have both the bandwidth and the human capacity to respond to a large number of inquiries. The spokesperson for the Department of Health and Human Services/and or the office of Homeland Security should regularly remind the public and health professionals that they can get such reliable information at the CDC website.

There is an important lesson in this experience for other government agencies. We do not know where other terrorist events may occur. Does the Department of Energy have the capacity to respond to inquiries from the public and professionals with high-quality, rapidly updated information should there be an incident involving radiological materials or a nuclear event? Can the Department of Agriculture respond with the type of credible expert to whom I refer and have a website with the capacity required for all inquiries, should there be any problems in the areas of agriculture or animal husbandry? Where will the appropriate website be located for information about an episode involving terrorism using a chemical agent? There are many agencies involved in these issues, but if information for the public is crucial the principles that I have outlined for the Department of Health and Human Services and the office of Homeland Security should also apply in each of these other areas. A single preferred information source should be assigned now to a single government agency in each case, and resources must be dedicated by this agency to maintaining this capability on high alert.

Mr. Chairman, many individuals in both the public and private sector work very long hours, seven days a week in coping with the anthrax episodes. They deserve enormous credit for their efforts in this regard. As the Institute of Medicine reports have emphasized for 13 years, the public health infrastructure at the state, federal, and local levels requires substantial upgrading. Strengthening the size and configuration of the Epidemic Intelligence Service, the facilities at the CDC, and the surveillance capacity of the federal, state, and local public health and medical entities are crucial. The Gilmore Commission has recommended that the Associate Director for Health in the Office for Homeland Security have an advisory panel consisting of representatives from a wide variety of hospitals, medical organizations, and first responders who can develop methodologies to rapidly communicate throughout the country the information required about how to meet emergencies in a timely way. The Institute of Medicine has published a preliminary report (to be followed by a full report next year) on the methodologies by which we can assess the capacity of local communities to respond to an episode of terrorism. The American Medical Association has developed an excellent plan to create educational programs and disaster planning efforts through their state and local societies. The Joint Commission on Accreditation of Healthcare Organizations has developed a plan for hospitals to improve their capabilities to cope with disasters. Resources will be necessary to make these happen, some of which will require federal help.

Additional research to deal with biological agents is essential. For example, the current anthrax vaccine requires six doses over 18 months. While studies are underway to determine the efficacy of fewer doses we desperately need a much better, purer, and more effective vaccine against anthrax. The Council of the Institute of Medicine has called for the establishment of a national vaccine authority or its equivalent to ensure supplies of vaccines that are not available through the market or that require public/private collaborations to ensure adequate supplies, as exemplified by shortages of anthrax and smallpox vaccines. This should include vaccines for childhood diseases, adult infections, and, in the case of preventing the spread of hoof and mouth disease, for animals. Improved diagnostic and therapeutic options are also required.

Central to all of these efforts is information and communication: information, which the American people can understand, and information about the concepts of risk and how to apply them. In the case of anthrax, less than 20 cases resulted in thousands of people taking antibiotics that were not indicated. Perhaps 20 percent of these individuals experienced some side affects from these drugs. These antibiotics changed the bacteriological environment and may have rendered some organisms resistant to the antibiotics employed. Several effective antibiotics were available and better early information might have prevented the exhaustion of stores of ciprofloxacin. A clear recommendation that one not take ciprofloxacin unless one is a member of a specifically defined high risk group, for example, postal workers or those with potential exposures on Capitol Hill, would also have been very helpful in this situation.

The debate about smallpox vaccination will be much more straightforward if the American public understands the concept of risk/benefit. Smallpox was controlled throughout the world by vaccination of the populations who had been potentially exposed to a case—that is, by surrounding the cases as soon as they were observed with vaccinations. Even two or three days after exposure, vaccination will prevent the disease. The public needs to be informed that this represents an excellent alternative to mass vaccination—which is likely to kill hundreds of people and seriously damage many more. We also need additional research to determine how many Americans who were vaccinated years ago have persistent immunity. This would help further refine the risk/benefit analysis and the needs for vaccine.

In summary, I have emphasized the critical role of a credible medical/public health expert spokesperson, knowledgeable about the current events, who speaks for the Department of Health and Human Services and stands side by side with the secretary in his or her communications. If a similarly qualified spokesperson on bioterrorism is to be designated in the Office of Homeland Security, the credible medical/public health expert spokesperson(s) must carefully coordinate their statements so that they are accurate, authoritative, and understandable, and consistent. Much more serious attention should be paid to the role of well-organized, well-presented, and technologically sophisticated websites for providing information to the public, health professionals, the media, and others. Such sites should be developed and be on alert (and when needed be well advertised) for each of the areas relevant to a potential terrorist attack.

Thank you Mr. Chairman, for this opportunity and I would be happy to answer any questions.

SOURCE:
http://www4.nationalacademies.org/ocga/testimon.nsf/By+Congress/2d04c080a441cbbf85256b130064a736?OpenDocument

Appendix K

Glossary and Acronyms

GLOSSARY

This glossary is intended to define terms encountered throughout this report as well as some terms that are commonly used in the public health arena. This glossary is not all-inclusive. New terms and new usages of existing terms will emerge with time and with advances in technology. The definitions for the terms presented here were compiled from a multitude of sources.

Antibiotic: Class of substances or chemicals that can kill or inhibit the growth of bacteria. Originally antibiotics were derived from natural sources (e.g., penicillin was derived from molds), but many currently used antibiotics are semi-synthetic and are modified by the addition of artificial chemical components.

Antibiotic resistance: Property of bacteria that confers the capacity to inactivate or exclude antibiotics or a mechanism that blocks the inhibitory or killing effects of antibiotics.

Antimicrobial agents: Class of substances that can destroy or inhibit the growth of pathogenic groups of microorganisms, including bacteria, viruses, parasites, and fungi.

Antiphagocytic: Counteracting or opposing phagocytosis, the process by which a cell engulfs particles such as bacteria, aged red blood cells, or foreign matter.

ALT (alanine aminotransferase): An enzyme normally present in liver and heart cells that is released into the bloodstream when the liver or heart is damaged.

AST (aspartate aminotransferase): An enzyme normally present in liver and heart cells that is released into blood when the liver or heart is damaged.

ATR (anthrax toxin receptor): A type I membrane protein with an extracellular Von Willebrand factor A domain that binds directly to PA.

Attenuated: To reduce the severity of (a disease) or virulence or vitality of a pathogenic agent.

Bacteremia: The presence of bacteria in the bloodstream.

Bacteria: Microscopic, single-celled organisms that have some biochemical and structural features different from those of animal and plant cells.

Basic research: Fundamental, theoretical, or experimental investigation to advance scientific knowledge, with immediate practical application not being a direct objective.

Benchmark: For a particular indicator or performance goal, the industry measure of best performance. The benchmarking process identifies the best performance in the industry (health care or non-health care) for a particular process or outcome, determines how that performance is achieved, and applies the lessons learned to improve performance.

Benign prostatic hypertrophy: Nonmalignant (noncancerous) enlargement of the prostate gland, a common occurrence in older men.

Blepharospasm: Tonic spasm of the orbicularis oculi muscle, producing more or less complete closure of the eyelids.

Broad-spectrum antibiotic: An antibiotic effective against a large number of bacterial species. It generally describes antibiotics effective against both gram-positive and gram-negative classes of bacteria.

BSL (biosafety level): Specific combinations of work practices, safety equipment, and facilities designed to minimize the exposure of workers and the environment to infectious agents. Biosafety level 1 applies to agents that do not ordinarily cause human disease. Biosafety level 2 is appropriate for agents that can cause human disease, but whose potential for transmission is limited. Biosafety level 3 applies to agents that may be transmitted by the respiratory route, which can cause serious infection. Biosafety level 4 is used for the diagnosis of exotic agents that pose a high risk of life-threatening disease, which may be transmitted by the aerosol route and for which there is no vaccine or therapy.

BT (bioterrorism): Terrorism using biological agents. Biological diseases and the agents that might be used for terrorism have been listed by the CDC and comprise viruses, bacteria, rickettsiae, fungi, and biological toxins. These agents have been classified according to the degree of danger each agent is felt to pose into one of three categories: A, B, and C (see definitions below).

Category A Biological Disease: High-priority agents include organisms that pose a risk to national security because they can be easily disseminated or transmitted person-to-person, cause high mortality, with potential for major public health impact, might cause public panic and social disruption, and require special action for public health preparedness. These diseases include: anthrax, botulism, plague, smallpox, tularemia, and viral hemorrhagic fevers.

Category B Biological Disease: Second-highest priority agents include those that are moderately easy to disseminate, cause moderate morbidity and low mortality, and require specific enhancements of CDC's diagnostic capacity and enhanced disease surveillance. These agents/diseases include: Q fever, brucellosis, glanders, ricin toxin, epsilon toxin, and staph toxin.

Category C Biological Disease: Third-highest priority agents include emerging pathogens that could be engineered for mass dissemination in the future because of availability, ease of production and dissemination, and potential for high morbidity and mortality and major health impact. These agents/diseases include Nipah virus, hantavirus, tickborne hemorrhagic fever viruses, tickborne encephalitis viruses, yellow fever, and tuberculosis.

CBER (Center for Biologics Evaluation and Research): A center of the FDA which regulates biological products including blood, vaccines, therapeutics and related drugs, and devices according to statutory authorities.

CDC (Centers for Disease Control and Prevention): A public health agency of the U.S. Department of Health and Human Services whose mission is to promote health and quality of life by preventing and controlling disease, injury, and disability.

CDER (Center for Drug Evaluation and Research): A center of the FDA whose mission it is to promote and protect public health by assuring that safe and effective drugs are available to Americans.

CDRH (Center for Devices and Radiological Health): A center of the FDA whose mission it is to provide reasonable assurance of the safety and effectiveness of medical devices and by eliminating unnecessary human exposure to radiation emitted from electronic products.

Chimeric: Relating to, derived from, or being a genetic chimera or its genetic material.

Clinical practice guidelines: Systematically developed statements that assist practitioners and patients with decision making about appropriate health care for specific clinical circumstances.

Clinical research: Investigations aimed at translating basic, fundamental science into medical practice.

Clinical trials: As used in this report, research with human volunteers to establish the safety and efficacy of a drug, such as an antibiotic or a vaccine.

Clinicians: One qualified or engaged in the clinical practice of medicine, psychiatry, or psychology, as distinguished from one specializing in laboratory or research techniques in the same fields.

CMV (cytomegalovirus): One of a group of highly host-specific herpesviruses that infect humans, monkeys, or rodents with the production of unique large cells bearing intranuclear inclusions.

Cutaneous: Related to the skin.

Cyanosis: The bluish color of the skin and the mucous membranes due to insufficient oxygen in the blood.

Cyclic AMP: A very close structural relative of adenosine monophosphate (AMP) containing an additional ester linkage between the phosphate and ribose units. It can act as a secondary messenger for several hormones and also plays a role in the transcription of some genes.

Cynomolgus monkeys: A macaque (*Macaca fascicularis* synonym *M. cynomolgus*) of southeastern Asia, Borneo, and The Philippines that is often used in medical research.

Cytokines: A small protein released by cells that has a specific effect on the interactions between cells, on communications between cells, or on the behavior of cells. The cytokines includes the interleukins, lymphokines, and cell signal molecules, such as tumor necrosis factor and the interferons, which trigger inflammation and respond to infections.

Cytosol: The liquid medium of the cytoplasm.

Dark Winter: An exercise portraying a fictional scenario depicting a covert smallpox attack on U.S. citizens which was held at Andrews Air Force Base, near Washington, D.C., on June 22–23, 2001.

DARPA (Defense Advanced Research Projects Agency): The DoD's central research and development organization which manages and directs selected projects, and pursues research and technology where risk and payoff are both very high and where success may provide dramatic advances for traditional military roles and missions.

DNIs (dominant negative inhibitors): Mutant forms of the protective antigen that block translocation of the virulence factors across the plasma membrane.

DoD (Department of Defense): DoD trains and equips the armed forces through three military departments—the Army, Navy, and Air Force whose primary job is to train and equip their personnel to perform warfighting, peacekeeping, and humanitarian/disaster assistance tasks.

Dyspnea: Difficult or labored breathing; shortness of breath.

Dystonia: Involuntary movements and prolonged muscle contraction, resulting in twisting body motions, tremor, and abnormal posture. These movements may involve the entire body, or only an isolated area.

EF (edema factor): One of two enzymes making up the anthrax toxin. After being cleaved by a protease, protective antigen (PA) binds to this toxic enzyme and mediates its transportation into the cytosol where it exerts its pathogenic effect.

EIS (Epidemic Intelligence Service): A unique, two-year postgraduate program of service and on-the-job training for health professionals interested in the practice of epidemiology.

ELISA (enzyme-linked immunosorbent assay): A rapid immunochemical test utilized to detect substances that have antigenic properties, primarily proteins. ELISA tests are generally highly sensitive and specific.

Emerging infections: Any infectious disease that has come to medical attention within the last two decades or for which there is a threat that its prevalence will increase in the near future (IOM, 1992). Many times, such diseases exist in nature as zoonoses and emerge as human pathogens only when humans come into contact with a formerly isolated animal population, such as monkeys in a rain forest that are no longer isolated because of deforestation. Drug-resistant organisms could also be included as the cause of emerging infections since they exist because of human influence. Some recent examples of agents responsible for emerging infections include human immunodeficiency virus, Ebola virus, and multidrug-resistant *Mycobacterium tuberculosis*.

Endemic: Disease that is present in a community or common among a group of people; said of a disease continually prevailing in a region.

Endocytosis: The uptake by a cell of particles that are too large to diffuse through its wall.

Epi-X (Epidemic Information Exchange): A system developed by the CDC which enables federal, state, and local epidemiologists, laboratorians, and other members of the public health community to instantly notify colleagues and experts of urgent public health events, review information on outbreaks and unusual health events through an easily searchable database, and rapidly communicate with colleagues through e-mail, Internet, and telecommunications capabilities.

Etiology: Science and study of the causes of diseases and their mode of operation.

FDA (U.S. Food and Drug Administration): A public health agency of the U.S. Department of Health and Human Services charged with protecting American consumers by enforcing the Federal Food, Drug, and Cosmetic Act and several related health laws.

FEMA (Federal Emergency Management Agency): An independent agency of the federal government founded in 1979. Its mission is to reduce loss of life and property and protect the nation's critical infrastructure from all types of hazards through a comprehensive, risk-based, emergency management program of mitigation, preparedness, response, and recovery.

Formulary: List of drugs approved for the treatment of various medical indications. It was originally created as a cost-control measure, but it has been used more recently to guide the use of antibiotics on the basis of information about resistance patterns.

GEIS (Global Emerging Infections Surveillance and Response System): A system designed to strengthen the prevention of, surveillance of, and response to infectious diseases that are a threat to military personnel and families, reduce medical readiness, or present a risk to U.S. national security. Its purpose is to create a centralized coordination and communication hub to help organize DoD resources and link with U.S. and international efforts.

Glycoprotein: A molecule that consists of a carbohydrate plus a protein.

Gram-negative: Gram-negative bacteria lose the crystal violet stain (and take the color of the red counterstain) in Gram's method of staining.

Gram-positive: Gram-positive bacteria, such as anthrax, retain the color of the crystal violet stain in the Gram stain. This is characteristic of bacteria that have a cell wall composed of a thick layer of a particular substance (called peptidologlycan).

HAN (Health Alert Network): A project of the CDC intended to ensure communications capacity at all local and state health departments, ensure capacity to receive distance learning offerings from CDC, and ensure capacity to broadcast and receive health alerts at every level.

Hemolytic: Referring to hemolysis, the destruction of red blood cells which leads to the release of hemoglobin from within the red blood cells into the blood plasma.

Hybridoma: A cell hybrid resulting from the fusion of a cancer cell and a normal lymphocyte (a type of white blood cell).

Hypoxia: Low oxygen content or tension; deficiency of oxygen in the inspired air.

Immunogenicity: The property that endows a substance with the capacity to provoke an immune response or the degree to which a substance possesses this property.

Immunomodulator: A chemical agent (as methotrexate or azathioprine) that modifies the immune response or the functioning of the immune system (as by the stimulation of antibody formation or the inhibition of white blood cell activity).

Incidence: The frequency of new occurrences of disease within a defined time interval. Incidence rate is the number of new cases of a specified disease divided by the number of people in a population over a specified period of time, usually one year.

IND (Investigational New Drug) Application: An application submitted by a sponsor to the FDA prior to human testing of an unapproved drug or of a previously approved drug for an unapproved use.

Infection: The invasion of the body or a part of the body by a pathogenic agent, such as a microorganism or virus. Under favorable conditions the agent develops or multiplies, the results of which may produce injurious effects. Infection should not be confused with disease.

IPV (inactivated polio vaccine): A vaccine for polio given as a shot in the arm or leg. The polio virus in IPV has been inactivated (killed). Also called the Salk vaccine.

LD50: The amount of a toxic agent that is sufficient to kill 50 percent of a population of animals usually within a certain time. Also called median lethal dose.

Leukocytosis: Increase in the number of white blood cells.

Level A Laboratory: Early detection of covert release. Primarily hospital laboratories with certified biological safety cabinet as a minimum biosafety requirement. These laboratories have the ability to rule out specific agents from clinical specimens and to forward organisms or specimens to higher-level laboratories.

Level B Laboratory: Core Capacity. State and county public health agency laboratories with BSL-2 biosafety facilities but which incorporate BSL-3 practices and maintain the proficiency to adequately perform confirmatory testing and characterize drug susceptibility. These laboratories have the ability to rule in specific agents, perform environmental testing and to forward organisms or specimens to higher-level laboratories. BSL-3 facilities are recommended but not required.

Level C Laboratory: Advanced Capacity. Rapid detection using nucleic acid amplification technology, molecular typing for comparison, and toxicity testing. Advanced capacity laboratories with BSL-3 facilities and proficiency sufficient to amplify, type, and perform toxicity testing. These laboratories will evaluate reagents and tests in order to facilitate transfer for use in Level B laboratories. Can conduct all tests performed in levels A and B laboratories.

Level D Laboratory: Can conduct all tests performed in levels A, B, and C laboratories. In addition, can detect genetic recombinants and bank isolate, and possesses BSL-3 and BSL-4 bio-containment facilities. These are highly specialized Federal laboratories with unique experience, ability to develop new tests and methods, and capability to securely maintain a bank of biological and threat agents.

LF (lethal factor): One of two enzymes making up the anthrax toxin. After being cleaved by a protease, protective antigen (PA) binds to this toxic enzyme and mediates its transportation into the cytosol where it exerts its pathogenic effect. Lethal factor is the crucial pathogenic enzyme and is the killer in the toxin.

***Listeria monocytogenes*:** A bacteria that can cause encephalitis, meningitis, blood-borne infection, and death. It is especially hazardous for pregnant women (posing a threat of miscarriage or stillbirth), newborn babies, the elderly, and immune-deficient patients. It causes about 28% of deaths due to food poisoning.

Lymph adenitis: Inflammation of lymph nodes.

Macrophages: A type of white blood cell that ingests foreign material. Macrophages are key players in the immune response to foreign invaders such as infectious microorganisms.

MAD (mutual assured destruction): The ability to kill 25 percent of a country's population and destroy 50 percent of its industry with nuclear weapons. The theory holds that countries would not strike with nuclear weapons if they knew their opponent could strike back and destroy them.

Mediastinitis: Inflammation of the mediastinum, a median septum or partition.

Monkeypox: A viral disease similar to smallpox still seen as a sporadic disease in parts of Central and West Africa.

Monoclonal antibodies: Immunoglobulin molecules secreted from a population of identical cells (i.e., cloned cells). They are homogeneous in structure and binding specificity.

Motile: Having spontaneous but not conscious or volitional movement.

NCID (National Center for Infectious Diseases): Its mission is to prevent illness, disability, and death caused by infectious diseases in the United States and around the world. NCID conducts surveillance, epidemic investigations, epidemiological and laboratory research, training, and public education programs to develop, evaluate, and promote prevention and control strategies for infectious diseases.

NDMS (National Disaster Medical System): Established in partnership with DOD, VA, FEMA, and the Public Health Service Commissioned Corps Readiness Force, a group of more than 7,000 volunteer health and support professionals who can be deployed anywhere in the country to assist communities in providing needed services to disaster victims.

NEDSS (National Electronic Disease Surveillance System): A public health initiative to provide a standards-based, integrated approach to disease surveillance and to connect public health surveillance to the burgeoning clinical information systems infrastructure.

NIAID (National Institute of Allergy and Infectious Diseases): A division of NIH that provides the major support for scientists conducting research aimed at developing better ways to diagnose, treat, and prevent the many infectious, immunological, and allergenic diseases that afflict people worldwide.

NIH (National Institutes of Health): A public health agency of the U.S. Department of Health and Human Services whose goal is to acquire new knowledge to help prevent, detect, diagnose, and treat disease and disability, from the rarest genetic disorder to the common cold.

Nosocomial infection: An infection that is acquired during hospitalization but that was neither present nor incubating at the time of hospital admission, unless it is related to a prior hospitalization, and that may become clinically manifest after discharge from the hospital.

NPS (National Pharmaceutical Stockpile) Program: Its mission is to ensure the availability and rapid deployment of life-saving pharmaceuticals, antidotes, other medical supplies, and equipment necessary to counter the effects of nerve agents, biological pathogens, and chemical agents. It stands ready for immediate deployment to any U.S. location in the event of a terrorist attack using a biological toxin or chemical agent directed against a civilian population.

Office of Homeland Security: Established by the president in 2001, its mission is to develop and coordinate the implementation of a comprehensive national strategy to secure the United States from terrorist threats or attacks. It coordinates the executive branch's efforts to detect, prepare for, prevent, protect against, respond to, and recover from terrorist attacks within the United States.

OPHP (Office of Public Health Preparedness): A newly created office, within the Department of Health and Human Services, which will coordinate the national response to public health emergencies.

OPV (oral polio vaccine): A vaccine for polio, given by mouth, and preferred for most children.

PA (protective antigen): A protein of anthrax toxin which binds to surface receptors on the host's cell membranes.

PhRMA (Pharmaceutical Research and Manufacturers of America): An association representing leading research-based pharmaceutical and biotechnology companies, which are devoted to inventing medicines that allow patients to live longer, healthier, happier, and more productive lives.

Plasmapheresis: A procedure in which the blood is filtered through a special machine to separate the plasma, the liquid portion of the blood, from the blood cells.

Plasmids: A self-replicating (autonomous) circle of DNA distinct from the chromosomal genome of bacteria. A plasmid contains genes normally not essential for cell growth or survival. Some plasmids can integrate into the host genome, be artificially constructed in the laboratory, and serve as vectors (carriers) in cloning.

Prions: A newly discovered type of disease-causing agent, neither bacterial nor fungal nor viral, and containing no genetic material. A prion is a protein that occurs normally in a harmless form. By folding into an aberrant shape, the normal prion turns into a rogue agent. It then co-opts other normal prions to become rogue prions.They have been held responsible for a number of degenerative brain diseases, including Mad Cow disease, Creutzfeldt Jakob disease, and possibly some cases of Alzheimer's disease.

Prophylactic antibiotics: Antibiotics that are administered before evidence of infection with the intention of warding off disease.

Pulmonary edema: Fluid in the lungs.

PulseNet: A national network of public health laboratories that perform DNA "fingerprinting" on bacteria that may be foodborne. The network permits rapid comparison of these "fingerprint" patterns through an electronic database at CDC.

Push Packages: Caches of pharmaceuticals, antidotes, and medical supplies designed to address a variety of biological or chemical agents. They are positioned in secure regional warehouses ready for immediate deployment to the airfield closest to the affected area following a federal decision to release NPS assets.

PVIs (polyvalent inhibitors): Chemically synthesized inhibitors that block toxin assembly.

Salmonella: A group of bacteria that cause typhoid fever, food poisoning, and enteric fever from contaminated food products.

Serotype: The kind of microorganism as characterized by serological typing (testing for recognizable antigens on the surface of the microorganism).

Sporulate: To form spores.

Strabismus: A condition in which the visual axes of the eyes are not parallel and the eyes appear to be looking in different directions.

Stridor: A harsh, high-pitched respiratory sound such as the inspiratory sound often heard in acute laryngeal obstruction.

Surveillance systems: Used in this report to refer to data collection and recordkeeping to track the emergence and spread of disease-causing organisms such as antibiotic-resistant bacteria.

Tachycardia: A rapid heart rate, usually defined as greater than 100 beats per minute.

TOPOFF: An exercise conducted by the Department of Justice which engaged key personnel in the management of mock chemical, biological, or cyberterrorist attacks. So named because it involved the participation of top officials of the U.S. government.

Toxoplasma: A genus of sporozoa that are intracellular parasites of many organs and tissues of birds and mammals, including humans.

USAMRIID (U.S. Army Medical Research Institute of Infectious Diseases): It is the lead medical research laboratory for the U.S. Biological Defense Research Program which conducts research to develop strategies, products, information, procedures, and training programs for medical defense against biological warfare threats and naturally occurring infectious diseases that require special containment. It is an organization of the U.S. Army Medical Research and Materiel Command (USAMRMC).

VA (Department of Veterans Affairs): A cabinet-level department that has the care of veterans as its primary mission and is composed of three administrations: Veterans Health Administration, Veterans Benefit Administration, and National Cemetery Administration.

Vaccine: A preparation of living, attenuated, or killed bacteria or viruses, fractions thereof, or synthesized or recombinant antigens identical or similar to those found in the disease-causing organisms, that is administered to raise immunity to a particular microorganism.

VHF (viral hemorrhagic fevers): A group of illnesses that are caused by viruses of four distinct families: arenaviruses, filoviruses, Bunyaviruses, and flaviviruses.

Virulence: The ability of any infectious agent to produce disease. The virulence of a microoganism (such as a bacterium or virus) is a measure of the severity of the disease it is capable of causing.

Zoonotic disease or infection: An infection or infectious disease that may be transmitted from vertebrate animals (e.g., a rodent) to humans.

ACRONYMS

ASM: American Society for Microbiology
AVA: Anthrax Vaccine Adsorbed
AVIP: Anthrax Vaccine Immunization Program
BW: biological warfare
BWC: Biological Weapons Convention
CFR: Code of Federal Regulations
CSF: cerebral spinal fluid
DIC: disseminated intravascular coagulation
EMSHG: Emergency Management Strategic Healthcare Group
ESSENCE: Electronic Surveillance System for the Early Notification of Community-Based Epidemics
GIS: Geographic Information Systems
GOCO: government-owned, contractor-operated
ICS: Incident Command System
JCAHO: Joint Commission on Accreditation of Healthcare Organizations
JVAP: Joint Vaccine Acquisition Program

LRN: Laboratory Response Network
MAV: Multiagent Vaccines
MBDRP: Medical Biological Defense Research Program
NDA: New Drug Application
NLS: National Laboratory System
NVSL: National Veterinary Services Laboratory
OTSG: Office of the Army Surgeon General
PCR: polymerase chain reaction
PFU: plaque forming units
PHLS: Public Health Laboratory Service
RHA: recombinant human antitoxin
RSVP: Rapid Syndrome Validation Project
TED: troop equivalent dose
VADAR: Virtually Assured Detection and Response

VEE: Viral Equine Encephalitis
VHA: Veterans Health Administration
VIG: vaccinia immune globulin

Appendix L

Forum Member, Speaker, and Staff Biographies

FORUM MEMBERS

ADEL A.F. MAHMOUD, M.D., Ph.D., *(chair),* is President of Merck Vaccines at Merck & Co., Inc. He formerly served Case Western Reserve University and University Hospitals of Cleveland as Chairman of Medicine and Physician-in-Chief from 1987 to 1998. Prior to that, Dr. Mahmoud held several positions, spanning 25 years, at the same institutions. Dr. Mahmoud and his colleagues conducted pioneering investigations on the biology and function of eosinophils. He prepared the first specific anti-eosinophil serum, which was used to define the role of these cells in host resistance to helminthic infections. Dr. Mahmoud also established clinical and laboratory investigations in several developing countries including Kenya, Egypt, and The Philippines to examine the determinants of infection and disease in schistosomiasis and other infectious agents. This work led to the development of innovative strategies to control those infections, which has been adopted by the World Health Organization as selective population chemotherapy. In recent years, Dr. Mahmoud turned his attention to developing a comprehensive set of responses to the problems associated with emerging infections in the developing world. He was elected to membership of the American Society for Clinical Investigation in 1978, the Association of American Physicians in 1980 and the Institute of Medicine of the National Academy of Sciences in 1987. He received the Bailey K. Ashford Award of the American Society of Tropical Medicine and Hygiene in 1983, and the Squibb Award of the Infectious Diseases Society of America in 1984. Dr. Mahmoud currently serves as Chair of the Forum on Emerging Infections and is a member of the Board on Global Health, both of the Institute of Medicine. He also chairs the U.S. Delegation to the U.S.-Japan Cooperative Medical Science Program.

STANLEY M. LEMON, M.D., *(vice-chair),* is Dean of the School of Medicine at the University of Texas Medical Branch at Galveston. He received his undergraduate degree in biochemical sciences from Princeton University summa cum laude, and his M.D. with honor from the University of Rochester. He completed postgraduate training in internal medicine and infectious diseases at the University of North Carolina at Chapel Hill, and is board certified in both. From 1977 to 1983, he served with the U.S. Army Medical Research and Development Command, directing the Hepatitis Laboratory at the Walter Reed Army Institute of Research. He joined the faculty of the University of North Carolina School of Medicine in 1983, serving first as Chief of the Division of Infectious Diseases, and then Vice Chair for Research of the Department of Medicine. In 1997, Dr. Lemon moved to the University of Texas Medical Branch as Professor and Chair of the Department of Microbiology & Immunology. He was subsequently appointed Dean *pro tem* of the School of Medicine in 1999, and permanent Dean of Medicine in 2000. Dr. Lemon's research interests relate to the molecular virology and pathogenesis of the positive-stranded RNA viruses responsible for hepatitis C and hepatitis A. He is particularly interested in the molecular mechanisms controlling cap-independent viral translation, and the replication of these RNA genomes. He has published over 180 papers, and numerous textbook chapters related to hepatitis and other viral infections, and has a longstanding interest in vaccine development. He has served previously as chair of the Anti-Infective Drugs Advisory Committee and the Vaccines and Related Biologics Advisory Committee of the U.S. Food and Drug Administration, and is past chair of the Steering Committee on Hepatitis and Poliomyelitis of the World Health Organization's Programme on Vaccine Development. He presently serves as Chairman of the U.S. Hepatitis Panel of the U.S.-Japan Cooperative Medical Sciences Program, and chairs an Institute of Medicine study committee related to vaccines for the protection of the military against naturally occurring infectious disease threats.

STEVEN J. BRICKNER, Ph.D., is Research Advisor, Antibacterials Chemistry, at Pfizer Global Research and Development. He received his Ph.D. in organic chemistry from Cornell University and was a NIH Postdoctoral Research Fellow at the University of Wisconsin-Madison. Dr. Brickner is a medicinal chemist with nearly 20 years of research experience in the pharmaceutical industry, all focused on the discovery and development of novel antibacterial agents. He is an inventor/co-inventor on 21 U.S. patents, and has published numerous scientific papers, primarily within the area of the oxazolidinones. Prior to joining Pfizer in 1996, he led a team at Pharmacia and Upjohn that discovered and developed linezolid, the first member of a new class of antibiotics to be approved in the last 35 years.

GAIL H. CASSELL, Ph.D., is Vice President, Infectious Diseases Research, Drug Discovery Research, and Clinical Investigation at Eli Lilly & Company. Previously, she was the Charles H. McCauley Professor and (since 1987) Chair, Department of Microbiology, University of Alabama Schools of Medicine and Dentistry at Birmingham, a department which, under her leadership, has ranked first in research funding from the National Institutes of Health since 1989. She is a member of the Director's Advisory Committee of the Centers for Disease Control and Prevention. Dr. Cassell is past president of the American Society for Microbiology (ASM) and is serving her third three-year term as chairman of the Public and Scientific Affairs Board of ASM. She is a former member of the National Institutes of Health Director's Advisory Committee and a former member of the Advisory Council of the National Institute of Allergy and Infectious Diseases. She has also served as an advisor on infectious diseases and indirect costs of research to the White House Office on Science and Technology and was previously chair of the Board of Scientific Counselors of the National Center for Infectious Diseases, Centers for Disease Control and Prevention. Dr. Cassell served eight years on the Bacteriology-Mycology-II Study Section and served as its chair for three years. She serves on the editorial boards of several prestigious scientific journals and has authored over 275 articles and book chapters. She has been intimately involved in the establishment of science policy and legislation related to biomedical research and public health. Dr. Cassell has received several national and international awards and an honorary degree for her research on infectious diseases.

GORDON DEFRIESE, Ph.D., is Professor of Social Medicine and Professor of Medicine (in the Division of General Medicine and Clinical Epidemiology) at the UNC-CH School of Medicine. In addition, he holds appointments as Professor of Epidemiology and Health Policy and Administration in the UNC-CH School of Public Health and as Professor of Dental Ecology in the UNC-CH School of Dentistry. From 1986–2000, he served as Co-Director of the Robert Wood Johnson Clinical Scholars Program, co-sponsored by the UNC-CH School of Medicine and the Cecil G. Sheps Center for Health Services Research. He received his Ph.D. from the University of Kentucky College of Medicine. Some of his research interests are in the areas of health promotion and disease prevention, medical sociology, primary health care, rural health care, cost-benefit analysis, and cost-effectiveness. He is a past president of the Association for Health Services Research and a fellow of the New York Academy of Medicine. He is founder of the Partnership for Prevention, a coalition of private-sector business and industry organizations, voluntary health organizations, and state and federal public health agencies based in Washington, D.C. that have joined together to work toward the elevation of disease prevention among the nation's health policy priorities. He is an at-large member of the National Board of Medical Examiners. Since 1994 he has served as President and CEO of the North Carolina Institute of Medicine. He is Editor-in-Chief and Publisher of the North Carolina Medical Journal.

CEDRIC E. DUMONT, M.D., is Medical Director for the Office of Medical Services (MED) at the U.S. Department of State. Dr. Dumont graduated from Columbia University with a B.A. in 1975 and obtained his medical degree from Tufts University School of Medicine in 1980. Dr. Dumont is a board-certified internist with subspecialty training in infectious diseases. He completed his internal medicine residency in 1983 and infectious diseases fellowship in 1988 at Georgetown University Hospital in Washington, D.C. Dr. Dumont has been a medical practitioner for over 19 years, 2 of which included service in the Peace Corps. Since joining the Department of State in 1990, he has had substantial experience overseas in Dakar, Bamako, Kinshasa, and Brazzaville. For the past 3 years, as the Medical Director for the Department of State, Dr. Dumont has promoted the health of all United States Government employees serving overseas by encouraging their participation in a comprehensive health maintenance program and by facilitating their access to high-quality medical care. Dr. Dumont is a very strong supporter of the professional development and advancement of MED's highly qualified professional staff. In addition, he has supported and encouraged the use of an electronic medical record, which will be able to monitor the health of all its beneficiaries, not only during a specific assignment but also throughout their career in the Foreign Service.

JESSE L. GOODMAN, M.D., M.P.H., was professor of medicine and Chief of Infectious Diseases at the University of Minnesota, and is now serving as Deputy Director for the U.S. Food and Drug Administration's (FDA) Center for Biologics Evaluation and Research, where he is active in a broad range of scientific, public health, and policy issues. After joining the FDA commissioner's office, he has worked closely with several centers and helped coordinate FDA's response to the antimicrobial resistance problem. He was co-chair of a recently formed federal interagency task force which developed the national Public Health Action Plan on antimicrobial resistance. He graduated from Harvard College and attended the Albert Einstein College of Medicine followed by internal medicine, hematology, oncology, and infectious diseases training at the University of Pennsylvania and University of California Los Angeles, where he was also chief medical resident. He received his master's of public health from the University of Minnesota. He has been active in community public health activities, including creating an environmental health partnership in St. Paul, Minnesota. In recent years, his laboratory's research has focused on the molecular pathogenesis of tickborne diseases. His laboratory isolated the etiological intracellular agent of the emerging tickborne infection, human granulocytic ehrlichiosis, and identified its leukocyte receptor. He has also been an active clinician and teacher and has directed or participated in major multi-center clinical studies. He is a Fellow of the Infectious Diseases Society of America and, among several honors, has been elected to the American Society for Clinical Investigation.

RENU GUPTA, M.D., is Vice President and Head, U.S. Clinical Research and Development at Novartis Pharmaceuticals. Previously, she was Vice President, Medical, Safety, and Therapeutics at Covance. Dr. Gupta is a board certified Pediatrician, with subspeciality training in Infectious Diseases from Children's Hospital of Philadelphia, and the University of Pennsylvania. She was also Postdoctoral Research Fellow in Microbiology at the University of Pennsylvania and the Wistar Institute of Anatomy and Biology, where she conducted research on the pathogenesis of infectious diseases. Dr. Gupta received her M.B.,Ch.B with distinction from the University of Zambia, where she examined the problem of poor compliance in the treatment of tuberculosis in rural and urban Africa. She is currently active in a number of professional societies, including the Infectious Diseases Society of America and the American Society of Microbiology. She is a frequent presenter at the Interscience Conference on Antimicrobial Agents and Chemotherapy and other major congresses, and has been published in leading infectious diseases periodicals. From 1989 to mid-1998, Dr. Gupta was with Bristol-Myers Squibb Company, where she directed clinical research as well as strategic planning for the Infectious Diseases and Immunology Divisions. For the past several years, her work has focused on a better understanding of the problem of emerging infections. This has led to her pioneering efforts in establishing the Global Antimicrobial Surveillance Program, SENTRY, a private-academic-public sector partnership. Dr. Gupta chaired the steering committee for the SENTRY Antimicrobial Surveillance Program. She remains active in women and children's health issues, and is currently furthering education and outreach initiatives. More recently Dr. Gupta has been instrumental in the formation of the Harvard-Pharma Management Board, of which she is a member, to further the educational goals of the Scholars in Clinical Science Program at the Harvard Medical School.

MARGARET A. HAMBURG, M.D., is Vice President for Biological Programs, Nuclear Threat Initiative, Washington, D.C. The Nuclear Threat Initiative (NTI) is a new organization whose mission is to strengthen global security by reducing the risk of use of nuclear and other weapons of mass destruction and preventing their spread. Dr Hamburg is in charge of the biological program area. Before taking on her current position, Dr. Hamburg was the Assistant Secretary for Planning and Evaluation, U.S. Department of Health and Human Services, serving as a principal policy advisor to the Secretary of Health and Human Services with responsibilities including policy formulation and analysis, the development and review of regulations and/or legislation, budget analysis, strategic planning, and the conduct and coordination of policy research and program evaluation. Prior to this, she served for almost six years as the Commissioner of Health for the City of New York. As chief health officer in the nation's largest city, Dr. Hamburg's many accomplishments included: the design and implementation of an internationally recognized tuberculosis control program that produced dramatic declines in tuberculosis cases; the development of initiatives that raised childhood immunization rates to record levels; and the creation of the first public health bioterrorism preparedness program in the nation. She completed her internship and residency in Internal Medicine at the New York Hospital/Cornell University Medical Center and is certified by the American Board of Internal Medicine. Dr. Hamburg is a graduate of Harvard College and Harvard Medical School. She currently serves on the Harvard University Board of Overseers. She has been elected to membership in the Institute of Medicine, the New York Academy of Medicine, the Council on Foreign Relations, and is a Fellow of the American Association of the Advancement of Science.

CAROLE A. HEILMAN, Ph.D., is Director of the Division of Microbiology and Infectious Diseases (DMID) of the National Institute of Allergy and Infectious Diseases (NIAID). Dr. Heilman received her bachelor's degree in biology from Boston University in 1972, and earned her master's degree and doctorate in microbiology from Rutgers University in 1976 and 1979. Dr. Heilman began her career at the National Institutes of Health as a postdoctoral research associate with the National Cancer Institute where she carried out research on the regulation of gene expression during cancer development. In 1986, she came to NIAID as the influenza and viral respiratory diseases program officer in DMID and, in 1988, she was appointed chief of the respiratory diseases branch where she coordinated the development of acellular pertussis vaccines. She joined the Division of AIDS as deputy director in 1997 and was responsible for developing the Innovation Grant Program for Approaches in human immunodeficiency virus vaccine research. She is the recipient of several notable awards for outstanding achievement. Throughout her extramural career, Dr. Heilman has contributed articles on vaccine design and development to many scientific journals and has served as a consultant to the World Bank and the World Health Organization in this area. She is also a member of several professional societies, including the Infectious Diseases Society of America, the American Society for Microbiology, and the American Society of Virology.

DAVID L. HEYMANN, M.D., is currently the Executive Director of the World Health Organization (WHO) Communicable Diseases Cluster. From October 1995 to July 1998 he was Director of the WHO Programme on Emerging and other Communicable Diseases Surveillance and Control. Prior to becoming director of this programme, he was the chief of research activities in the Global

Programme on AIDS. From 1976 to 1989 prior to joining WHO, Dr Heymann spent thirteen years working as a medical epidemiologist in sub-Saharan Africa (Cameroon, Cote d' Ivoire, the former Zaire and Malawi) on assignment from the Centers for Disease Control and Prevention (CDC) in CDC-supported activities aimed at strengthening capacity in surveillance of infectious diseases and their control, with special emphasis on the childhood immunizable diseases, African haemorrhagic fevers, pox viruses, and malaria. While based in Africa, Dr Heymann participated in the investigation of the first outbreak of Ebola in Yambuku (former Zaire) in 1976, then again investigated the second outbreak of Ebola in 1977 in Tandala, and in 1995 directed the international response to the Ebola outbreak in Kikwit. Prior to 1976, Dr Heymann spent two years in India as a medical officer in the WHO Smallpox Eradication Programme. Dr Heymann holds a B.A. from the Pennsylvania State University, an M.D. from Wake Forest University, a Diploma in Tropical Medicine and Hygiene from the London School of Hygiene and Tropical Medicine, and completed practical epidemiology training in the Epidemic Intelligence Service (EIS) training program of the CDC. He has published 131 scientific articles on infectious diseases in peer-reviewed medical and scientific journals.

JAMES M. HUGHES, M.D., received his B.A. in 1966 and M.D. in 1971 from Stanford University. He completed a residency in internal medicine at the University of Washington and a fellowship in infectious diseases at the University of Virginia. He is board-certified in internal medicine, infectious diseases, and preventive medicine. He first joined CDC as an Epidemic Intelligence Service officer in 1973. During his CDC career, he has worked primarily in the areas of foodborne disease and infection control in healthcare settings. He became Director of the National Center for Infectious Diseases in 1992. The center is currently working to address domestic and global challenges posed by emerging infectious diseases and the threat of bioterrorism. He is a fellow of the American College of Physicians, the Infectious Diseases Society of America, and the American Association for the Advancement of Science. He is an Assistant Surgeon General in the Public Health Service.

SAMUEL L. KATZ, M.D., is Wilburt C. Davison Professor and chairman emeritus of pediatrics at Duke University Medical Center. He has concentrated his research on infectious diseases, focusing primarily on vaccine research, development and policy. Dr. Katz has served on a number of scientific advisory committees and is the recipient of many prestigious awards and honorary fellowships in international organizations. He attained his M.D. from Harvard Medical School and completed his residency training at Boston hospitals. He became a staff member at Children's Hospital, working with Nobel laureate John Enders, during which time they developed the attenuated measles virus vaccine now used throughout the world. He has chaired the Committee on Infectious Diseases of the American Academy of Pediatrics (the Redbook Committee), the Advisory Committee on Immunization Practices (ACIP) of the Centers for Disease Control and Prevention, the Vaccine Priorities Study of the Institute of Medicine (IOM), and several World Health Organization (WHO) and Children's Vaccine Initiative panels on vaccines. He is a member of many scientific advisory committees including those of the NIH, IOM, and WHO. Dr. Katz's published studies include abundant original scientific articles, chapters in textbooks, and many abstracts, editorials, and reviews. He is the coeditor of a textbook on pediatric infectious diseases and has given many named lectures in the United States and abroad. Currently he co-chairs the Indo-US Vaccine Action Program as well as the National Network for Immunization Information (NNii).

COLONEL PATRICK KELLEY, M.D., M.P.H., Dr.P.H., is Director of the Department of Defense Global Emerging Infections System and the Director of the Division of Preventive Medicine at the Walter Reed Army Institute of Research (WRAIR), Silver Spring, Maryland. He obtained his M.D. from the University of Virginia and a Dr.P.H. in infectious disease epidemiology from the Johns Hopkins Bloomberg School of Public Health. He is board-certified in general preventive medicine and a fellow of the American College of Preventive Medicine. For many years he directed the Army General Preventive Medicine Residency at WRAIR. Colonel Kelley has extensive experience leading military infectious disease studies and in managing domestic and international public health surveillance efforts. He has spoken before professional audiences in over 15 countries and has authored or co-authored over 40 scientific papers and book chapters on a variety of infectious disease and preventive medicine topics. He serves as the specialty editor for a textbook entitled, *Military Preventive Medicine: Mobilization and Deployment.*

MARCELLE LAYTON, M.D., is the Assistant Commissioner for the Bureau of Communicable Diseases at the New York City Department of Health. This bureau is responsible for the surveillance and control of 51 infectious diseases and conditions reportable under the New York City Health Code. Current areas of concern include antibiotic resistance; foodborne, waterborne, and tickborne diseases; hepatitis C and biological disaster planning for the potential threats of bioterrorism and pandemic influenza. Dr. Layton received her medical degree from Duke University. She completed an internal medicine residency at the University Health Science Center in Syracuse, New York, and an infectious disease fellowship at Yale University. In addition, Dr. Layton spent two years with the Centers for Disease Control and Prevention as a fellow in the Epidemic Intelligence Service, where she was assigned to the New York City Department of Health. In the past, she has volunteered or worked with the Indian Health Service, the Alaskan Native Health Service, and clinics in northwestern Thailand and central Nepal.

JOSHUA LEDERBERG, Ph.D., is Professor emeritus of Molecular Genetics and Informatics and Sackler Foundation Scholar at The Rockefeller University, New York, New York. His lifelong research, for which he received the Nobel Prize in 1958, has been in genetic structure and function in microorganisms. He has a keen interest in international health and was co-chair of a previous Institute of Medicine Committee on Emerging Microbial Threats to Health (1990–1992) and currently is co-chair of the Committee on Emerging Microbial Threats to Health in the 21st Century. He has been a member of the National Academy of Sciences since 1957 and is a charter member of the Institute of Medicine.

CARLOS LOPEZ, Ph.D., is Research Fellow, Research Acquisitions, Eli Lilly Research Laboratories. He received his Ph.D. from the University of Minnesota in 1970. Dr. Lopez was awarded the NTRDA postdoctoral fellowship. After his fellowship he was appointed assistant professor of pathology at the University of Minnesota, where he did his research on cytomegalovirus infections in renal transplant recipients and the consequences of those infections. He was next appointed assistant member and head of the Laboratory of Herpesvirus Infections at the Sloan Kettering Institute for Cancer Research, where his research focused on herpes virus infections and the resistance mechanisms involved. Dr. Lopez's laboratory contributed to the immunological analysis of the earliest AIDS patients at the beginning of the AIDS epidemic in New York. He is co-author of one of the seminal publications on this disease, as well as many scientific papers and co-editor of six books. Dr. Lopez has been a consultant to numerous agencies and organizations including the National Institutes of Health, the Department of Veterans Affairs, and the American Cancer Society.

LYNN MARKS, M.D., is board certified in internal medicine and infectious diseases. He was on faculty at the University of South Alabama College of Medicine in the Infectious Diseases department focusing on patient care, teaching and research. His academic research interest was on the molecular genetics of bacterial pathogenicity. He subsequently joined SmithKline Beecham's (now GlaxoSmithKline) anti-infectives clinical group and later progressed to global head of the Consumer Healthcare division Medical and Regulatory group. He then returned to pharmaceutical research and development as global head of the Infectious Diseases Therapeutic Area Strategy Team for GlaxoSmithKline.

STEPHEN S. MORSE, Ph.D., is Director of the Center for Public Health Preparedness at the Mailman School of Public Health of Columbia University, and a faculty member in the Epidemiology Department. Dr. Morse recently returned to Columbia from 4 years in government service as Program Manager at the Defense Advanced Research Projects Agency (DARPA), where he co-directed the Pathogen Countermeasures program and subsequently directed the Advanced Diagnostics program. Before coming to Columbia, he was Assistant Professor (Virology) at The Rockefeller University in New York, where he remains an adjunct faculty member. Dr. Morse is the editor of two books, *Emerging Viruses* (Oxford University Press, 1993; paperback, 1996) (selected by "American Scientist" for its list of "100 Top Science Books of the 20th Century"), and *The Evolutionary Biology of Viruses* (Raven Press, 1994). He currently serves as a Section Editor of the CDC journal "Emerging Infectious Diseases" and was formerly an Editor-in-Chief of the Pasteur Institute's journal "Research in Virology". Dr. Morse was Chair and principal organizer of the 1989 NIAID/NIH Conference on Emerging Viruses (for which he originated the term and concept of emerging viruses/infections); served as a member of the Institute of Medicine-National Academy of Sciences' Committee on Emerging Microbial Threats to Health (and chaired its Task Force on Viruses), and was a contributor to its report, *Emerging Infections* (1992); was a member of the IOM's Committee on Xenograft Transplantation; currently serves on the Steering Committee of the Institute of Medicine's Forum on Emerging Infections, and has served as an adviser to WHO (World Health Organization), PAHO (Pan-American Health Organization), FDA, the Defense Threat Reduction Agency (DTRA), and other agencies. He is a Fellow of the New York Academy of Sciences and a past Chair of its Microbiology Section. He was the founding Chair of ProMED (the non-profit international Program to Monitor Emerging Diseases) and was one of the originators of ProMED-mail, an international network inaugurated by ProMED in 1994 for outbreak reporting and disease monitoring using the Internet. Dr. Morse received his Ph.D. from the University of Wisconsin-Madison.

MICHAEL T. OSTERHOLM, Ph.D., M.P.H., is Director of the Center for Infectious Disease Research and Policy at the University of Minnesota where he is also Professor at the School of Public Health. Previously, Dr. Osterholm was the state epidemiologist and Chief of the Acute Disease Epidemiology Section for the Minnesota Department of Health. He has received numerous research awards from the National Institute of Allergy and Infectious Diseases and the Centers for Disease Control and Prevention (CDC). He served as principal investigator for the CDC-sponsored Emerging Infections Program in Minnesota. He has published more than 240 articles and abstracts on various emerging infectious disease problems and is the author of the best selling book, Living Terrors: What America Needs to Know to Survive the Coming Bioterrorist Catastrophe. He is past president of the Council of State and Territorial Epidemiologists. He currently serves on the National Academy of Sciences, Institute of Medicine (IOM) Forum on Emerging Infections. He has also served on the IOM Committee, Food Safety, Production to Consumption, the IOM Committee on the Department of Defense Persian Gulf Syndrome Comprehensive Clinical Evaluation Program and as a reviewer for the IOM report on chemical and biological terrorism.

GARY A. ROSELLE, M.D., received his M.D. from Ohio State University School of Medicine in 1973. He served his residency at Northwestern University School of Medicine and his Infectious Diseases fellowship at the University of Cincinnati School of Medicine. Dr. Roselle is the Program Director for Infectious Diseases for VA Central Office in Washington, D.C., as well as the Chief of the Medical Service at the Cincinnati VA Medical Center. He is a professor of medicine in the Department of Internal Medicine, Division of Infectious Diseases at the University of Cincinnati College of Medicine. Dr. Roselle serves on several national advisory committees. In addition, he is currently heading the Emerging Pathogens Initiative for the Department of Veterans Affairs. Dr. Roselle has received commendations from the Cincinnati Medical Center Director, the Under Secretary for Health for the Department of Veterans Affairs, and the Secretary of Veterans Affairs for his work in the infectious diseases program for the Department of Veterans Affairs. He has been an invited speaker at several national and international meetings, and has published over 80 papers and several book chapters.

DAVID M. SHLAES, M.D., Ph.D., is Vice President and Therapeutic Area Co-Leader for Infectious Diseases at Wyeth. Before joining Wyeth, Dr. Shlaes was professor of medicine at the Case Western Reserve University School of Medicine and chief of the Infectious Diseases Section and the Clinical Microbiology Unit at the Veterans Affairs Medical Center in Cleveland, Ohio. His major research interest has been the mechanisms and epidemiology of antimicrobial resistance in bacteria where he has published widely. He has recently become more involved in the area of public policy as it relates to the discovery and development of antibiotics. He has served on the Institute of Medicine's Forum on Emerging Infections since 1996.

JANET SHOEMAKER, is director of the American Society for Microbiology's Public Affairs Office, a position she has held since 1989. She is responsible for managing the legislative and regulatory affairs of this 42,000-member organization, the largest single biological science society in the world. She has served as principal investigator for a project funded by the National Science Foundation (NSF) to collect and disseminate data on the job market for recent doctorates in microbiology and has played a key role in American Society for Microbiology (ASM) projects, including the production of the ASM *Employment Outlook in the Microbiological Sciences* and *The Impact of Managed Care and Health System Change on Clinical Microbiology*. Previously, she held positions as Assistant Director of Public Affairs for ASM, as ASM coordinator of the U.S./U.S.S.R. Exchange Program in Microbiology, a program sponsored and coordinated by the National Science Foundation and the U.S. Department of State, and as a freelance editor and writer. She received her baccalaureate, cum laude, from the University of Massachusetts, and is a graduate of the George

Washington University programs in public policy and in editing and publications. She has served as commissioner to the Commission on Professionals in Science and Technology, and as the ASM representative to the ad hoc Group for Medical Research Funding, and is a member of Women in Government Relations, the American Society of Association Executives, and the American Association for the Advancement of Science. She has co-authored published articles on research funding, biotechnology, biological weapons control, and public policy issues related to microbiology.

P. FREDERICK SPARLING, M.D., is J. Herbert Bate Professor emeritus of Medicine, Microbiology and Immunology at the University of North Carolina (UNC) at Chapel Hill and is Director of the North Carolina Sexually Transmitted Infections Research Center. Previously, he served as chair of the Department of Medicine and chair of the Department of Microbiology and Immunology at UNC. He was president of the Infectious Disease Society of America in 1996–1997. He was also a member of the Institute of Medicine's Committee on Microbial Threats to Health (1991–1992). Dr. Sparling's laboratory research is in the molecular biology of bacterial outer membrane proteins involved in pathogenesis, with a major emphasis on gonococci and meningococci. His current studies focus on the biochemistry and genetics of iron-scavenging mechanisms used by gonococci and meningococci and the structure and function of the gonococcal porin proteins. He is pursuing the goal of a vaccine for gonorrhea.

I. KAYE WACHSMUTH, Ph.D., serves as Deputy Administrator of the Office of Public Health and Science in USDA's Food Safety and Inspection Service. Before joining USDA, she was the Deputy Director for Programs at the Food and Drug Administration's Center for Food Safety and Applied Nutrition. Dr. Wachsmuth was with the Centers for Disease Control and Prevention in Atlanta from 1972–1994, where she was Deputy Director, Division of Bacterial and Mycotic Diseases from 1991–1994, and Chief, Enteric Diseases Laboratory Section from 1985–1991. While at CDC, she developed programs and conducted studies in the areas of molecular epidemiology and bacterial pathogenesis. She also worked extensively in Southeast Asia and South America to establish laboratory-based diarrheal disease surveillance programs. In addition to her positions at FDA and CDC, Dr. Wachsmuth chairs the National Advisory Committee on Microbiological Criteria for Foods and the Codex Committee for Food Hygiene, and is a member of the World Health Organization (WHO) Expert Advisory Panel on Food Safety. She has been director of WHO's International Collaborating Center for Shigella. Through adjunct faculty appointments at the University of North Carolina School of Public Health, Emory University, and Georgia State University, Dr. Wachsmuth was doctoral research advisor to students in the microbial sciences. She also mentored postdoctoral students through the National Research Council, WHO, and Fogarty and Fulbright fellowship programs. At CDC in the early 1990s, she directed the summer research program for students enrolled in the University of Tuskegee Veterinary School and Morehouse University Medical School. Dr. Wachsmuth received her B.S. degree from Stetson University, Deland, Florida and Ph.D. degree in Microbiology from the University of Tennessee. She is a fellow of the Infectious Diseases Society of America and the American Academy of Microbiology. She has received awards for benchmark epidemiological investigations of Legionnaire's disease, cholera in Latin America, drug-resistant tuberculosis, hantavirus in the western United States, and diphtheria in the former Soviet Union. Author of more than 160 scientific papers, she is on the editorial board of scientific journals and is editor of a book on cholera.

C. DOUGLAS WEBB, Jr., Ph.D., received his bachelor's degree in biology from Emory University and his master's and doctoral degrees in microbiology from the University of Georgia. He served in the Public Health Service at the Centers for Disease Control and Prevention (CDC) as both a research microbiologist and supervisory microbiologist. After the CDC, Dr. Webb went to Pfizer Pharmaceuticals and was involved in the development of ampicillin-sulbactam, carbenicillin, cefoperazone, fluconazole, azithromycin, and trovafloxacin. Dr. Webb is Senior Medical Director in Infectious Diseases in U.S. Medicines at Bristol-Myers Squibb, working on the strategy and development for the anti-infective portfolio.

MICHAEL ZEILINGER, D.P.M, M.P.H., is Infectious Disease Team Leader at the Office of Health and Nutrition, Environmental Health Division at the U.S. Agency for International Development. Dr. Zeilinger serves as the senior advisor and manager of the infectious disease strategic objective team which encompasses four sub-teams: malaria, tuberculosis, antimicrobial resistance, and surveillance. He is also the Cognizant Technical Officer for USAID's umbrella Inter-Agency Agreement with the Centers for Disease Control and Prevention. Prior to his work at USAID/W, Dr. Zeilinger was the Program Director for Project HOPE in the Central Asian republics, which included public and private sector-funded TB control projects in Kazakhstan, Uzbekistan, Kyrgyzstan and Turkmenistan. His work in Central Asia also included humanitarian assistance and child survival programs. Prior to that, he worked with Birch and Davis on the Department of Defense's Military Health Service System, and the Public Health Foundation on a U.S. Department of Health and Human Services funded Empowerment Zones/Enterprise Communities Health Benchmarks Demonstration Project. Dr. Zeilinger is a Doctor of Podiatric Medicine and managed several community health programs during his surgical residency. He also has a master's degree in public health (International Health Promotion) from George Washington University and is currently an Adjunct Professor at their School of Public Health.

SPEAKERS

KENNETH ALIBEK, M.D., Ph.D., Sc.D., is President of Hadron Advanced Biosystems, and corporate vice president of its publicly held parent company, Hadron. Dr. Alibek also holds the position of Distinguished Professor at George Mason University. He leads medical and scientific research programs dedicated to developing new forms of medical protection against biological weapons and other infectious diseases. Dr. Alibek served as first deputy chief of the civilian branch of the Soviet Union's offensive biological weapons program and has more than 20 years of experience in development, management, and supervision of high containment (BL-4) pathogen laboratories. He has extensive knowledge of biotechnology, including large-scale bioprocessing and production of biological weapons; antibacterial and antiviral drug development; development of regimens for urgent prophylaxis and treatment of the diseases caused by biological weapons; and mass casualty handling. He is a former Soviet Army colonel. Dr. Alibek has published more than 80 articles in classified journals on the development of new types of biological weapons and on medical aspects of biodefense. Dr. Alibek defected to the United States from the Soviet Union in 1992 and subsequently served as a consultant to numerous U.S. government agencies in the areas of industrial technology, medical microbiology, biological weapons defense, and biological weapons nonproliferation. He has worked with the National Institutes of Health, testified extensively before the U.S. Congress on nonproliferation of biological weapons and is the author of *Biohazard*, published by Random House. Dr. Alibek holds an M.D. in epidemiology and infectious diseases, a Ph.D. in medical microbiology, and an Sc.D. in industrial biotechnology.

STEPHEN S. ARNON, M.D., is the co-discoverer of infant botulism who in 1976 founded the State of California's research program into this and pathophysiologically related diseases that is now known as the Infant Botulism Treatment and Prevention Program. For the past 13 years he has led the California effort to develop the first specific treatment for infant botulism, the orphan drug known as botulism immune globulin intravenous (human), for which he is one of the volunteer donors of hyperimmune source plasma. Dr. Arnon trained in internal medicine at the University of Colorado Hospitals in Denver, after which he joined the Epidemic Intelligence Service of the Centers for Disease Control and Prevention (CDC) and served in the Enteric Disease Branch. During medical school and residency he did medical work and mountaineering in Afghanistan, India, and Nepal. Dr. Arnon received his undergraduate, public health, and medical degrees from Harvard University and in 1989 was elected directly to fellowship in the Infectious Diseases Society of America.

MICHAEL S. ASCHER, M.D., FACP, graduated from Dartmouth and Harvard medical schools and trained in internal medicine, infectious disease, and immunology at Bellevue Hospital Center in New York City. He served in the U.S. Army as chief of medicine and in the Bacteriology Division at the U.S. Army Medical Research Institute of Infectious Diseases (USAMRIID) and as a traveling fellow at the Royal College of Surgeons of the United Kingdom. After working with the Division of Infectious Disease at the U.C. Irvine College of Medicine, he was a public health medical officer in the Viral and Rickettsial Disease Laboratory of the California Department of Health Services and was appointed chief of the laboratory in 1995. He is a lecturer in the School of Public Health/U.C. Berkeley and a colonel in the U.S. Army Reserve as well as commander of the 1488th Medical Team in Oakland. He chaired the Disease Control Subcommittee of the Armed Forces Epidemiological Board and served on an interagency advisory panel on Biological Warfare Preparedness for the 21st Century. He has 28 years of biological defense experience and currently consults to the Department of Defense, Department of Health and Human Services, the Centers for Disease Control and Prevention, Mitre Corporation, the National Domestic Preparedness Office of the FBI, the Association of Public Health Laboratories, the Nuclear Threat Initiative, the National Research Council, and the Lawrence Livermore National Laboratory. He is a founding member of the Working Group on Civilian Biodefense of the Center for Civilian Biodefense Studies of the Johns Hopkins Bloomberg School of Public Health. He is a member of the bioterrorism subcommittee of the Emerging Infections Committee of the Infectious Diseases Society of America and the Infectious Disease Committee of the Association of Public Health Laboratories. He is a founding member of the board of Epi-X, the electronic epidemiology information exchange of the CDC. He is the lead medical officer for biological defense activities in the California Department of Health Services and Principal Investigator of the CDC cooperative agreement to the state for preparedness and response to bioterrorism. Dr. Ascher's research interests include mechanisms of protective immunogenicity of microbial vaccines, advanced methods for diagnosis of infectious diseases, and fundamental issues of HIV pathogenesis. He is a member of numerous scientific societies and has over 90 publications.

RONALD M. ATLAS, Ph.D., is currently a professor of biology, dean of the graduate school, and Co-Director of the Center for the Deterrence of Biowarfare and Bioterrorism at the University of Louisville. He is president-elect of the American Society for Microbiology and co-chair of the ASM task force on biological weapons. Dr. Atlas' recent studies have focused on the application of molecular techniques to environmental problems. His studies have included the development of "suicide vectors" for the containment of genetically engineered microorganisms and the use of gene probes and the polymerase chain reaction for environmental monitoring, including the detection of pathogens in water, soil, and air. He received a B.S. degree from the State University of New York at Stony Brook in 1968, an M.S. from Rutgers University in 1970, and a Ph.D.

from Rutgers University in 1972. After one year as a National Research Council Research Associate at the Jet Propulsion Laboratory he joined the faculty of the University of Louisville in 1973. He is a member of the American Academy of Microbiology and was the recipient of the American Society for Microbiology award in Applied and Environmental Sciences.

RUTH L. BERKELMAN, M.D., is currently with the Rollins School of Public Health, Emory University. An Assistant Surgeon General, she has served as a Senior Adviser to the Director, Centers for Disease Control and Prevention (CDC) and is a member of the Board of Trustees of Princeton University. She was formerly the deputy director of the National Center for Infectious Diseases and led CDC's efforts to respond to the threat of emerging infectious diseases. Board-certified in pediatrics and internal medicine, she also serves on the American Society of Microbiology's Committee on Public Health and is a member of the American Epidemiological Society and a fellow of the Infectious Diseases Society of America. She has extensive experience in disease surveillance and is a member of the Institute of Medicine Committee to Review the DoD Global Emerging Infectious Surveillance and Response System (DoD) Strategic Plan.

DONALD S. BURKE, M.D., is professor of international health and epidemiology and Director of the Center for Immunization Research at the Johns Hopkins Bloomberg School of Public Health. Previously, he served 23 years at the Walter Reed Army Institute of Research, including six years at the Armed Forces Research Institute of Medical Sciences in Bangkok. His research focuses on the epidemiology and prevention of human epidemic virus diseases including HIV/AIDS, dengue, flavivirus encephalitis, and hepatitis. He is past president of the American Society of Tropical Medicine. He has served on the NRC Roundtable for the Development of Drugs and Vaccines Against AIDS, and the NRC Committee on Climate, Ecology, Infectious Diseases, and Human Health (as chairman), and is currently a member of the IOM Committee to Review the Department of Defense Global Emerging Infections Surveillance and Response System and the IOM Committee on Emerging Microbial Threats to Health in the 21st Century.

R. JOHN COLLIER, Ph.D., is currently Maude and Lillian Presley Professor of Microbiology and Molecular Genetics at the Harvard Medical School where he has been since 1984. His career has been devoted to understanding at a molecular level how bacterial toxins cause the symptoms of disease. After graduating from Rice University summa cum laude, he earned his doctorate in biology from Harvard University. He completed post-doctoral studies at the Molecular Biology Institute in Geneva. Subsequently, he was professor of microbiology at the University of California, Los Angeles. At Harvard Medical School, he has held the positions of Faculty Dean for Graduate Education and Chairman of the Division of Medical Sciences. In recent years, he has served on the Advisory Panel for Defense Advanced Research Projects Agency (DARPA) Biological Warfare Defense Unconventional Pathogen Countermeasures. Dr. Collier has been a consultant to Cetus Corporation, Virus Research Institute, AVANT Immunotherapeutics and is co-founder of PharmAthene, Inc. Dr. Collier was initiated as a member of the National Academy of Sciences in 1991 which awarded him the Selman A. Waksman Award in Microbiology in 1999. In addition, he is a fellow at the American Academy of Arts and Sciences and the American Academy of Microbiology and a member of the Norwegian Academy of Science and Letters. One of his research articles was chosen for inclusion in *Microbiology, a Centenary Perspective*, a compilation of one hundred landmark research articles in microbiology over the past century. Dr. Collier has authored nearly 200 publications, was associate editor of *Protein Science* for several years and has served on the editorial boards of several scientific journals.

DAVID T. DENNIS, M.D., M.P.H., is currently Chief of the Bacterial Zoonoses Branch at the Division of Vector-Borne Infectious Diseases at the Centers for Disease Control and Prevention (CDC). He graduated from Whitman College and attended the Cornell University Medical College followed by training in internal medicine at San Francisco General Hospital and Tulane Charity Hospital and in infectious diseases at New York Hospital-Cornell Medical Center. Dr. Dennis received his master's of public health from Harvard University and was a resident in preventive medicine and Epidemic Intelligence Service officer at CDC. Prior positions include that of research medical officer in Ethiopia and Indonesia for the U.S. Naval Medical Research Detachment as well as team leader for the Regional Center for Research and Training in Tropical Diseases, Malaysia. Dr. Dennis has served as medical epidemiologist to the Pennsylvania Department of Health and the New Hampshire Division of Public Health Services, where he was also Chief of the Bureau of Disease Control. His international activities also include consultancies in Mongolia, Kosovo, China, India, Vietnam, Saudi Arabia, Bahrain, United Arab Emirates, Brazil, Sudan, Nicaragua, and South Africa (Zululand). His scientific interests include research into the epidemiology and control of communicable diseases as well as the ecology of vector-borne infectious diseases. Dr. Dennis has authored or co-authored nearly 160 scientific publications.

ERIC EISENSTADT, Ph.D., joined the Defense Advanced Research Projects Agency (DARPA) as a program manager in 1999. He is an active program officer for the Office of Naval Research and also serves as a Navy representative to the Joint Service Technical Panel for Chemical and Biological Warfare Defense. Prior to government service, Dr. Eisenstadt was on the faculty of the Harvard University School of Public Health in the Department of Microbiology and the

Laboratory of Toxicology. Dr. Eisenstadt earned his bachelor's degree and doctorate in biology from Washington University in St. Louis. Following his graduate studies, he completed a one-year NSF-NATO Postdoctoral Fellowship at the Université de Paris, Orsay, and a two-year Deutsche Forschungsgemeinschaft Postdoctoral Fellowship at the Universität zu Köln. Dr. Eisenstadt was also a NIH Staff Fellow at the Laboratory of Molecular Biology, NINDS. In addition to authoring and co-authoring numerous journal articles, books, and monographs, Dr. Eisenstadt is a member of the American Association for the Advancement of Science and the American Society for Microbiology.

COLONEL EDWARD M. EITZEN, Jr., M.D., M.P.H., is currently Commander at the U.S. Army Medical Research Institute of Infectious Diseases (USAMRIID). He graduated from Auburn University as a ROTC Distinguished Military Graduate and with honors from the University of Alabama School of Medicine in Birmingham. After completing training in pediatrics at Fitzsimons Army Medical Center, he became chief of pediatrics at the 121st Evacuation Hospital in Seoul, Korea. Since then, he has taught residents at Brooke Army Medical Center and Madigan Army Medical Center, completed an M.P.H. degree at the University of Washington in Seattle, and completed a preventive medicine residency at Madigan Army Medical Center. He served in Operation Desert Storm as the DCCS and Surgeon of the 62nd Medical Group, 18th Airborne Corps, before coming to USAMRIID. Before taking command of USAMRIID, Colonel Eitzen served as chief of operational medicine, a division he established at USAMRIID in 1991. Colonel Eitzen is board-certified in pediatrics, emergency medicine, and preventive medicine, and is subspecialty board-certified in pediatric emergency medicine and in tropical medicine and traveler's health. Colonel Eitzen holds the Bronze Star Medal for his service in the Gulf War, the Meritorious Service Medal with four oak leaf clusters, the Army Commendation Medal, and the Army Achievement Medal. He is a member of the Order of Military Medical Merit and a recipient of the Army's "A" Proficiency Designator in Pediatrics. Colonel Eitzen has served on several committees for both the American Academy of Pediatrics and the American College of Emergency Physicians and has appeared as a speaker and consultant at numerous conferences. He has published extensively on the medical effects of biological agents and the medical management of biological casualties. He also maintains clinical appointments in pediatrics, emergency medicine, and preventive medicine at Walter Reed Army Medical Center. Colonel Eitzen holds a dual appointment as Adjunct Associate Clinical Professor of Pediatrics and Emergency Medicine at the Uniformed Services University of Health Sciences. In 1998 he was selected as one of ten finalists for the Frank Brown Berry Prize in Federal Medicine awarded by *U.S. Medicine.*

ANTHONY S. FAUCI, M.D., received his medical degree from Cornell University Medical College and subsequently completed an internship and residency at the New York Hospital-Cornell Medical Center. Dr. Fauci came to the NIH as a clinical associate in the Laboratory of Clinical Investigation (LCI) at NIAID. Dr. Fauci became Head of the Clinical Physiology Section, LCI, and in 1980, he was appointed Chief of the Laboratory of Immunoregulation, a position he still holds. Dr. Fauci became Director of NIAID in 1984. Dr. Fauci has made many contributions to clinical research on the pathogenesis and treatment of immune-mediated and infectious diseases. He has pioneered the field of human immunoregulation by making observations that serve as the basis for current understanding of the regulation of the human immune response. Dr. Fauci is widely recognized for delineating the precise mechanisms whereby immunosuppressive agents modulate the human immune response. He has developed effective therapies for several formerly fatal diseases. Dr. Fauci has contributed to the understanding of how the AIDS virus destroys the body's defenses leading to its susceptibility to deadly infections. He has also delineated the mechanisms of induction of HIV expression by endogenous cytokines. Furthermore, he has been instrumental in developing strategies for the therapy and immune reconstitution of patients with this serious disease as well as for a vaccine to prevent HIV infection. He continues to devote much of his research time to identifying the nature of the immunopathogenic mechanisms of HIV infection and the scope of the body's immune responses to the AIDS retrovirus. Dr. Fauci has served as visiting professor at major medical centers nationwide. He is the recipient of numerous prestigious awards for his scientific accomplishments, including 22 honorary doctorate degrees. Dr. Fauci is a member of the National Academy of Sciences, the American Philosophical Society, the Institute of Medicine (Council Member), the American Academy of Arts and Sciences, and the Royal Danish Academy of Science and Letters, as well as a number of other professional societies. He serves on the editorial boards of many scientific journals; as an editor of *Harrison's Principles of Internal Medicine*; and as author, co-author, or editor of more than 1,000 scientific publications.

ARTHUR M. FRIEDLANDER, M.D., is currently Senior Military Scientist at the U.S. Army Medical Research Institute of Infectious Diseases (USAMRIID). He received his medical degree from the University of Pittsburgh after graduating from Harvard College. After completing his medical training at SUNY in Brooklyn, and a postdoctoral fellowship at the National Cancer Institute, Dr. Friedlander was a NIH Postdoctoral Research Fellow in Infectious Diseases at the University of California School of Medicine in San Diego where he subsequently became Assistant Professor. He then joined USAMRIID as a Principal Research Investigator, and in the following years was Chief of Airborne Diseases Division; Chief of Department of Pathobiology, Pathology Division; and Chief of the Bacteriology Division. He is currently Adjunct Professor of Medicine, Uniformed Services University of the Health Sciences as well as Chairman of the Human Scientific Review Committee at USAMRIID. Dr. Friedlander's research interests are in the pathogenesis of infectious diseases and more recently in vaccine development. He has patents filed for both the development of an attenuated strain of *Bacillus anthracis* for production of a recombinant anthrax vaccine and an improved recombinant F1-V fusion protein vaccine against plague. He is a fellow of the Infectious Diseases Society of America and has served as a member of their Committee on Emerging Infections and Bio-Defense. Dr. Friedlander has authored or co-authored nearly 70 articles and 13 book chapters on anthrax and other biological agents.

WILLIAM FRIST, M.D., has been a U.S. Senator from Tennessee since 1994 and is the first practicing physician elected to the Senate since 1928. After graduating from Princeton University, he earned his medical degree with honors from Harvard Medical School and spent the next several years in surgical training at Massachusetts General Hospital; Southampton General Hospital, Southampton, England; and Stanford University Medical Center. He is board-certified in both general surgery and heart surgery. Senator Frist taught at Vanderbilt University Medical Center where he founded and subsequently directed the multidisciplinary Vanderbilt Transplant Center, which under his leadership became an internationally renowned center of multi-organ transplantation. In addition to performing 200 heart and lung transplant procedures, he has written more than 100 articles on medical research and three books: *Transplant,* which examines the social and ethical issues of transplantation and organ donation; *Grand Rounds in Transplantation*, which he co-authored with J. H. Helderman, and *Tennessee Senators, 1911–2001: Portraits of Leadership in a Century of Change,* which he wrote with Lee J. Annis. In the Senate, Senator Frist serves on the Budget; Foreign Relations; and Health, Education, Labor and Pensions committees, and chairs both the Subcommittee on Public Health and Safety and the Subcommittee on African Affairs. In 1999, he was named a Deputy Whip of the Senate; in 2000 he was tapped to head the National Republican Senatorial Committee; and in 2001 he was named one of two congressional representatives to the United Nations General Assembly. U.S. Senate Liaison to the George W. Bush for President Committee, Senator Frist is now the Senate's Liaison to the White House.

JULIE L. GERBERDING, M.D., M.P.H., is the Acting Director, National Center for Infectious Diseases, Centers for Disease Control and Prevention (CDC), an associate clinical professor of medicine (infectious diseases) at Emory University, and an associate professor of medicine (infectious diseases) and epidemiology and biostatistics at the University of California, San Francisco (UCSF). She earned her B.A. degree magna cum laude in chemistry and biology and M.D. degree at Case Western Reserve University and then completed her internship and residency in internal medicine at UCSF, where she also served as chief medical resident before completing her fellowship in clinical pharmacology and infectious diseases at UCSF. She earned her M.P.H. at the University of California, Berkeley in 1990. Dr. Gerberding is a member of Phi Beta Kappa, Alpha Omega Alpha, the American Society for Clinical Investigation (ASCI), and the American College of Physicians, and is a fellow in the Infectious Diseases Society of America (IDSA). She has served as chair and co-chair of the IDSA's Committee on Professional Development and Diversity and co-chair of the Annual Program Committee, and was elected to serve as a member of the nominations committee. Dr. Gerberding is also a member of the Society for Healthcare Epidemiology of America and has served as a member of the AIDS/Tuberculosis Committee and as Academic Counselor on the SHEA Board, and will be president of SHEA in 2003. In the past, she served as a member of NCID/CDC Board of Scientific Counselors, the CDC HIV Advisory Committee, and the Scientific Program Committee, National Conference on Human Retroviruses. She has also been a consultant to NIH, AMA, CDC, OSHA, the National AIDS Commission, Office of Technology Assessment, and WHO. Her editorial activities have included appointments to the Editorial Board, *Annals of Internal Medicine*; Associate Editor, *American Journal of Medicine*, and service as a peer-reviewer for numerous journals. She has authored/co-authored more than 120 publications. Her scientific interests encompass infection prevention/healthcare quality promotion among patients and their healthcare providers and emerging infectious diseases threats. Currently, she is actively engaged in CDC's response to the recent bioterrorist anthrax attacks through the U. S. mail delivery system.

JAMES J. GIBSON, M.D., M.P.H., is State Epidemiologist and Director of the Bureau of Disease Control of the South Carolina Department of Health and Environmental Control. In that job he is responsible for communicable and other acute disease surveillance, epidemiology, and control programs including planning, management, and evaluation of HIV, STD, TB, and vaccine-preventable disease control programs. In his prior career he represented the Centers for Disease Control at the U.S. Agency for International Development as medical officer for child survival programs, and as a tenured associate professor of preventive and internal medicine at the University of South Carolina School of Medicine. He has published on the epidemiology of herpes simplex infection, syphilis, and other infectious diseases, as well as complications of therapeutic abortions. His training included service in the Epidemic Intelligence Service of the Centers for Disease Control and Prevention, and a fellowship in infectious diseases. He is board-certified in internal medicine and in preventive medicine.

MARY J. R. GILCHRIST, Ph.D., was named the director of the University Hygienic Laboratory on July 1, 1995. She holds a bachelor's degree in microbiology from the University of Iowa and an M.S. and Ph.D. in microbiology from the University of Illinois at Champaign-Urbana. She is a Diplomate of the American Board of Medical Microbiology. After a fellowship in clinical and public health microbiology at the Mayo Clinic, Dr. Gilchrist served in the state public health laboratories of Minnesota and Iowa and in two hospitals in Ohio. She was Director of Clinical Microbiology at the Children's Hospital Medical Center and at the Veteran's Affairs Medical Center in Cincinnati and Associate Professor at the University of Cincinnati. In 1991, after the Persian Gulf War, she was nominated as Federal Employee of the Year for her contributions to the bioterrorism response and planning for the Department of Veterans Affairs. In 1994, Dr. Gilchrist was named the Eagleson Institute Lecturer of the American Biological Safety Association. Dr. Gilchrist is an active member of the American Society for Microbiology and currently serves as president of the Association of Public Health Laboratories. She is a member of the CDC's National Center for Infectious Diseases, Board of Scientific Counselors. Dr. Gilchrist is very active in the public health response to bioterrorism on the local, state, and national levels and has several committee appointments related to bioterrorism. She has recently been appointed to the Secretary of Health and Human Services' Advisory Council on Public Health Preparedness.

DONALD A. HENDERSON, M.D., currently is director of the newly created Office of Public Health Preparedness, which coordinates national response to public health emergencies. Dr. Henderson directed the World Health Organization's global smallpox eradication campaign and was instrumental in 1974 in initiating WHO's global program of immunization, which is now vaccinating 80 percent of the world's children against six major diseases and has a goal of eradicating of poliomyelitis. Dr. Henderson is a Johns Hopkins University Distinguished Service Professor with appointments in the departments of Epidemiology and International Health at the Bloomberg School of Public Health. For the past four years, he has directed the Johns Hopkins Center for Civilian Biodefense Studies, of which he is a founding director. The center was established to increase awareness of the medical and public health threats posed by biological weapons. From 1977 through August 1990, Dr. Henderson was dean of the Johns Hopkins School of Public Health. He rejoined the Hopkins faculty in June 1995 after five years of federal government service in which he served initially as Associate Director, Office of Science and Technology Policy, Executive Office of the President and later as Deputy Assistant Secretary and Senior Science Advisor in the Department of Health and Human Services. Dr. Henderson has been recognized for his work by many institutions and governments. In 1986, he received the National Medal of Science, presented by the President of the United States. He is the recipient of the National Academy of Sciences' highest award, the Public Welfare Medal, and, with two colleagues, he shared the Japan Prize. Most recently he received from the Royal Society of Medicine the Edward Jenner Medal. In all, 13 universities have conferred honorary degrees and 14 countries have honored him with awards and decorations.

PETER B. JAHRLING, Ph.D., is Principal Scientific Advisor at the U.S. Army Medical Research Institute of Infectious Diseases (USAMRIID) based in Fort Detrick, Maryland, where he advises the Commander of USAMRIID on the development and coordination of research programs directed at prevention, treatment, and surveillance of infectious disease threats. He also conducts research to evaluate countermeasures, especially vaccines, against viral infectious disease and biological warfare threats. Dr. Jahrling is a consultant to the World Health Organization (WHO), the Department of Health and Human Services, and the National Research Council. He is also head of the WHO Collaborating Center for Arbovirus and Hemorrhagic Fever Reference and Research at USAMRIID. Since 1996, Dr. Jahrling has served as a consultant to the National Academy of Sciences/Institute of Medicine's Russian/U.S. Collaborative Program for Research and Monitoring of Pathogens of Global Importance. He has authored more than 140 scientific papers and chapters on viruses, biological warfare agents, and vaccines. Dr. Jahrling received his A.B. in biology from Cornell University and his Ph.D. in microbiology from Cornell University Graduate School of Medical Sciences.

ANNA JOHNSON-WINEGAR, Ph.D., is the Deputy Assistant to the Secretary of Defense, Chemical and Biological Matters (DATSD(CBM)). She serves as the single focal point within OSD responsible for oversight, coordination, and integration of the chemical/biological defense, counterproliferation support, chemical demilitarization, and Assembled Chemical Weapons Assessment (ACWA) programs. She is a member of the OSD Steering Committee for Chemical-Biological Defense, and represents the DoD on numerous interagency and international groups addressing CB issues. Before joining the Pentagon staff, Dr. Johnson-Winegar was head of the Human Systems Department at the Office of Naval Research (ONR), where she was responsible for the direction, program planning, management, and oversight of their programs in biomedical, cognitive and neural sciences, human factors, and training technologies. Prior to that, she served as Director of Environmental and Life Sciences in the Office of the Director of Defense Research and Engineering (DDR&E) and Director of Medical, Chemical, and Biological Defense Research Programs at the United States Army Medical Research and Materiel Command at Fort Detrick, Maryland. Her previous positions included product manager at the U. S. Army Medical Materiel Development Activity, and research investigator at the U. S. Army Medical Research Institute of Infectious Diseases. She also participated as a biological weapons inspector in Iraq for UNSCOM. Dr. Johnson-Winegar received a B.A. in biology from Hood College, as well as an M.S. and Ph.D in microbiology from Catholic University of America. She has published numerous technical manuscripts, and authored/co-authored several book chapters. She is a long-standing member of many professional societies, serves as a member of the National Board of Directors of the American Cancer Society (ACS), and is President of the ACS Mid-Atlantic Division. In 1998, she received the lifetime achievement award from Women in Science and Engineering.

KRISTI L. KOENIG, M.D., FACEP, a board-certified emergency physician and Director of Emergency Management Strategic Healthcare Group, serves as the principal advisor on emergency management and disaster medicine to the Office of the Under Secretary for Health, Veterans Health Administration, Department of Veterans Affairs. Prior to joining the department, she served as the Director of Prehospital and Disaster Medicine at Alameda County Medical Center in Oakland, California, and was associate professor on the Emergency Medicine Faculty at the University of California at San Francisco. She was invited on sabbatical to be the Co-Director of the Accident and Emergency Department at St. George's Hospital National Health Service Trust in London, England where she concurrently served as the Director of Undergraduate Medical Student Education and Honorary Senior Lecturer at the University of London. Dr. Koenig has held appointments on multiple committees and boards including the Joint Commission on Accreditation of Healthcare Organizations' (JCAHO) Committee on Healthcare Safety, chair of the Society for Academic Emergency Medicine Disaster Medicine Task Force, the American College of Emergency Physicians liaison to the Federal Emergency Management Agency, the National Association of EMS Physicians Standards and Practice Committee, a California Senate appointment as the California Medical Association representative to the State Emergency Medical Services Commission, the California Health Policy and Data Advisory Commission, Advance Directive Subcommittee, the California Association of Hospitals and Health Systems EMS/Trauma Committee, and the London Ambulance Service Accreditation Ambulance Standards Working Group. Dr. Koenig is an honors graduate in applied mathematics from the University of California at San Diego, received her medical degree from Mount Sinai School of Medicine in New York, and completed an emergency medicine residency at Highland Hospital in Oakland, California, serving as chief resident in her final year. She is a fellow of the American College of Emergency Physicians and a clinical professor of emergency medicine at the George Washington University School of Medicine and Health Sciences. Dr. Koenig has also been an associate editor for *Academic Emergency Medicine* and serves on the editorial board for *Journal Watch Emergency Medicine*. She has authored numerous articles on emergency and disaster management.

SCOTT P. LAYNE, M.D., Ph.D., is a tenured associate professor of epidemiology at the UCLA School of Public Health. He is board-certified in internal medicine and infectious diseases and trained in applied physics. Before joining

UCLA, Dr. Layne was a staff member at the Los Alamos and Lawrence Livermore national laboratories. There, his work in infectious diseases utilized mathematical models and epidemiological data to understand the spread of HIV/AIDS in the United States. His further work in virology utilized mathematical models and laboratory experiments to understand the kinetics of HIV infection and biological blocking activities of immunoglobulins against HIV. In 1999, Dr. Layne organized a meeting under the auspices of the Institute of Medicine and National Academy of Engineering to discuss infectious disease threats such as influenza, multidrug-resistant tuberculosis, and HIV and consider new approaches against them. What emerged was a plan to create new kinds of high-throughput laboratory and database resources that expedite disease surveillance and intervention efforts on an international scale. Dr. Layne is currently organizing a new effort to build a global lab against influenza in collaboration with the World Health Organization Influenza Program. He also is proposing a plan to build a high-throughput automated lab and database against bioweapons agents to strengthen homeland security and facilitate compliance and verification procedures for the Biological Weapons Convention. Dr. Layne has authored many publications and is editor of the book, *Firepower in the Lab: Automation in the Fight Against Infectious Diseases and Bioterrorism*.

SCOTT R. LILLIBRIDGE, M.D., is Special Assistant to the Secretary of Health and Human Services for National Security and Emergency Management. Dr. Lillibridge works with HHS Secretary Tommy G. Thompson to enhance national preparedness for bioterrorism and other health emergencies. These efforts have included support to the various departmental initiatives in response to the recent anthrax crimes following the September 11 attacks. Until July 2001, Dr. Lillibridge was the first director of the CDC Bioterrorism Preparedness and Response Program. This program was charged with enhancing CDC's capacities to assist states and other partners in responding to bioterrorism. In addition to infectious disease concerns, the CDC program included consideration of chemical terrorism, the development of a National Pharmaceutical Stockpile, and support for a National Laboratory Response Network for bioterrorism. The CDC's program was initiated in 1999. Dr. Lillibridge's career has focused on emergency public health issues. He was the lead physician during the initial U.S. Public Health Service (PHS) response to the Oklahoma City bombing and also led the U.S. medical delegation to Tokyo following the Sarin release in 1995. During the 1996 Olympics, he served as the PHS science advisor to the multiagency task force that was assembled to protect the public against biological and chemical terrorism. From 1990–1992, Dr. Lillibridge was a member of CDC's Epidemic Intelligence Service (EIS).

JAMES D. MARKS, M.D., Ph.D., currently is Professor of Anesthesia and Pharmaceutical Chemistry at the University of California, San Francisco. Dr.

Marks received his undergraduate training at UC Berkeley, majoring in Biochemistry, and received his M.D. from UCSF. He completed residencies in internal medicine and anesthesia and a fellowship in critical care medicine. He is board-certified in all three specialties. He was a graduate student at the Medical Research Council Laboratory of Molecular Biology under the supervision of Dr. Greg Winter and received his Ph.D. in 1992 for a dissertation titled, *Making Human Antibody Fragments in Bacteria and Bacteriophage*. From 1996–2001 he was the medical director of the Medical-Surgical Intensive Care Unit at San Francisco General Hospital and continues to be an attending physician in the intensive care unit and operating rooms at San Francisco General Hospital. Dr Marks is a world recognized pioneer in the field of antibody engineering. He directs a research group using antibody diversity libraries and molecular evolution to dissect the molecular basis of infectious diseases and cancer and develop novel antibody based therapeutic approaches for these diseases. For the last 8 years he has worked on understanding the requirements for potent antibody neutralization of the botulinum neurotoxins under funding from the Department of Defense. He has 86 relevant publications in the field and is a co-inventor on 7 issued patents and 6 patent applications.

ANDREA MEYERHOFF, M.D., M.Sc., DTMH, is the Director, Antiterrorism Programs, U.S. Food and Drug Administration (FDA). FDA antiterrorism activities refer to the agency's dual missions in public health and law enforcement. These include the protection of regulated products from terrorist tampering, and the availability of safe and effective medical products. FDA antiterrorism preparedness and response planning to meet these public health needs is developed and coordinated by the director, who serves as the point of contact for anti-terrorism issues at FDA. Dr. Meyerhoff joined the FDA in 1996 as a medical officer in the Division of Special Pathogens, and assumed her present position in July 2001. She is board-certified in infectious disease and internal medicine, and holds a M.Sc. in clinical tropical medicine. She is a Clinical Assistant Professor of Medicine at Georgetown University.

THOMAS L. MILNE, is the executive director of the National Association of City and County Health Officials (NACCHO), a position he has held since January 1998. NACCHO serves the 3,000 local health departments in the country, providing a broad range of membership services, national policy advocacy, and cutting-edge tools and services for local public health practitioners. Mr. Milne reports to a thirty-member board of health officials who are elected by their member peers. He is a member of several national committees and boards addressing such issues as workforce development, bioterrorism, public health infrastructure, academics, and leadership, and was recently appointed to the HHS secretary's Advisory Council on Public Health Preparedness. Prior to his current position, Mr. Milne served for 15 years as the executive director of the

Southwest Washington Health District, a three county public health department in Washington State. While there, he conceived and helped lead a dynamic healthy community process that has received national attention. Mr. Milne was a member of the steering committee that developed the innovative Public Health Improvement Plan in the state, and served on a variety of state and local boards relating to HIV/AIDS, managed care, and higher education. He has also served on a number of national boards, including those for NACCHO, the American Public Health Association, and the Public Health Leadership Society. Mr. Milne was a scholar in the inaugural year of the Public Health Leadership Institute.

GARY J. NABEL, M.D., Ph.D., is currently Director of the Vaccine Research Center at the National Institutes of Health. He came from the University of Michigan in Ann Arbor, where he was the Henry Sewall Professor of Internal Medicine and professor of biological chemistry, as well as a Howard Hughes Medical Institute investigator. Dr. Nabel graduated magna cum laude from Harvard College, and then earned his M.D. and Ph.D. from Harvard University. He completed his postdoctoral fellowship in the laboratory of David Baltimore at the Whitehead Institute, MIT. Dr. Nabel is well known as a molecular virologist and immunologist for his work in the fields of HIV, cancer, and Ebola virus research. Dr. Nabel's laboratory has studied mechanisms by which cells coordinate in the regulation of the expression of genes during viral infection and development. Specifically, they have examined the molecular basis of HIV transcriptional activation. In late 1997, Dr. Nabel led a group of researchers who demonstrated in guinea pigs that a DNA-based vaccine could generate protective immune responses to Ebola virus. He and his colleagues were also the first to use direct gene transfer to introduce therapeutic proteins into patients with melanoma, showing the feasibility and safety of this approach. Dr. Nabel is a member of the Institute of Medicine and his honors include the James Tolbert Shipley Prize for Research for Harvard Medical School, the Midwest American Federation for Clinical Research Young Investigator Award, and the ASBMB-Amgen Scientific Achievement Award. Dr. Nabel currently is associate editor of the *Journal of Virology* and the *Journal of Clinical Investigation* and serves on the editorial boards of several other journals.

C.J. PETERS, M.D., is a professor in the Department of Microbiology, Immunology and Pathology at the University of Texas Medical Branch in Galveston. Dr. Peters has recently been named director of the Center for Biodefense at UTMB, which will serve as a catalyst for research and development efforts on effective medical countermeasures against bioterrorism and biological warfare. He had been Chief of Special Pathogens at the Centers for Disease Control and Prevention in Atlanta. Formerly chief of the Disease Assessment Division at USAMRIID, he has worked in the field of infectious diseases for three decades with the CDC, the U.S. Army, and the U.S. Public Health Service. He was the head of the unit that contained the outbreak of Ebola virus at Reston, Virginia. He was also called in to contain an outbreak of deadly hemorrhagic fever in Bolivia. He received his M.D. from Johns Hopkins University and has more than 275 publications in the area of virology and viral immunology. Dr. Peters is currently a member of the National Research Council Committee on Occupational Health and Safety in Care of Nonhuman Primates and the Institute of Medicine Committee on Emerging Microbial Threats to Health in the 21st Century.

STANLEY PLOTKIN, M.D., is currently a medical and scientific consultant, Aventis Pasteur, after seven years as Medical and Scientific Director, Pasteur Merieux Connaught Vaccines, Paris. He is also Emeritus Professor of Pediatrics at the University of Pennsylvania and Emeritus Professor of Virology at the Wistar Institute. Over the course of his career he has served as Senior Assistant Surgeon, Epidemic Intelligence Service, USPHS, director of the Division of Infectious Diseases at Children's Hospital of Philadelphia, and as associate chairman, Department of Pediatrics, University of Pennsylvania. Dr. Plotkin has developed many vaccines, including the Rubella vaccine, RA27/3 strain, now exclusively used in the United States and throughout the world. He has held editorial positions with many scholarly journals and is a member of numerous professional and scientific societies, including the American Academy for the Advancement of Science, the Society for Pediatric Research, the American Society for Microbiology, the Infectious Diseases Society of America, and the American Epidemiologic Society. Dr. Plotkin has received several professional awards including the French Legion Medal of Honor (1998); the Clinical Virology Award, Pan American Group for Rapid Viral Diagnosis (1995); the Distinguished Physician Award, Pediatric Infectious Disease Society (1993); and the Bruce Medal of the American College of Physicians.

DAVID A. RELMAN, M.D., is associate professor of medicine (infectious diseases) and of microbiology and immunology at Stanford University, and Acting Chief of Infectious Diseases at the Veterans Affairs Palo Alto Health Care System. Dr. Relman received his clinical training at Harvard Medical School and Massachusetts General Hospital, and his postdoctoral research training at Stanford. He joined the Stanford faculty in 1992. Since then, his research activities have focused on the molecular mechanisms of bacterial pathogenesis, and on the discovery of previously unrecognized microbial pathogens and commensals. He has described a number of novel disease-causing infectious agents for the first time, and expanded our understanding of human microbial ecology. Recent work includes efforts to employ human and microbial genomic approaches for detection and classification of infectious diseases. Dr. Relman serves on advisory panels for federal agencies, such as the departments of Defense and Energy, CDC, and NIH, and committees for professional societies, such as the Infectious Diseases Society of America and the American Society for Microbiology.

PHILIP K. RUSSELL, M.D., received his bachelor's degree from Johns Hopkins University and his medical degree from the University of Rochester. He is board-certified in internal medicine and has authored or co-authored over 100 publications on infectious diseases. He is a professor emeritus at Johns Hopkins School of Hygiene and Public Health. Prior to joining the university in 1990, Dr. Russell served in the U.S. Army Medical Department where he pursued a career in infectious disease research, retiring as a major general. Military assignments included Director, Walter Reed Army Institute of Research; Commander, Fitzsimons Army Medical Center; and Commander, U.S. Army Medical Research and Development Command. Overseas tours included Pakistan, Thailand, and Vietnam. His military awards include the Legion of Merit and the Distinguished Service Medal. Academic appointments included professor of preventive Medicine at the Uniformed Services University of the Health Sciences. Dr. Russell is a past president of the American Society of Tropical Medicine and Hygiene, and a fellow of the Infectious Disease Society of America. He served as special adviser to the international Children's Vaccine Initiative. He was a member of the Board of Scientific Counselors of the National Center for Infectious Diseases and served on the President's Advisory Committee on Human Radiation Experiments. He has served on numerous boards and advisory committees for national and international agencies and now serves on the boards of directors of the International AIDS Vaccine Initiative and the Albert B. Sabin Vaccine Institute. He is member of the Strategic Advisory Committee of the Bill and Melinda Gates Children's Vaccine Program. and a consultant to the Bill and Melinda Gates Foundation. He chairs the Malaria Vaccine Task Force of the National Institute of Allergy and Infectious Diseases. He currently serves as special adviser in the Office of Public Health Preparedness, Department of Health and Human Services.

JOHN SIMPSON, M.B.B.S., M.F.P.H.M., received his medical degree from University College, London. He trained as a general practitioner and was a principal in general practice in Croydon, London. He then trained in public health and was a consultant in communicable disease control in Surrey (where he set up the countywide service), and Wiltshire, England. These posts entailed considerable involvement in emergency planning and response. Since October 2000 he has been Regional Epidemiologist, Communicable Disease Surveillance Centre, South East, covering a population of 8 million in southern England. As part of this post he has recently coordinated a major study of the health effects of the flooding in Lewes, Sussex, England, in October/November 2000. He was seconded part-time to be head of the Emergency Planning Co-Ordination Unit at the Department of Health in London and subsequent to this has been working at the

Department of Health since September as part of the team coordinating the U.K. Health Response to the September 11 and deliberate anthrax release incidents in the United States. He is also a senior research fellow at the University of Bath.

KATHRYN E. STEIN, Ph.D., is the Director of the Division of Monoclonal Antibodies (DMA), Office of Therapeutics Research and Review, Center for Biologics Evaluation and Research (CBER), FDA, and Acting Chief, Laboratory of Molecular and Developmental Immunology (LMDI), DMA. Dr. Stein received her B.A. in chemistry from Bard College and her Ph.D. in microbiology and immunology from Albert Einstein College of Medicine. She received a National Research Service Award from the National Institutes of Health (NIH) for post-doctoral studies at Harvard with Dr. Harvey Cantor and at the NIH with Dr. William Paul prior to her joining CBER's Division of Bacterial Products in 1980. She has been the director of DMA since 1992 and prior to that time was chief of the LMDI in the Division of Bacterial Products, Office of Vaccines Research and Review. She has an active research laboratory at CBER and is an expert in the field of immune responses to polysaccharide antigens, including the polysaccharide capsules of human pathogens.

LIEUTENANT COMMANDER DONALD C. WETTER, P.A.-C, M.P.H., is an Emergency Coordinator with the Office of Emergency Preparedness in the U.S. Public Health Service. He has been involved in emergency services since 1968, primarily emergency medicine hospital, prehospital, and urgent care. He has had experience in disaster response since 1995. Mr. Wetter received his bachelor's in microbiology from Arizona State University and his physician assistant degree from the University of Alabama in Birmingham. He has a master's of public health degree from the University of Washington.

KEVIN YESKEY, M.D., an active duty 0-6 in the U.S. Public Health Service, is the director of the Bioterrorism Preparedness and Response Program, National Center for Infectious Diseases, Centers for Disease Control and Prevention. He has served as the deputy director, Emergency Public Health in the Division of Emergency and Environmental Health Services, National Center for Environmental Health, CDC. Dr. Yeskey is board-certified in emergency medicine. His previous assignments include associate professor and vice chair, Department of Military and Emergency Medicine, Uniformed Services University School of Medicine, and chief medical officer, PHS Office of Emergency Preparedness. His disaster response experience includes deployments to hurricanes, earthquakes, floods, mass migrations, and terrorist bombings.

FORUM STAFF

STACEY L. KNOBLER, is Director of the Forum on Emerging Infections at the Institute of Medicine (IOM). She previously served as the co-director of the IOM Board on Global Health's study, *Neurological, Psychiatric, and Developmental Disorders in Developing Countries,* and research associate for the *Assessment of Future Scientific Needs for Live Variola Virus.* Ms. Knobler is actively involved in program research and development for the Board on Global Health. Previously, she has held positions as a Research Associate at the Brookings Institution, Foreign Policy Studies Program and as an Arms Control and Democratization Consultant for the Organization for Security and Cooperation in Europe in Vienna and Bosnia-Herzegovina. Ms. Knobler has also worked as a research and negotiations analyst in Israel and Palestine. She is currently a member of the CBACI Senior Working Group for Health, Security, and U.S. Global Leadership. Ms. Knobler has conducted research and co-authored published articles on biological and nuclear weapons control, foreign aid, health in developing countries, poverty and public assistance, and the Arab-Israeli peace process.

MARJAN NAJAFI, M.P.H., is a research associate for the Forum on Emerging Infections in the Board on Global Health. She has also worked with the IOM committee that produced *Veterans and Agent Orange: Update 2000.* She received her undergraduate degrees in chemical engineering and applied mathematics from the University of Rhode Island. Ms. Najafi served as a public health engineer with the Maryland Department of Environment and, later, the Research Triangle Institute. After obtaining a master's of public health from the Bloomberg School of Public Health at Johns Hopkins University, she managed a lead poisoning prevention program in Micronesia with a grant from the U.S. Department of Health and Human Services. Prior to joining IOM, she worked on a study researching the effects of cellular phone radiation on human health.

LAURIE A. SPINELLI is a project assistant for the Forum on Emerging Infections in the Board on Global Health. Ms. Spinelli joined the IOM in July 2000 and has worked with the IOM committee that generated the *Neurological, Psychiatric, and Developmental Disorders: Meeting the Challenge in the Developing World* report. Currently, she is working on two forthcoming reports: *Reducing the Impact of Birth Defects in Developing Countries* and *Improving Birth Outcomes in Developing Countries*. Prior to joining IOM, she graduated from Syracuse University with a bachelor of arts degree in speech communications. Ms. Spinelli also teaches an interpersonal communication course at the College of Southern Maryland.